Chemical Reactivity and Reaction Paths

Chemical Reactivity and Reaction Paths

EDITED BY
Gilles Klopman

A Wiley-Interscience Publication
JOHN WILEY & SONS
New York · London · Sydney · Toronto

To my wife Malvina

Library of Congress Cataloging in Publication Data:

Klopman, Gilles, 1933–
   Chemical reactivity and reaction paths.

   "A Wiley-Interscience publication."
   Includes bibliographical references.
   1. Reactivity (Chemistry) I. Title.

QD505.5.K55 1974   541'.393   73–17325
ISBN 0-471-49355-4

Printed in the United States of America

10 9 8 7 6 5 4 3 2 1

Contributors

R. C. Dunbar
Case Western Reserve University

J. Feuer
Texas Tech University at Lubbock

H. Fujimoto
Kyoto University

K. Fukui
Kyoto University

W. B. Giles
Texas Tech University at Lubbock

W. C. Herndon
University of Texas at El Paso

R. F. Hudson
University of Kent at Canterbury

G. Klopman
Case Western Reserve University

J. Michl
University of Utah

G. A. Olah
Case Western Reserve University

D. Otteson
Texas Tech University at Lubbock

E. Silber
Texas Tech University at Lubbock

M. Simonetta
University of Milan

Preface

The aim of the present volume is to present the actual state of thinking about chemical reactivity problems. In the following pages, several leading contributors in the field present their viewpoints, sometimes on parallel subjects, in chapters that are self-containing. It is by comparing the different approaches or interpretations that the reader will best be able to judge the merits of the various theories and their applicability to his own field of interest.

Several unavoidable duplications will be encountered particularly in such general areas as free energy correlations, perturbation theory, and orbital correlation diagrams. This simply reflects the widely held opinion that these techniques provide the most powerful tools for the study of chemical reactivity.

No attempt has been made to include sophisticated quantum mechanical calculations of energy profiles as this approach represents merely an application of complex computations to specific cases rather than the development of a general theory of chemical reactivity. The main emphasis is on general theories and their application to a wide variety of cases without the assistance of complex computer routines. In this way, it is hoped that the methods will be more appealing to the chemist who seeks a rationale for understanding the development of reactions along specific routes. Most of the chapters have been arranged so that they present the general theories believed to be the most appropriate for the problem, together with key applications in various fields of chemistry.

July 1973   *Gilles Klopman*

Contents

Chemical Reactivity and Reaction Paths

CHAPTER 1

Chemical Reactivity and Reaction Paths: General Introduction

G. Klopman
Case Western Reserve University
Cleveland, Ohio

I. INTRODUCTION

The prediction of chemical reactivity presents a constant challenge to the chemist who desires to define optimum conditions for performing specific reactions. The basic aims of the theories of reactivity are to find an explanation of how chemical transformations come about and to go further in the prediction of properties and reactions of interest to organic chemists.

Although we speak of theoretical chemistry, we must realize that from the point of view of a theoretical physicist this is an applied subject and to a purist largely an empirical one. The laws of chemical reaction must be regarded as provisional and in no way immutable. Because of the extreme electronic complexity of molecular systems, wide and frequently crude assumptions have to be made in most theoretical treatments, and empirical factors introduced into theoretical expressions except in the simplest of systems. One must not, however, underestimate the enormous progress in theoretical understanding of molecular structure that has been made in recent years. The advent of computer techniques means that the most difficult problems are, in principle, capable of solution within a reasonable period of time.

An investigation into the factors responsible for the chemical behavior of molecules must start with the study of the basic laws that apply to such

1

phenomena. These laws may be considered as of two kinds: (1) thermo-dynamic and statistical laws, and (2) quantum-mechanical laws. Thermo-dynamics and statistics apply essentially to groups of objects (molecules) and hence lead to quantitative descriptions of the physical properties of chemical substances. For example, the overall rate with which one chemical substance is transformed into another is based on statistical laws, that is, on the be-havior of a large number of molecules. Thus for a given reaction, proceeding by a bimolecular mechanism,

$$A + B \xrightarrow{k_2} C + D$$

the rate given by change in molar concentration of A (or B) is determined by the probability of molecule A meeting molecule B (whatever their electronic structures). This probability is given by the product of concentration,* leading to the classical rate equation

$$-\frac{d[A]}{dt} = k_2[A][B]$$

The *principle* of activation, that is, the postulate that only those molecules with combined energy equal to or greater than some critical value can react, is likewise a statistical concept, but the actual value of this critical energy (which depends on electronic changes in the reacting molecules) cannot be found without a knowledge of quantum laws.

The introduction of the concept of an activation energy leads to a con-sideration as to how a system consisting of two molecules passes over the potential energy barrier. By taking a statistical view of the energy distribu-tion of an assembly of a large number of molecules, the origin of such a critical energy can be explained. The passage over the barrier, however, involves a complete change in the positions and behavior of the electrons associated with nuclei A and B. The overall rate, however, is determined only by the difference between the ground-state energy and the energy at the top of the barrier, that is, of the transition state. We need consider, therefore, only ground-state and transition-state energies and structures in order to derive reaction rates theoretically. A description of the transition state is therefore necessary so that its properties may be computed. The great success of the transition state (or absolute rate) theory was in providing such a description in terms of the bond vibrations of the reactant molecules.

According to this theory, reactants and transition states are supposed to be in thermodynamic equilibrium, namely,

$$AB \underset{}{\overset{K}{\rightleftharpoons}} AB^* \longrightarrow \text{products}$$

and provided the transition state is defined in a particular way, it may be

* Neglecting specific interactions, and volumes, which require activity corrections.

treated in the same way as an ordinary molecule. This means that the ordinary laws of thermodynamics may be applied to such a system, and the transition state may be given an enthalpy and entropy.

In principle, the former can be calculated by the standard methods of statistical thermodynamics, in terms of partition functions of reactants and transition state. In practice, however, it is impossible to evaluate vibrational partition functions for the transition state without a detailed consideration of the potential energy surface for the reaction. This of course requires quantum-mechanical methods, but the solutions of the appropriate energy equations even for only three- or four-electron systems are extremely difficult.

It follows, therefore, that the calculation of absolute rate constants other than for the simplest systems is impossible, and one has to turn to more empirical methods of treating reaction rates.

The first task confronting us in the search for such a theory is to identify the reactive sites as a function of the molecular structure, and to determine their relative reactivity. Chemistry is concerned with the properties and reactions of an enormous number of different compounds, which for the purpose of expediency may be classified into groups of certain similarity, in various ways. For example, in organic chemistry the conventional way is to recognize groups of compounds with the same functional group. Then within each group the properties of a particular compound may be inferred from the behavior of any other member. This means that quantitative and qualitative relations exist between the properties, physical and chemical, of the members of each group, and also to a greater approximation between the members of different groups.

This comparative element in chemistry, although in many ways intellectually dissatisfying, has led to considerable clarification and organization in the interpretation of chemical effects. In particular in the field of reaction kinetics the selective comparison of rate constants has led to considerable advances in the molecular interpretation of chemical reactions. It is this comparison of rate constants which has led to the concept of chemical reactivity, and the consideration of chemical reactivity in molecular terms which has led to reaction mechanisms.

Let us take a simple example in order to clarify these points. Consider one of the first reactions to be studied kinetically, the saponification of esters. Ethyl benzoate reacts with sodium hydroxide in aqueous solution as follows, found by chemical analysis,

I

The rate constant can be measured by conventional means, and the reaction is readily found to be of the second order. Similarly a rate constant, k_2', can be measured for the hydrolysis of ethyl *p*-nitrobenzoate.

$$NO_2-\bigcirc-C\begin{matrix}O\\OEt\end{matrix} + HO^{(-)} \xrightarrow{k_2'} NO_2-\bigcirc-C\begin{matrix}O\\O^{(-)}\end{matrix} + EtOH$$

II

This reaction leads to *p*-nitrobenzoic acid, and the two reactions are obviously very similar. Now it is found that $k_2' > k_2$, and we may say that II is more reactive than I. This fact alone is not particularly interesting. It is only when we find that the NO_2 group increases the reactivity of a wide range of electrophiles that we search for an explanation, by relating the change in reactivity to some action or property of the nitro group.

The simple postulate that the NO_2 group withdraws electrons from σ and π bonds enables these observations to be interpreted. Such interpretations, however, involve knowledge of reaction mechanisms, that is, the position of attack, the bond broken, and the electronic changes in the molecule resulting from the interaction with the reagent. Providing that these have been reasonably established, the interpretation can be carried further. Thus other groups show effects similar to those attributed to the NO_2 group. Most halogens increase the reactivity of benzoic esters and may equally be assigned an electron withdrawing effect. Alkane groups, methoxy, hydroxy, and amino groups, on the other hand, retard such reactions and are tentatively assigned an electron-donating effect.

A more quantitative picture of these effects can even be obtained by attributing to each substituent an index derived from the relative reactivity of a substituted standard compound engaged in a reference reaction. This is done, as in the Hammett equation, by characterizing each substituent by a number equal to the ratio of the ionization of the substituted benzoic acid and that of the unsubstituted benzoic acid in water at 25°.

An experimental correlation is thus being set, such as

$$\log\left(\frac{k_s}{k_0}\right) = \rho\sigma_s$$

where σ_s characterizes the substituent, and ρ, equal by definition to unity for the ionization of benzoic acids, is in other cases a parameter specific to the reaction under investigation. Its value then has to be determined for each type of reaction (and each different set of conditions) by running a couple of kinetic measurements of compounds substituted by a group of known σ.

The Hammett relationship, as well as other "free-energy relationships," have been extremely valuable to the chemist in helping him to predict the re-activity of compounds.

Owing to the complexity of organic molecules, a particular reagent may contain several alternative nucleophilic and electrophilic centers, and hence the competition for these alternative reaction sites is a very important general problem. Basically, the problem is the same as that encountered in the comparison of the reactivity of specific functional groups attached to struc-turally different compounds, except that in this case, the various reactive sites are located on the same molecule. However, contrary to the former case, where only the rate of the reaction was of consequence, here not only the rate but also the nature of the product formed in the reaction depends on the factors influencing reactivity.

The electrophilic substitution of aromatic derivatives presents the elements of such a situation. Thus, naphthalene is more reactive than benzene toward electrophilic reagents, say nitration. This by itself is an interesting observa-tion that may help us define the conditions under which these compounds should be nitrated. But whereas for benzene the problem is closed by this observation, it is not at all the case for naphthalene. Here a crucial problem remains as to where is the nitration going to take place. Naphthalene has two possible sites susceptible of undergoing electrophilic attack, the α and the β positions:

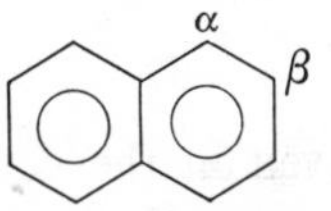

The identification of the reactive site presents, in this case, an acute im-portance and a serious challenge that can be met only if we understand how the molecular structure affects the reactivity of the various active centers of the molecule. In the particular case of naphthalene, the α position is known to be more reactive than the β position, whatever the nature of the electro-philic reagent.

Since naphthalene is also always more reactive than benzene, it is clear that a consideration of the structure·of the aromatic compound (nucleophile) should be sufficient to account for the observed differences in reactivity.

This is not always the case, as in most "ambient" reactions, the relative reactivity of active sites varies with the nature of the attacking reagent. For example, the methyl group in toluene is known to be ortho para directing in electrophilic substitutions:

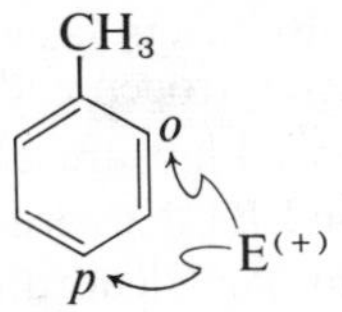

However, the product ratio determined by the relative reactivity of the ortho and para positions varies widely with the nature of the electrophilic reagent.

Thus chlorination yields up to 75% of the ortho derivative, whereas mercuration produces mostly the para derivative, 80%.

Another example of ambient reactivity where the nature of the product critically depends on the reagent includes the nucleophilic attack on alkyl halides. Here the two electrophilic centers competing for the nucleophile

$$H-\underset{|}{\overset{|}{C}}-\underset{|}{\overset{|}{C}}-SR + Cl^{(-)} \underset{RS^{(-)}}{\longleftarrow} H-\underset{|}{\overset{|}{C}}-\underset{|}{\overset{|}{C}}-Cl \underset{HO^{(-)}}{\longrightarrow} H_2O + \underset{|}{\overset{|}{C}}=\underset{|}{\overset{|}{C}} + Cl^{(-)}$$

are of a different nature, H versus C. In discussing the yield of olefin, the nucleophile must be considered in addition to the structure of the organic substrate.

In a similar way, the competition between C and O alkylation of enolate ions gives rise to the two following possibilities:

$$\underset{H_3C}{\overset{R'}{\diagdown}}CH-C\overset{\diagup O}{\underset{\diagdown R}{}} \underset{\text{C-alk.}}{\overset{CH_3X}{\longleftarrow}} \left[R'-CH\cdots C\overset{\diagup\diagup O}{\underset{\diagdown R}{}} \right]^{(-)} \underset{\text{O-alk.}}{\overset{MeOCH_2Cl}{\longrightarrow}}$$

$$R-CH=C\overset{\diagup OCH_2OMe}{\underset{\diagdown R}{}}$$

The situation is complex even in the case of relatively simple organic molecules as, for example, in the substitution reactions of α-halogenated ketones, where different nucleophiles react on different reaction sites.

$$(RO)_3P \longrightarrow \underset{\underset{\underset{H}{|}}{\overset{\overset{O}{\|}}{R-C-CH-Br}}}{} \overset{R_3N, I^{(-)}, RS^{(-)}}{\nearrow}$$

$$\longleftarrow I^{(-)}, RS^{(-)}, R_3P$$

$$\longleftarrow \text{strong bases, } RO^{(-)}$$

This observation reflects the preference manifested by nucleophiles for specific reactive sites. For example, sulfur nucleophiles are extremely active toward saturated carbon atoms, whereas oxygen and nitrogen nucleophiles prefer carbonyl carbons. Thus $R-CH_2Br$ reacts faster with SH^- than with OH^-, and SCN^- reacts with it through its sulfur end.

On the other hand, RCOBr reacts faster with OH^- than with SH^- and SCN^- attacks it through its nitrogen end.

Marked preferences of this kind are also found in inorganic reactions where, for example, Fe^{3+} and Ba^{2+} coordinate with halogens as F > Cl > Br > I, whereas Hg^{2+} and Ag^+ coordinate as I > Br > Cl > F, even though all metals probably form their most stable bonds in the gas phase in the order $F > Cl > Br > I$.

In explaining such reactions, it is essential to differentiate between thermo-dynamic and kinetic control. When the reactions under consideration are reversible, the thermodynamically stable product is formed (rate factors permitting), for example,

$$AB + XY \underset{k_{-1}}{\overset{k_1}{\rightleftarrows}} AX + BY$$

$$k_2 \updownarrow k_{-2}$$

$$AY + BX$$

Thus if $k_1 > k_2$, the product AX will be isolated if k_{-1} is small, that is, if the reaction is effectively irreversible. One may then say that the competing reactions are subject to kinetic control. On the other hand, if k_{-1} is large, then AY will be isolated if this is the thermodynamically most stable product. It is particularly important to differentiate between these possibilities when ambident reagents are used.

Ambidency may also be exhibited as a result of experimental conditions, and its interpretation must be included in a satisfactory theory of chemical reactivity. A specific example is provided by the alkylation of the phenolate ions with 3-bromopropene.

ONa

+ $CH_2{=}CH{-}CH_2Br$

t-butyl alcohol → 100% O-alkylation 0% C-alkylation

phenol → 23% O-alkylation 77% C-alkylation

The reaction affords exclusive O-alkylation when it is conducted in *t*-butyl alcohol, but mostly C-alkylation in phenol.

This stresses the role of the solvent whose importance manifests itself by pro-ducing sometimes drastic mechanistic changes and enormous rate variations.

Finally, temperature effects and catalytic activities are important general problems and represent an additional challenge to the derivation of reactivity theories.

II. QUANTUM-MECHANICAL APPROACH TO CHEMICAL REACTIVITY

The variation in heat content or enthalpy which occurs during the course of a chemical reaction depends on dissociation energies, electron reorganization, and changes in solvation energies involved in the rate determining process.

Because of the uncertainty of the entropy term, the calculation of absolute rates of reaction can hardly be expected to be done, but the influence of structure on reactivity, which is one of the most important aspects of organic chemistry, has been largely studied and often quantitatively correlated by the widespread use of quantum-mechanical methods.

The problem consists, in principle, in calculating the energy profile, that is, a large number of possible configurations for the interacting species, and determining the easiest reaction path. The transition state is the structure corresponding to the highest energy when going from the reactants to the products; the heat of activation is the difference between this energy and that of the initial reactants. In order to be able to calculate successfully the energy of the various possible configurations, one must, however, be ensured that the accuracy of the quantum-mechanical method used to perform the calculations enables results of significant value to be obtained. As a matter of fact, only methods leading to a correct bond-length–bond-energy relationship would be suitable for such an approach.

Unfortunately, the usual approximations used in this context do not provide any guarantee that such relationship holds and therefore that the calculations would be of any validity.

Only the exact solution of the Schrodinger equation and possibly also its LCAO *ab initio* approximation would prove sufficiently accurate to be significant. For the time being, however, calculations based on the exact solution of the Schrödinger equation have been in practice impossible to realize except for the reaction $H_2 + H$, which was successfully calculated. There is little hope that such approach may even be extended to other systems in the near future.

Nevertheless, *ab initio* techniques in their recent development using Gaussian functions may well be our best chance to obtain an answer to some chemical problems. Until recently, they were still too complicated to be handled even for small interacting systems, but with the improvement in computers and computational techniques, they now appear to be of much broader interest. Preliminary calculations have already been made for small systems in the gas phase. Thus the reaction between NH_3 and HCl was investigated by Clementi and the existence of a charge transfer complex, stabilized at approximate distances of $N \cdot Cl$ equal to 2.9 Å and $H \cdot Cl$ of 1.65 Å with a stabilization energy of 19 kcal was established. The structure of carbocationic intermediates was also extensively studied with these methods. Such approach might well illustrate a viable route for further investigations.

But the procedure remains too complicated for practical use with large systems of common chemical interest. And even if it could be applied successfully to such systems, ninety-nine percent of the reactions would escape this investigation, because no method is available for the *ab initio* calculation of the influence of the solvent, and only gas-phase reactions could be studied. Another difficulty, which arises very quickly when the size of the reactants increase, is due to the increase of the degree of freedom of the system. The number of possible structure arrangements increases exponentially and owing to the length of time required to calculate one single configuration, it is practically beyond our actual possibilities to determine the exact reaction path.

For large systems, therefore, there is a need for approximate quantum-mechanical methods. With such methods, however, not only would the calculations of all possible intermediate configurations be tedious but probably also incorrect. Hence, when the reactivities of several similar reactants are investigated under the same experimental conditions, one usually simplifies the procedure by assuming a reasonable common structure for the transition state. By an adequate quantum-mechanical procedure, one then calculates the energy $\Delta E_0^{\ddagger}$ necessary to reach this hypothetical transition state for some standard compound and determines the relative rate of reaction of other compounds by the following equation:

$$\mathrm{RT} \log \frac{k}{k_0} = \Delta E^{\ddagger} - \Delta E_0^{\ddagger}$$

where k_0 is the experimentally determined rate of reaction of the standard compound, $\Delta E^{\ddagger}$ the calculated activation energy of the other compound, and k its unknown reaction rate.

Changes in reactivity produced by substituents or other structural differences of the reactants can hopefully be correlated in this manner.

Such procedure makes use of a rule known as the noncrossing rule, which states that for similar reactants the ratio of the energy necessary to reach any particular (but common) point on the respective reaction path curves is proportional to the ratio of the activation energies. This is illustrated in Fig. 1-1.

Although there is neither proof nor reason for such behavior, it has reasonably been verified experimentally and serves as a basis for most attempts to correlate chemical reactivity, particularly aromatic reactivity.

For reactions occurring in solution, an additional approximation is usually made which consists in assuming that changes in solvation energy are constant, irrespective of the structural differences between the reactants. It has been shown, however, that not all chemical properties can be accommodated with this approximation and, although it is reasonable for most of the organic reactions, it must be made sure that the structural changes do not affect the charge of ions or polar species involved in the rate-determining step.

Under these circumstances, the solvation energy remains approximately

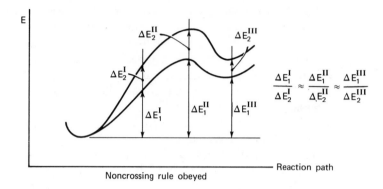

Noncrossing rule obeyed

$$\frac{\Delta E_1^I}{\Delta E_2^I} \approx \frac{\Delta E_1^{II}}{\Delta E_2^{II}} \approx \frac{\Delta E_1^{III}}{\Delta E_2^{III}}$$

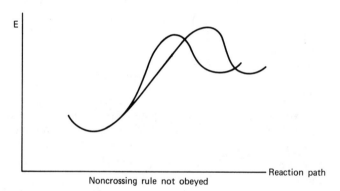

Noncrossing rule not obeyed

Fig. 1–1. The noncrossing rule

constant and does not affect the correlation between the calculated values and the experimental rates of reaction.

Several quantum-mechanical methods have been suggested for the calculation of the activation energies. These methods, though all derived from approximations to the Schrödinger equation, have been of various degrees of sophistication. Among them, the one-electron methods, neglecting the interaction between the electrons and providing some kind of topological description of the molecules, have been by far the most extensively used.

The Hückel method, for example, which allows one to calculate electron properties, was found to be particularly well adapted to this problem and has shown considerable success in dealing with aromatic reactivity. In these cases, quantitative correlations have been systematically observed between the various reactivity indices and experimentally determined rate constants.

More recently also, several applications have been studied by more involved methods such as the Pariser-Parr, Pople, or Dewar methods, which take into account the electron-electron interactions, but neglect the differential overlaps. These procedures are all restricted to the calculation of changes in conjugation energy.

The reactivity of nonconjugated species or of species which react without involving a change in resonance energy have also been investigated mostly by one-electron methods such as that of Fukui and Hoffman. No systematic procedure, like those utilized for aromatic reactivity, have been described, however, and each case needs specific study. The agreement with experiment is only fair and the extent of the calculations can by no means be compared with that on aromatic reactivity.

One of the dangers of such procedures is that it requires an accurate knowledge of the small difference between two large quantities calculated by quantum-mechanical approximate methods, that is, the heat of formation of the two reactants and that of the transition state. It is clear, therefore, that small nonsystematic errors in estimating total heats of formation will affect to a fatal extent the prediction of the value of the heats of activation.

For this reason, much effort is now being drawn in order to obtain more accurate heats of formation of the ground state for both σ- and π-bonded species.

There is thus now a general tendency to provide more sophisticated methods allowing the calculations of both σ and π electrons to be made simultaneously with the final aim of obtaining accurate charge distributions, heats of formation, and ultimately heats of activation of reactions. Such methods, which were not available for large molecules until very recently, have only begun to be applied to reactivity problems.

The approximations used in these methods, in turn, provide another approach particularly well adapted for dealing with the problem of chemical reactivity—the perturbation method. This method, indeed, takes advantage of the fact that in most effectively occurring chemical reactions, the transition-state structure remains close to that of the starting reactants. The free-energy difference produced by the modification of only one or two bonds remains a small fraction of the total energy and can thus be found by a perturbational approach based on the known molecular properties of the isolated reactants.

CHAPTER 2

Empirical and Semiempirical Calculations of Chemical Reactivity and Reaction Paths

M. Simonetta
Institute of Physical Chemistry
University of Milan
Milan, Italy

I. INTRODUCTION

In recent years a number of reviews have appeared in the literature devoted to the subject of quantum-mechanical interpretation of chemical reactivity.[1-6] In most of these papers, however, attention is focused on the old concepts of reactivity index[1-4], or special consideration is given to a particular approach, like the frontier orbital theory[2] or the perturbational molecular orbital method.[3] In others[5,6] the impact of orbital symmetry rules and of molecular orbital methods in which all valence electrons are included is fully recognized. In all these papers the theoretical approach is represented as a counterpart of the experimental approach.

There are some numbers, like equilibrium or rate constants, entropy or enthalpy of activation, ratios of different products, and so on; the theoretical chemist builds a model of the system under study, makes a choice of a suitable method of calculation, and produces his numbers, to be compared with those obtained by experiment. This does not seem to be the right way of presenting the situation.

Chemists today are interested in a deep knowledge of what is going on during a chemical reaction. Each molecule undergoes a series of changes in which the different atoms move in many different ways and there is usually an enormous number of paths along which the reaction can occur.

We can distinguish groups of paths as belonging to different mechanisms and make experiments to find out which mechanism is favored, the ratio of molecules following different mechanisms when more than one is feasible, the energy and free-energy changes involved, and so on.

Since the experiments are almost exclusively performed on macroscopic samples, the methods of thermodynamic and statistical mechanics are extensively used in the interpretation of results. The more precise the questions raised by the chemist, the more difficult it is to find the way to the answers. The number of different techniques to be used in making significant experiments is bound to increase at the same time. Furthermore, the interpretation becomes more difficult, requiring the introduction of new concepts and models of increasing sophistication. Quantum mechanics, and to a lesser extent molecular mechanics, are simply tools for the mechanistic chemist to use. Theoretical chemistry is at its best when it is merged in experiment.

Theoretical calculations can be used to investigate properties of molecular species not accessible to experiment, but still postulated as necessary intermediates along a probable reaction path. Quantum mechanics at its highest levels can be used to calculate energy differences between simple molecules, or the rules of conservation of orbital symmetry can be used to predict if a sigmatropic reaction obtains with retention or inversion of configuration. All the different methods offered by quantum chemistry can be useful in different problems or at different stages in the same problem. In this chapter a number of examples will be considered in which theoretical methods have proved to be useful in understanding chemical equilibria or the mechanism and rate of chemical reactions. I shall try to choose the examples as much as possible from my own experience. Before going through the examples, I would like to clarify exactly what is meant in the title by the terms "empirical" and "semiempirical" methods. In the first method the energy of a molecular system is calculated as the sum of a limited set of energy functions obtained by fitting a large set of experimental data; the second method implies the solution by means of some more or less satisfactory approximation of the Schrödinger equation.

II. THE H_3 PROBLEM

One of the simplest chemical reactions, namely,

$$H_2 + H \longrightarrow H + H_2 \qquad (1)$$

has been the subject of a large number of experimental and theoretical investigations. For this system the energy has been calculated as a function of geometry in a quite accurate way by means of self-consistent field molecular orbital theory, using an extended orbital set and including complete configuration interaction on that basis.[7,8] These calculations give good agreement with experimental results for the activation energy. Using a semiempirical potential energy surface and a quasiclassical procedure, the collision dynamic of reaction (1) has been examined[9] and the calculated rate constants have been compared with values obtained from absolute rate theory. Quantum mechanics here contributes an accurate potential energy surface, and multiconfigurational MO wavefunctions as well as VB wavefunctions can be used. The same potential energy surfaces are used in the interpretation of crossed-beam scattering experiments. In the field of the simple gas-phase reactions (e.g., atom + diatomic molecule), the interaction of high quality *ab initio* calculations and crossed molecular beam experiments can lead to detailed understanding of the mechanism and perhaps to a better formulation of the theory of chemical kinetics.

III. BASICITY OF THE AZULENES

It is well known that azulene and methylazulenes are soluble in 50% sulfuric acid. While the solutions of azulene in organic solvents have a blue color, the sulfuric acid solution is yellow. Plattner[10] has suggested that in acid solution a complex is formed by proton addition and a conductimetric investigation has shown that it is a 1:1 complex; that is, one proton is added to one azulene molecule. The next question is at which position the addition takes place. This problem was solved theoretically.[11] If the reactions

$$Az + H \rightleftharpoons (AzH^+)_i \qquad (2)$$

are considered, where $i = 1, \ldots, 6$ corresponds to six different possible positions of attack, the equilibrium constants

$$K_i = \frac{|(AzH^+)_i|}{|Az|\,|H^+|}$$

are related to the enthalpies and entropies of reactions as follows: $-RT \ln K_i = \Delta H_i - T\Delta S_i$ and since ΔS_i can be taken as independent from position i, $-RT \ln K_i = \Delta H_i + \text{const.}$ ΔH_i is given by $H_{(AzH^+)_i} - H_{Az} - H_{H^+}$, where the H are the enthalpies of formation of the different species. In order to compare two different reactions we must compare the corresponding ΔH_i; that is we have to calculate the difference $H_{(AzH^+)_k} - H_{(AzH^+)_l}$.

This difference is measured, with the reasonable assumption of a constant σ bond energy in all complexes, by the difference in localization energy, that is the difference in π-electron energy in complexes k and l.

The Hückel method allows a prompt, reliable calculation of such energies. It was found that the most stable cation was the one with the methylene group in position 2. It was also found that only for such a cation was there an hypsochromic shift of the first π-π^* transition; for all other cations the calculations predict a bathochromic shift. The influence of a methyl substituent at different positions was also correctly predicted. For example, it turns out, correctly, that a methyl group in the seven-membered ring has a greater stabilizing effect that one in the five-membered ring. The fact that azulene is more basic than naphthalene was rationalized. It is gratifying that good agreement with experimental values of such properties as charge distribution and electronic spectra was obtained for simple cations by means of a quite independent semiempirical valence bond calculation.[12]

IV. RATE OF SOLVOLYSIS AND HOMOALLYLIC RESONANCE

The rate-enhancing effect of unsaturated β substituents in solvolysis has been rationalized with the assumption of a stabilization of the intervening cation due to homoconjugation.[13] To support this assumption, quantum-mechanical calculations have been carried out for a few simple cations.[14] The difference in energy between different cations has been calculated taking into account both the additional resonance stabilization related to 1–3 interactions and the strain energy involved in the distortion of nuclear arrangement to make this resonance possible. Differences in stability of the cations correlate quite well with solvolysis rate constants. These calculations have been heavily criticized[15] since they are based on the Hückel theory, in which electronic and core repulsion are not taken explicitly into account. The same criticism has been raised[15] against the more sophisticated extended Hückel method.[16]

The same calculations for cations as those of Ref. 14 have been performed using the all valence electron CNDO/2 method[17]; the new results completely confirm those obtained by the earlier treatment.[18] For an appreciation of the impact of the extended Hückel theory on the theory of reaction mechanism and determination of reaction paths I suggest the reading of Ref. 19 and literature cited therein.

Experimental rates of solvolysis have been satisfactorily correlated with calculated quantities for bridgehead substrates.[20,21] This time molecular mechanics (see below) was used as the basis of the calculations; activation free energies were compared with the difference in energy of the related hydrocarbon (taken as a model of the ground state of the bridgehead substrate) and the carbonium ion (taken as a model of the transition state). It seems well established that the usual assumptions of constant entropy and solvent effects are reasonable.

V. THE COPE AND FRITSCH REARRANGEMENTS

The Cope rearrangement is a reaction which obtains in biallylic systems,

$$\tag{3}$$

and is particularly amenable to theoretical investigations. It is an intramolecular reaction that occurs in the gas phase without secondary reactions, insensitive to external catalytic influence, and following first-order kinetics.

A small part of the potential energy surface has been explored by the "molecular mechanics" approach.[22] Two paths have been followed, via a chair-like transition state and via a boat-like transition state. The energy along these paths has been calculated as the sum of compression energy, bending energy, torsional energy, nonbonded interactions, variation of σ- and π-bond energy. The chair-like path was found to be more favorable, in agreement with experimental results. Also the increased ease of reaction in *cis*-1,2-divinylcyclopropane and *cis*-1,2-divinylcyclobutane has been rationalized. The results for hexa-1,5-diene have been almost exactly reproduced by an all electron molecular orbital calculation.[23] It has been pointed out[24] that other pathways in which a plane of symmetry is not present in the transition state might be energetically competitive. A different situation is present when the diallylic system is favorably preoriented, as in cyclopentadiene dimers:

$$\tag{4}$$

A kinetic study of the Cope rearrangement of some tricyclo(5.2.1.0^{2,6})deca-4,8-dienols[25] has led to values for the activation energy not sensibly higher than those found for open-chain dienes, suggesting that a relief of steric strain in the reaction compensates for the high energy of the boat-like transition state, while the activation entropy was found slightly positive, in sharp contrast with the values for the open-chain systems and in agreement with predictions obtained by means of statistical mechanic methods. This data plus the geometry of the cyclopentadiene dimer nucleus obtained by an x-ray investigation of the crystal and molecular structure of tricyclo(5.2.1.0^{2,6})deca-4,8-dienyl-*p*-bromobenzoate,[26] allowed the allocation of the transition state in a small region of the potential energy surface for the reaction. The geometry and electronic structure of the transition state were then calculated by means of the CNDO/2 method.[27]

The cooperation of the kinetic and molecular orbital approach again was a major item in the elucidation of the mechanism of the Fritsch rearrangement[28]:

$$\begin{matrix} Ar \\ \\ Ar' \end{matrix} C{=}C \begin{matrix} X \\ \\ H \end{matrix} \xrightarrow{\text{EtO}^-} Ar{-}C{\equiv}C{-}Ar' \quad (X = Cl, Br) \qquad (5)$$

Under the conditions necessary to obtain the rearrangement, the direct substitution reaction may also occur:

$$\begin{matrix} Ar \\ \\ Ar' \end{matrix} C{=}C \begin{matrix} X \\ \\ H \end{matrix} + B^- \longrightarrow \begin{matrix} Ar \\ \\ Ar' \end{matrix} C{=}C \begin{matrix} B \\ \\ H \end{matrix} + X^- \qquad (6)$$

For a given reagent, different substrates can show a range of results that go from 100% rearrangement to 100% substitution. A model for the transition state was assumed, as shown in Fig. 2-1. The influence of one OCH$_3$ substituent in the *p* position of one phenyl group is predicted to increase the rate

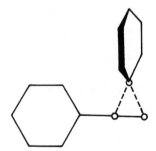

Fig. 2-1. Transition-state model for the Fritsch rearrangement.

of the rearrangement, with the OCH$_3$ group more effective when in the phenyl trans to the halogen than when in the phenyl cis to the halogen. These predictions are fully confirmed by experiment.

VI. NUCLEOPHILIC AROMATIC SUBSTITUTION

Nucleophilic aromatic substitution has been the subject of numerous experimental and theoretical investigations.[29] Energy profiles for these reactions obtained by different theoretical approaches have shown, in agreement with kinetics results, that in most cases the reaction goes in two steps, and a stable intermediate with the well-known Wheland structure has been postulated (anionic σ complex). In a few examples such intermediates could be isolated and the crystal structure of one of them, namely the picryl ether adduct with ethoxide, has been determined.[30] The geometry found for this compound agrees completely with that suggested by the Wheland model. The carbon to which the two alkoxy groups are bonded has paraffinic character, as shown by bond lengths and bond angles afferent to it. A quinoid structure is evident from the fact that the C—N distance for the nitro-group in para position is significantly shorter than the same distance for the two ortho nitro-groups. The two ortho groups are coplanar with the ring, at variance with the situation in the parent compound, 2,4,6-trinitrophenetole, where the two ortho nitro-groups are rotated, with respect to the ring plane, by 32° and 61°.[31] This fact suggests that steric factors as well as electronic factors contribute to the stability of the adduct and to the variation of energy along the reaction path. Since the Wheland model fits so well with the intermediate, it cannot be retained as a model for the transition state and an alternative representation is needed for it.[32] The leaving and attacking groups are considered together as a "pseudo-atom." The orbitals of these groups are combined to give a "quasi-σ" bonding orbital, symmetric with respect to the plane of the ring, and a "quasi-π" antibonding orbital, antisymmetric with respect to the same plane. The first orbital can combine with the sp^2 orbital of the carbon atom site of the substitution, and the second one with its p_z orbital. This description allows for a large degree of freedom in the activated complex and is compatible with a synchronous one-stage as well as with a two-stage mechanism. The model has been applied in the calculation of activation energies in nucleophilic aromatic substitution in benzene and naphthalene derivatives[33] and in the related aza compounds.[34] The agreement with experimental activation energies and free energies is very satisfactory.

VII. ELECTROPHILIC AND NUCLEOPHILIC SUBSTITUTION AT SATURATED CARBON

The simplest model of a transition state for electrophilic substitution at alkanes is the CH_5^+ ion. Three configurations have been suggested for this ion, which have D_{3h}, C_{4v}, C_s symmetry. A calculation according to the CNDO/2 method has shown the configuration with C_s symmetry to be the most stable.[35] This finding has been substantiated by the results of other calculations carried out with different techniques.[36] It turns out also that rotation of the CH_2 group around the axis through the carbon atom and perpendicular to the plane of the other three hydrogen atoms obtains at no energy expense, so that entropy factors as well as the energetic contribution act in favor of the mechanism of electrophilic substitution with configuration retention. A potential energy surface calculation has shown that the approach of H_2 along the ternary axis of CH_3^+ leads to a stable configuration with no activation energy along the path, while the formation of a C_{2v} ion (of which the configuration of symmetry C_{4v} and D_{3h} are limiting forms) requires a high activation energy.[37]

On the other hand all the calculations performed for the CH_5^- ion have given the D_{3h} configuration as the most stable one, suggesting configuration inversion in S_{N2} reactions. As an example let us consider the reaction

$$CH_3F + F^- \longrightarrow FCH_3 + F^- \tag{7}$$

A CNDO/2 calculation of the energy along the reaction path predicts the transition state to be more stable than the reactants.[38] However, if a sufficient number of water molecules is reasonably arranged around the reacting molecules and included in the calculation, an activation energy of 10 to 20 kcal/mole is calculated. This is because solvation stabilizes the transition state much less than the reactants. This is perhaps the first example of a calculation in which solvent effect on a reaction path is explicitly included, on a molecular basis. This opens the field of the quantum-mechanical study of reactions in solution.

References

1. R. D. Brown, in *Molecular Orbitals in Chemistry, Physics and Biology*, P. O. Lowdin and B. Pullman, Eds., Academic, New York, 1964, p. 495.
2. K. Fukui, in *Molecular Orbitals in Chemistry, Physics and Biology*, Ref. 1, p. 513.
3. M. J. S. Dewar, *Adv. Chem. Phys.* **8**, 65 (1965).
4. H. H. Greenwood and R. McWeeny, *Adv. Phys. Org. Chem.* **4**, 73 (1966).
5. R. Zahradnik, Colloques Internationaux du CNRS, Menton, July 1970.
6. M. Simonetta, Twenty-third International Congress of Pure and Applied Chemistry, Butterworths, London, 1971, Vol. I, p. 127.
7. I. Shavitt, M. R. Stevens, F. L. Minn, and M. Karplus, *J. Chem. Phys.* **48**, 2700 (1968).
8. E. Gianinetti, G. F. Majorino, E. Rusconi, and M. Simonetta, *Int. J. Quantum Chem.* **3**, 45 (1969).
9. M. Karplus, R. N. Porter, and R. D. Sharma, *J. Chem. Phys.* **43**, 3259 (1965).
10. Pl. A. Plattner, E. Heilbronner, and S. Weber, *Helv. Chim. Acta* **35**, 1036 (1952).
11. E. Heilbronner and M. Simonetta, *Helv. Chim. Acta* **35**, 1049 (1952).
12. M. Simonetta and E. Heilbronner, *Theor. Chim. Acta* **2**, 228 (1964).

13. S. Winstein, *Bull. Soc. Chim.* **18**, 55C (1951).
14. See, e.g., M. Simonetta, S. Winstein, *J. Am. Chem. Soc.* **76**, 18 (1954).
15. M. J. S. Dewar, *The Molecular Orbital Theory of Organic Chemistry*, McGraw-Hill, New York, 1969, p. 361.
16. R. Hoffmann, *J. Chem. Phys.* **39**, 1397 (1963).
17. J. A. Pople and G. A. Segal, *J. Chem. Phys.* **44**, 3289 (1966).
18. A. Gavezzotti and M. Simonetta, unpublished results.
19. R. Hoffmann, C. C. Wan, and V. Neager, *Mol. Phys.* **19**, 113 (1970).
20. G. J. Gleicher and P. von Schleyer, *J. Am. Chem. Soc.* **89**, 582 (1967).
21. R. C. Bingham and P. von Schleyer, *J. Am. Chem. Soc.* **93**, 3189 (1971).
22. M. Simonetta, G. Favini, C. Mariani, and P. Gramaccioni, *J. Am. Chem. Soc.* **90**, 1280 (1968).
23. A. Brown, M. J. S. Dewar, W. Schoeller, *J. Am. Chem. Soc.* **92**, 5516 (1970).
24. L. Salem, *Acc. Chem. Res.* **4**, 322 (1971).
25. I. R. Bellobono, P. Beltrame, M. G. Cattania, and M. Simonetta, *Tetrahedron* **26**, 4407 (1970).
26. I. R. Bellobono, R. Destro, C. M. Gramaccioli, and M. Simonetta, *J. Chem. Soc.* (B) 710 (1969).
27. P. Beltrame, A. Gamba, and M. Simonetta, *Chem. Comm.* 1660 (1970).
28. M. Simonetta and S. Carrà, *Tetrahedron* **19**, Suppl. **22**, 467 (1963).
29. For a review, see, e.g., M. J. Strauss, *Chem. Rev.* **70**, 667 (1970).
30. R. Destro, C. M. Gramaccioli, and M. Simonetta, *Acta Cryst.* **24B** 1369 (1968).
31. C. M. Gramaccioli, R. Destro, and M. Simonetta, *Acta Cryst.* **24B**, 129 (1968).
32. M. Simonetta and S. Carrà, in *Nitro Compounds*, Pergamon, London, 1963, p. 383.
33. S. Carrà, M. Raimondi, and M. Simonetta, *Tetrahedron* **22**, 2673 (1966).
34. P. Beltrame, P. L. Beltrame, and M. Simonetta, *Tetrahedron* **24**, 3043 (1968).
35. A. Gamba, G. Morosi, and M. Simonetta, *Chem. Phys. Letters* **3**, 20 (1969).
36. See, e.g., W. A. Lathan, W. J. Hehre, and J. A. Pople, *J. Am. Chem. Soc.* **93**, 808 (1971).
37. H. F. Guest, J. N. Murrel, and J. B. Padley, *Mol. Phys.* **20**, 81 (1971).
38. A. Gamba, P. Cremaschi, and M. Simonetta, *Theor. Chim. Acta*, **25**, 237 (1972).

CHAPTER 3

Intermolecular Interactions and Chemical Reactivity

H. Fujimoto and K. Fukui
Faculty of Engineering
Kyoto University
Kyoto, Japan

I. INTRODUCTION

Chemical reactions have long been an attractive subject of quantum chemistry. Especially, molecular orbital (MO) theory has experienced a brilliant success in interpreting and predicting the orientation and the stereoselection of a large number of chemical reactions. The fundamental character of the various sorts of chemical reactions, intermolecular and intramolecular, have been discussed in terms of MOs. Chemical reactivity indices for π and σ electronic systems which have been developed based upon several reaction models seem now to be in common use by organic chemists.[1-5] The Woodward-Hoffmann selection rules for pericyclic reactions have disclosed the basic principles underlying apparently different kinds of reactions and have stimulated new experiments on the prediction of the rules.[6-9] Recent progress in the high-speed computer has supplied us with several semiempirical MO methods, which can conveniently be applied to complicated chemically interacting systems with tolerable accuracy in a chemical sense.[10-18] The paths of some reactions have been studied using semiempirical MO methods.[19-26] *Ab initio* MO calculations have also been attempted on some chemical interactions between sizable molecules.[27-31] On the other hand, some extensive studies on the modes of chemical interaction have been developed by the use of MOs which are obtained with respect to the reactant and reagent in their isolated states.[32-39] Application of these methods of representing chemically interacting systems is in general limited to the early stage of reaction, since the energy of interacting systems is usually expanded in power series of overlap integrals, interaction integrals, and so on. However, these approaches are expected to be able to give detailed information on the mechanisms of chemical reactions. In the following sections, we will make a brief survey of the chemical reactivity indices and the stereoselection rules, which have been established on the basis of simplified reaction models, in relation to more comprehensive studies of chemical reactions. We will not discuss developments in the theories of long-range forces the main interest of which are not directed toward chemical reactions. The variational calculations on reaction intermediates or activated complexes will also mostly be disregarded here.

A. Chemical Reactivity Indices

The chemical reactivity indices that have been proposed by the use of MOs are conveniently divided into three groups: the static, the localization, and the delocalization approaches.[2]

In the early stage of calculations on π electrons of planar conjugated systems, the π electron densities were regarded as a measure of chemical reactivity toward ionic reactions.[40-42] In the case of alternant hydrocarbons, the π electron densities of every carbon atom calculated by the simple Hückel MO method are unity.[43] The chemical reactivity index, self-atompolarizability, was proposed on the basis of the perturbation theory.[44] For the purpose of discussing the chemical reactivity toward radical reagents, the chemical reactivity index, free valence, was put forth.[45] These indices are classified as belonging to the static approaches.

The localization energy is a measure of the change in the π electron energy of a system by deleting the atom under attack by a reagent from the remainder of the π conjugated part.[46] The position of the smallest unstablization, obtained by putting 0, 1, and 2 electrons in the deleted atom, is the most reactive toward an attacking nucleophile, radical, and electrophile, respectively. Dewar's reactivity number is an approximation of the localization energy for alternant hydrocarbons, which can be obtained by much simpler calculations.[47] These indices belong to the localization approaches.

In 1952, the new reactivity index, frontier electron density, was proposed.[48] It was first pointed out that electrophilic aromatic substitutions took place most dominantly at the position where the partial electron density in the highest occupied (HO) MO was the largest. Afterwards, it was stated that the position of the largest partial electron density in the lowest unoccupied (LU) MO was usually the most favorable position of nucleophilic aromatic substitutions.[49] These particular MOs were termed frontier orbitals and the partial densities in these MOs were called frontier electron densities. In homolytic reactions both HOMO and LUMO serve as frontier orbitals. Assuming hyperconjugation between the aromatic π electron system of a molecule and the pseudo π orbital composed of the atomic orbital (AO) of the attacking reagent and the AO of hydrogen to be replaced, at the transition state of aromatic substitution, the chemical reactivity index, superdelocalizability, was derived.[50,51] While frontier electron density is employed for an intramolecular comparison of reactive positions of a given molecule, superdelocalizability can be used for the purpose of comparing the relative reactivities of different molecules for a given reaction. The chemical reactivity indices, frontier electron density, and superdelocalizability are regarded as the delocalization model of chemical interaction, together with Brown's Z value[52] and with the ion free valence by Nagakura and Tanaka.[53]

With the progress of MO methods for σ electrons, some theoretical treatments of the chemical reactivity of saturated compounds have been developed. Delocalizability was defined in order to discuss the reactivities of hydrogens in paraffin hydrocarbons and their derivatives.[3,54] Frontier electron density was also used as a reactivity measure of hydrogens in some halogenated hydrocarbons.[55-57] The relative ease of thermal cleavage of carbon–carbon bonds of alkanes has been elucidated by means of σ MO perturbation calculations.[2,58]

B. Stereoselection Rules

Woodward and Hoffmann derived significant selection rules of electrocyclic reactions, sigmatropic reactions, and concerted cycloaddition reactions, by an ingenious use of correlation diagrams.[6-9] The correlation of the state functions in electrocyclic reactions was discussed by Longuet-Higgins and

Abrahamson.[59] The discovery of Woodward-Hoffmann selection rules have attracted much attention from organic chemists and have quickly been applied to numerous cases.[60-62]

The chemical reactivity indices, free valence, and localization energy were applied to Diels-Alder reactions.[63,64] In 1964, the importance of nodal properties of the MOs of reactant and reagent in concerted multicentric processes was first pointed out by Fukui.[2] Diels-Alder additions of dienes and dienophiles were discussed by the use of perturbation MO method. The conjugation stabilization energy due to the concerted interaction between the two termini of diene and dienophile is given by[65-67]

$$\Delta E = 2 \sum_{i}^{occ} \sum_{l}^{uno} \frac{(c_t^{(i)}c_u^{(l)}\gamma_{tu} + c_{t'}^{(i)}c_{u'}^{(l)}\gamma_{t'u'})^2}{\varepsilon_{ai} - \varepsilon_{bl}} + 2 \sum_{k}^{occ} \sum_{j}^{uno} \frac{(c_t^{(j)}c_u^{(k)}\gamma_{tu} + c_{t'}^{(j)}c_{u'}^{(k)}\gamma_{t'u'})^2}{\varepsilon_{bk} - \varepsilon_{aj}} \quad (1)$$

where $c_t^{(i)}$, $c_{t'}^{(i)}$ are the coefficients of the AOs t, t' in the LCAO MO a_i of diene and $c_u^{(1)}$, $c_{u'}^{(1)}$ are those of the AOs u, u' in the MO b_l of dienophile, ε_{ai} and ε_{bl} are the MO energies and γ_{tu} and $\gamma_{t'u'}$ stand for the interaction integrals, as is illustrated in Figure 3-1. The first term on the right-hand side of Eq. (1) means the summation over all the combinations of the occupied MOs a_i of diene and the unoccupied MOs b_l of dienophile and the second term covers all the combinations of the unoccupied MOs a_j of diene and the occupied MOs b_k of dienophile. Here it should be noted that the interaction is still not too strong, thus still satisfying the condition that the numerators of Eq. (1) are small. Of the terms appearing in Eq. (1), the interaction between HOMO of diene and

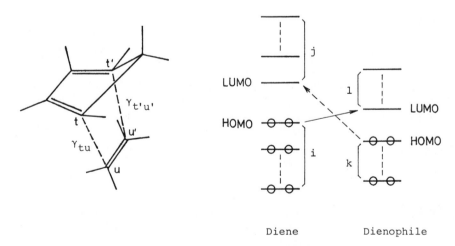

Fig. 3-1. Mode of orbital interaction in Diels-Alder reactions.

LUMO of dienophile usually plays a discriminative role. Then Eq. (1) can be approximated by the following one-term expression:

$$\Delta E \cong 2 \frac{(c_t^{(\text{HO})} c_u^{(\text{LU})} \gamma_{tu} + c_{t'}^{(\text{HO})} c_{u'}^{(\text{LU})} \gamma_{t'u'})^2}{\varepsilon_{a\text{HO}} - \varepsilon_{b\text{LU}}} \tag{2}$$

It was shown that the products of the AO coefficients of both reaction centers, $c_t^{(\text{HO})} c_u^{(\text{LU})}$ and $c_{t'}^{(\text{HO})} c_{u'}^{(\text{LU})}$, did have the same sign with respect to a number of combinations of diene and dienophile, making the concerted process favorable.[2]

The stereoselection rules for thermal pericyclic reactions were found to be governed by a single basic principle in which the number of electrons participating in the formation of the cycle had a key importance.[6-9,67-74] The selection rules for photochemical interactions seem to be not necessarily defined in a unique manner, although they are usually contrary to thermal ones.[6-9,67-74] Existence of singly occupied (SO) MOs makes it difficult to derive general selection rules.

The concept of stereoselections, represented by the Woodward-Hoffmann selection rules, has been applied to other interesting systems. Hoffmann and Woodward extended the basic concept of Eq. (2) to discuss the exo-endo selectivity in Diels-Alder reactions, regarding the interaction as three-centric with respect to both diene and dienophile.[7,8] Herndon and Hall pointed out the possible difference in the interaction integrals in the exo-endo processes.[75]

The importance of the secondary attractive force in Diels-Alder addition of butadiene, which was different from that proposed by Hoffmann and Woodward, was pointed out by Dewar.[71] Salem also disclosed from numerical calculations that the same kind of secondary interaction as was proposed by Dewar was operative in the dimerization of butadiene and of acrolein.[35] Sustmann and Binsch investigated the exo-endo selectivity in the interaction of dienes and dienophiles which had no extra π electrons available for the secondary interaction.[39] Goldstein obtained a selection rule for bicycloaromaticity.[76] Spironconjugation was discussed by Simmons and Fukunaga and by Hoffmann, Imamura, and Zeiss.[77,78] Mango and Schachtschneider showed how the unstabilization due to the thermal interaction between two olefinic double bonds was removed through the electron-donating and the electron-accepting interactions with the d orbitals of transition-metal catalysts.[79-81] While they took only the π electrons of olefins into account, the potential significance of the σ electrons of the carbon-carbon bonds was also pointed out.[5,82] These interactions can also be discussed by means of Eqs. (1) and (2).[5]

Although the selection rules are of profound significance in elucidating the basic principle of chemical interactions, they are qualitative in their nature. Some semiempirical and *ab initio* MO calculations have been carried out on electrocyclic reactions of conjugated systems.[83-88] Valence bond (VB) calculation of the photochemical ring-closing reaction of butadiene was also carried out by van der Lugt and Oosterhoff.[89]

II. DEVELOPMENT OF MO THEORIES OF CHEMICAL REACTIONS

As has been discussed, several chemical reactivity indices have been proposed using MOs, each depending upon an assumed reaction model. In some fortunate cases, like the substitution reactions of alternant hydrocarbons, any of the reactivity indices mentioned before was found to give satisfactory results. In some other unfortunate cases, some of them failed to give correct prediction. In recent years, several theories were presented which tried to treat chemical reactions more comprehensively.[33-39] In these, the chemical reactivity is not expressed by a single term, but by a combined sum of more than two different terms. Klopman discussed the interaction between an electrophile and a nucleophile, taking the two effects into account, neighboring effect and electron-transfer effect.[34] The interaction energy was represented by the sum of the three energy terms

$$\Delta E = -\Delta q_t \Delta q_u \frac{\Gamma}{\varepsilon} + \Delta solv + \sum_i^{occ} \sum_l^{uno} \frac{2(c_t^{(i)})^2 (c_u^{(l)})^2 \gamma_{tu}^2}{\varepsilon_{ai}^* - \varepsilon_{bl}^*} \tag{3}$$

where Δq_t is the total initial charge of the atom t, Γ is the Coulomb repulsion term between atoms t and u, ε is the local dielectric constant of the solvent, ε_{ai}^* is the energy of the ith occupied MO of the nucleophile under the influence of the electrophile and the possible effect of solvent, and ε_{bl}^* is associated with the lth unoccupied MO of the electrophile. The first term on the right-hand side of Eq. (3) stands for the Coulombic interaction between the charges of nucleophile and electrophile, the second term is the change in solvation energy, and the last represents the stabilization due to electron-transfer interaction. In the case of interactions between a hard base (with low ε_{ai}^* value) and a hard acid (with high ε_{bl}^* value), the first term dominates, while the third term plays the most dominant role in the interaction between a soft base (with high ε_{ai}^* value) and a soft acid (with low ε_{bl}^* value). Klopman called the former case "charge controlled" and the latter case "frontier controlled." He elegantly interpreted the changes in the position of additions and substitutions according to the differences in the softness and/or the hardness of attacking reagents.

In deriving Eq. (1), the overlap integrals between two interacting AOs were neglected. Including the overlap integrals explicitly, Salem obtained a new equation for the ground-state interaction of two closed-shell systems, A and B:

$$\Delta E = - \sum_t^A \sum_u^B (q_t + q_u)\gamma_{tu}s_{tu}$$
$$- 2 \sum_i^{occ} \sum_l^{uno} \left[\frac{(\sum_{t,u} c_t^{(i)} c_u^{(l)} \gamma_{tu})^2}{\varepsilon_{bl} - \varepsilon_{ai}} + \tfrac{1}{4}(\varepsilon_{bl} - \varepsilon_{ai})\left(\sum_{t,u} c_t^{(i)} c_u^{(l)} s_{tu} \right)^2 \right]$$
$$- 2 \sum_k^{occ} \sum_j^{uno} \left[\frac{(\sum_{t,u} c_t^{(j)} c_u^{(k)} \gamma_{tu})^2}{\varepsilon_{aj} - \varepsilon_{bk}} + \tfrac{1}{4}(\varepsilon_{aj} - \varepsilon_{bk})\left(\sum_{t,u} c_t^{(j)} c_u^{(k)} s_{tu}^- \right)^2 \right] \quad (4)$$

where q_t is the charge density of the AO t and s_{tu} is the overlap integral between the two AOs t of A and u of B. Salem estimated the second term in the brackets. This term was smaller than the first term in the brackets, having the same form as Eq. (1). The first term on the right-hand side of Eq. (4) behaves repulsively in the interaction and Salem ascribed this to the penetration of electrons into the "exclusion core." Applying the theory to Diels-Alder reactions of several dienes and dienophiles, Salem proved the important role of the HOMO-LUMO interaction.[35] An equation similar to Eq. (4) was also derived for the photochemical interactions of two conjugated systems. Devaquet and Salem tried to represent the energy of the combined system of two reactants applying the perturbation theory in the Fock-Roothaan SCF MO formalism[90] and introduced an additional energy term, $\sum_{t,u} \Delta q_t \Delta q_u / R_{tu}$, representing the Coulombic interaction between the net charges of two reactants.[37,38] They calculated the interaction energy of the dimerization of acrolein and found that the Coulombic interaction energy was important in determining the favorable reaction path. Although Devaquet and Salem put an approximation in constructing the matrix elements of the combined system of two reactants, Sustmann and Binsch got the interaction energy between two closed-shell systems by an iterative determination of the Fock perturbed matrix elements of the combined system.[39] Murrell, Randić, and Williams expanded the interaction energy between two systems in the power series of interaction potential and overlap integral, dividing it into Coulomb, exchange, polarization, exchange polarization and charge-transfer interaction energies.[32] The theory has been applied to rather static systems like hydrogen bonds, charge-transfer complexes, and so on.[91–93]

In order to clarify the basic principles governing chemical reactions, the interaction energy between two reactants was represented by the combined sum of four energies, Coulomb, exchange, delocalization, and polarization interactions, of which chemically graspable formulas were derived using LCAO SCF MOs of reactant and reagent.[36] When the theory is applied to π electron systems, the Coulomb interaction energy is related to the reactivity measure "total π electron density," the delocalization interaction energy to "super-delocalizability" and the polarization interaction energy to "self-atompolarizability." The exchange interaction energy emerges from the antisymmetrization of the wave functions of two reactants obtained for each of them in an isolated state. The examination of the magnitudes of these energy terms suggested the increasing importance of the delocalization interaction in stabilizing chemically interacting systems with the progress of the reactions. The delocalization interaction is also important as the major origin of bond interchange, that is, the formation of new bonds and the disappearance of old bonds, in chemical reactions.[94,95] In the following, we will discuss how the chemical interaction is represented by the MOs of two reactants in some details, with the aid of some illustrative examples.

A. Interaction Between Two Systems

Let us represent the wave function of the system composed of two independent systems A and B, interacting with each other, by a linear combination of the wave functions of adiabatic interaction, charge-transfer interaction and polarization interaction:

$$\Psi = C_0\Psi_0 + \sum_p C_p\Psi_p \qquad (5)$$

We assume that both of the two reactants A and B have closed-shell electronic structures. Although it is desirable to include as many electron con-

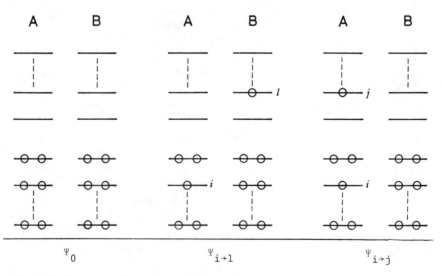

Fig. 3-2. Electron configurations of the combined system A-B.

figurations as possible, this is rather a tedious chore and, hence, we may here take only "one-electron transferred" and "one-electron excited" electron configurations with respect to the original state. These are schematically illustrated in Figure 3-2. The contributions of other electron configurations are small in the early stage of chemical interaction.[36] The wave functions of these states are represented by the Slater determinants composed of spin orbitals of A and B. The state of the adiabatic interaction is given simply by a single determinantal wave function

$$\Psi_0 = \mathcal{NA} |a_1 \bar{a}_1 a_2 \bar{a}_2 \cdots a_m \bar{a}_m b_1 \bar{b}_1 b_2 \bar{b}_2 \cdots b_n \bar{b}_n| \tag{6}$$

where $\mathcal{NA}$ is the normalization-antisymmetrization operator, barred and nonbarred MOs are the spin orbitals with spin functions α and β, respectively. Similarly, the wave function of a singlet one-electron transferred state is given by

$$\Psi_{i \to l} = (1/\sqrt{2})\{\mathcal{NA} |a_1 \bar{a}_1 \cdots a_i \cdots a_m \bar{a}_m b_1 \bar{b}_1 \cdots b_n \bar{b}_n \bar{b}_l| \\ - \mathcal{NA} |a_1 \bar{a}_1 \cdots \bar{a}_i \cdots a_m \bar{a}_m b_1 \bar{b}_1 \cdots b_n \bar{b}_n b_l|\} \tag{7}$$

Here, the spatial parts of the one electron wave functions, a_is and b_ks are assumed to be represented in LCAO forms:

$$a_i(1) = \sum_t c_t^{(i)} t(1) \quad \text{for system } A$$

and

$$b_k(1) = \sum_u c_u^{(k)} u(1) \quad \text{for system } B$$

The MOs a_i are orthonormalized with respect to the isolated system A. The MOs b_k also form an orthonormal set in an isolated state of B. The MOs a_i and the MOs b_k are not orthogonal in general, except for the case of infinite separation. We can take all the functions as real.

The Hamiltonian operator of the interacting system of A and B in the frame of the Born-Oppenheimer approximation is given by (in atomic units)[96]

$$H = \sum_\lambda \left(-\tfrac{1}{2}\Delta(\lambda) - \sum_\alpha^A \frac{Z_\alpha}{r_{\lambda\alpha}} - \sum_\beta^B \frac{Z_\beta}{r_{\lambda\beta}} \right) + \sum_{\lambda < \lambda'} \frac{1}{r_{\lambda\lambda'}} + \sum_{\gamma < \gamma'} \frac{Z_\gamma Z_{\gamma'}}{R_{\gamma\gamma'}} \tag{8}$$

where Δ is the Laplacian operator, Z_α and Z_β are the positive charges of the nuclei of the atoms α and β, belonging to the systems A and B, respectively, $r_{\lambda\alpha}$ is the distance between the electron λ and the nucleus α, $r_{\lambda\lambda'}$ is the separation between the two electrons λ and λ', and $R_{\gamma\gamma'}$ is the distance between the two nuclei γ and γ'.

The lowest energy of the interacting system can be obtained by solving the secular equation

$$
\begin{vmatrix}
H_{0,0} - E & H_{0,p} - S_{0,p}E & \cdots \\
H_{p,0} - S_{p,0}E & H_{p,p} - E & \cdots \\
\cdots\cdots\cdots\cdots\cdots\cdots\cdots\cdots
\end{vmatrix} = 0 \tag{9}
$$

where

$$
H_{p,p'} = \int \Psi_p^* H \Psi_{p'} \, d\tau, \qquad S_{p,p'} = \int \Psi_p^* \Psi_{p'} \, d\tau
$$

The interaction energy may be defined as the difference between the ground-state energy of the combined system and the sum of the energies of A and B in their isolated states,

$$
\Delta W = W - (W_{A0} + W_{B0}) \tag{10}
$$

If the interaction is assumed to be not yet strong, the interaction energy can be expanded in the following form:

$$
\Delta W \cong (H_{0,0} - W_{A0} - W_{B0}) - D - \Pi \tag{11}
$$

where

$$
D \cong \sum_i^{occ} \sum_l^{uno} \frac{|H_{0,i\to l} - S_{0,i\to l}H_{0,0}|^2}{H_{i\to l,i\to l} - H_{0,0}} + \sum_k^{occ} \sum_j^{uno} \frac{|H_{0,k\to j} - S_{0,k\to j}H_{0,0}|^2}{H_{k\to j,k\to j} - H_{0,0}} \tag{12}
$$

and

$$
\Pi \cong \sum_i^{occ} \sum_j^{uno} \frac{|H_{0,i\to j} - S_{0,i\to j}H_{0,0}|^2}{H_{i\to j,i\to j} - H_{0,0}} + \sum_k^{occ} \sum_l^{uno} \frac{|H_{0,k\to l} - S_{0,k\to l}H_{0,0}|^2}{H_{k\to l,k\to l} - H_{0,0}} \tag{13}
$$

We call the D and Π energy terms, the "delocalization interaction energy" and the "polarization interaction energy," respectively. The first-order interaction energy $(H_{0,0} - W_{A0} - W_{B0})$ may be divided into the two terms, the Coulomb interaction energy and the exchange interaction energy

$$
\varepsilon_Q \cong 2\sum_i^{occ} V_{Bii} + 2\sum_k^{occ} V_{Akk} + 4\sum_i^{occ}\sum_k^{occ} (ii|kk) + \sum_\alpha^A \sum_\beta^B \frac{Z_\alpha Z_\beta}{R_{\alpha\beta}} \tag{14}
$$

and

$$
\varepsilon_K \cong -2\sum_i^{occ}\sum_k^{occ}\left[(ik|ki) + S_{ik}V_{ik} + \sum_{i'}^{occ} S_{ik}\{2(ik|i'i') - (ii'|i'k)\} \right.
$$

$$
\left. + \sum_{k'}^{occ} S_{ik}\{2(ik|k'k') - (ik'|k'k)\} \right] \tag{15}
$$

where

$$V_{Bii} = \int a_i(1)\left(-\sum_{\beta}^{B}\frac{Z_\beta}{r_{1\beta}}\right)a_i(1)\,dv(1)$$

$$V_{ik} = \int a_i(1)\left(-\sum_{\alpha}^{A}\frac{Z_\alpha}{r_{1\alpha}} - \sum_{\beta}^{B}\frac{Z_\beta}{r_{1\beta}}\right)b_k(1)\,dv(1)$$

$$S_{ik} = \int a_i(1)b_k(1)\,dv(1)$$

$$(ii|kk) = \int\int a_i(1)b_k(2)\frac{1}{r_{12}}a_i(1)b_k(2)\,dv(1)\,dv(2)$$

When we employ the point-charge approximation for nuclear attractions and electron repulsions, then

$$\int t(1)\frac{-Z_\beta}{r_{1\beta}}t(1)\,dv(1) = -\frac{Z_\beta}{R_{\alpha\beta}} \qquad (t \text{ belongs to } \alpha)$$

$$\int\int t(1)u(2)\frac{1}{r_{12}}t(1)u(2)\,dv(1)\,dv(2) = \frac{1}{R_{\alpha\beta}} \qquad (u \text{ belongs to } \beta)$$

the Coulomb interaction energy is given in a chemically graspable form, by the aid of the Mulliken approximation[97]:

$$\varepsilon_Q \cong \sum_{\alpha}^{A}\sum_{\beta}^{B}\frac{(Z_\alpha - N_\alpha)(Z_\beta - N_\beta)}{R_{\alpha\beta}} \tag{16}$$

where N_α is the gross atomic population of the atom α[98]:

$$N_\alpha = 2\sum_{i}^{occ}\sum_{t}^{\alpha}\sum_{t'}^{A}c_t^{(i)}c_{t'}^{(i)}s_{tt'}$$

The exchange interaction energy represents a repulsive term in the usual chemical reactions of two closed-shell systems.

Neglecting exchange terms, the numerator of the delocalization interaction energy is given by

$$H_{0,i\to l} - S_{0,i\to l}H_{0,0} \cong \sqrt{2}\left[V_{Bil} + 2\sum_{k}^{occ}(il|kk) - S_{il}\left\{V_{Bii} + 2\sum_{k}^{occ}(ii|kk)\right\}\right] \tag{17}$$

Here it was assumed that the MOs satisfy the Fock equation for the isolated systems. When the interaction takes place through the orbital overlapping of paired AOs (t, u) (t', u') and so on, we have

$$H_{0,i\to l} - S_{0,i\to l}H_{0,0} \cong \sqrt{2}\sum_{t,u}c_t^{(i)}c_u^{(l)}\gamma_{tu}^{(i)} \tag{18}$$

where

$$\gamma_{tu}^{(i)} \cong -\int t(1)\left(\sum_{\beta}^{B}\frac{Z_\beta - N_\beta}{r_{1\beta}}\right)u(1)\,dv(1) + S_{tu}\sum_{\alpha}^{A}\sum_{\beta}^{B}\frac{n_\alpha^{(ii)}(Z_\beta - N_\beta)}{R_{\alpha\beta}} \qquad (19)$$

$$n_\alpha^{(ii)} = \sum_{t}^{\alpha}\sum_{t'}^{A} c_t^{(i)}c_{t'}^{(i)}s_{tt'}$$

Equation (18) has a form similar to the numerator of Eqs. (1), (3), and (4) derived from simpler calculations. We see that the interaction integral γ depends on the net charges of the atoms of the electron accepting molecule.

The denominator of the energy term D is given by

$$H_{i\to l,i\to l} - H_{0,0} \cong -(E_{Bl} + \Delta E_{Bl}) + (I_{Ai} + \Delta I_{Ai}) + a_{il} \qquad (20)$$

where

$$E_{Bl} = -\left[H_{Bll} + \sum_{k}^{occ}\{2(ll|kk) - (lk|kl)\}\right]$$

$$\Delta E_{Bl} = -\left\{V_{All} + 2\sum_{i'}^{occ}(ll|i'i')\right\} + \sum_{i'}^{occ}\{V_{li'}S_{li'} + (li'|i'l)\}$$

$$I_{Ai} = -\left[H_{Aii} + \sum_{i'}^{occ}\{2(ii|i'i') - (ii'|i'i)\}\right]$$

$$\Delta I_{Ai} = -\left\{V_{Bii} + 2\sum_{k}^{occ}(ii|kk)\right\} + \sum_{k}^{occ}\{V_{ik}S_{ik} + (ik|ki)\}$$

$$a_{il} = -(ii|ll) + 2\{V_{il}S_{il} + (il|li)\}$$

$$H_{Aii} = \int a_i(1)\left(-\tfrac{1}{2}\Delta(1) - \sum_{\alpha}^{A}\frac{Z_\alpha}{r_{l\alpha}}\right)a_i(1)\,dv(1)$$

Therefore, the denominator is not simply given by the difference between the energy of the MO a_i and that of the MO b_l of B as in the case of a simple Hückel calculation. Klopman took the effect of opponent reactant upon the MO energies in the interaction, still neglecting differential overlaps.[34] I_{Ai} is the vertical ionization potential of the electron in the MO a_i and ΔI_{Ai} is the change in the I_{Ai} value due to the interaction with B.[99] Similarly, E_{Bl} is the vertical electron affinity of the MO b_l of B and ΔE_{Bl} is the change in the E_{Bl} value due to the interaction with A, When both A and B are neutral nonpolar molecules, the absolute values of ΔI_{Ai} and ΔE_{Bl} may be relatively small in comparison with the absolute value of a_{il}, making $(H_{i\to l,i\to l} - H_{0,0})$ smaller than $(I_{Ai} - E_{Bl})$. By taking explicitly the electron repulsions, nuclear attractions, and overlap integrals into account, we see that the charge-transfer interaction between the HOMO of A and the LUMO of B is relatively more important over other terms than expected simply from $(I_{Ai} - E_{Bl})$. The same can be said for the charge-transfer from B to A. When one of the two, say A, is an anion, then $-\Delta E_{Bl}$ will have a large positive value which can barely be compensated by adding a_{ij}. However, in this case I_{Ai} is small in comparison with the cases of neutral A, making the interaction between the HOMO of A and the LUMO of B dominant over all other charge-transfer terms. In the case of the interaction between a neutral system A and a cationic system B, E_{Bl} will have a large positive value and the interaction between the HOMO of A and the LUMO of B will be important. In some cases, the energy difference between the electron transferred state from the HOMO of one reactant into the LUMO of another, $\Psi'_{HO \to LU}$, and the original state, Ψ'_0, is so small that the second-order perturbation expression like Eq. (12) is no more valid. Then it may be more appropriate to write the energy term D as

$$D \cong \sqrt{2}\,|H_{0,HO \to LU} - S_{0,HO \to LU}H_{0,0}| \tag{21}$$

When one of the charge-transferred states, say c, happens to have a lower energy than the original state, Eq. (11) should be replaced by[36,52,53]

$$\Delta W = H_{c,c} - (W_{A0} + W_{B0}) - \frac{|H_{0,c} - S_{0,c}H_{c,c}|^2}{H_{0,0} - H_{c,c}} + \cdots \tag{22}$$

The numerator of the polarization energy Π is given by

$$H_{0,i \to j} - S_{0,i \to j}H_{0,0} \cong \sqrt{2}\left\{ V_{Bij} + 2\sum_{k}^{occ}(ij|kk) \right\} \tag{23}$$

where

$$V_{Bij} = \int a_i(1)\left(-\sum_{\beta}^{B}\frac{Z_\beta}{r_{1\beta}} \right)a_j(1)\,dv(1)$$

The denominator of Π is written as

$$H_{i \to j,i \to j} - H_{0,0} \cong -(E_{Aj} + \Delta E_{Aj}) + (I_{Ai} + \Delta I_{Ai}) + a_{ij} \tag{24}$$

where

$$a_{ij} = -(ii|jj) + 2(ij|ji)$$

The overlap integral $S_{0,i \to j}$ is the second order of the overlap integrals between the MOs of A and B. The denominators of Π cannot usually be so small as those of D in the majority of chemical reactions. Therefore, the contribution of Π to the stabilization of the interacting system may become less important than that of D as chemical interaction gets strong through the orbital overlappings of the MOs of A and B. In addition to D and Π, the dispersion energy may be important in the interaction between two completely nonpolar systems.[32,91] However, in most chemical reactions, one or both of the two reactants have local dipoles. Thus the dispersion energy may tentatively be neglected here.

B. Application to Chemically Reacting Systems

As has been discussed above, the interaction energy between two interacting systems can be divided into several terms, in an approximate sense. In some cases, one or two terms are dominant and others can be neglected. It is one of the characteristic features of ionic reactions that the Coulomb interaction energy plays a dominant role in the early stage of interaction. On the other hand, the contribution of the delocalization interaction energy to the stabilization becomes increasingly important with the progress of the reaction, both in polar and in nonpolar cases. It may be suitable for a better understanding of the important role of the delocalization interaction to look at some numerical results. In Table 3-1 are shown the results of a simple Hückel MO perturbation calculation for some two-centric interactions, using Eq. (1) with $|\gamma_{tu}| = |\gamma_{t'u'}|$.[67] Although the contributions of the HOMO-LUMO interactions are underestimated in Eq. (1) in comparison with the estimate of Eq. (12), the values may still serve as measures for the purpose of discussing the favorable reaction paths. In two centric cases we may assume, in principle, two modes of interactions. One is the case in which γ_{tu} and $\gamma_{t'u'}$ have the same sign, and another is the case in which γ_{tu} and $\gamma_{t'u'}$ have different signs. We may call the

Table 3-1. The Favorable Mode of Cyclic Interactions

Reagent (u, u')	Reactant (t, t')	$\Delta E(\beta/\gamma^2)$ [a] syn	anti
Ethylene 1,2	Ethylene 1,2	0	2.000
	Allyl anion 1,3	2.414	0.414
	Butadiene 1,2	0.106	1.895
	Butadiene 1,4	1.789	0.422
	Hexatriene 1,6	0.622	1.657
	Benzene 1,2	0.333	1.444
	Benzene 1,4	1.333	0.444
	Anthracene 1,2	0.250	1.604
	Anthracene 1,4	1.475	0.424
	Anthracene 9,10	1.616	0.404
Butadiene 1,4	Butadiene 1,2	1.744	0.492
	Butadiene 1,4	0.716	1.789
	Naphthalene 1,2	1.454	0.573
	Naphthalene 1,4	0.658	1.427
Allyl cation 1,4	Ethylene 1,2	0.414	2.414
	Butadiene 1,4	2.789	0.789
Allyl anion 1,3	Butadiene 1,4	0.789	2.789

[a] $\gamma = |\gamma_{tu}| = |\gamma_{t'u'}|$, β: resonance integral of C=C double bond.

former case "syn" interaction and the latter "anti" interaction.[67,70,74] Here it is assumed that all the π AOs have the same sign on the same side of the molecular plane of each conjugated system. The modes of syn and anti interactions are schematically illustrated in Figure 3-3. The interaction integral γ is negative when the AOs overlap mainly with the lobes of the same

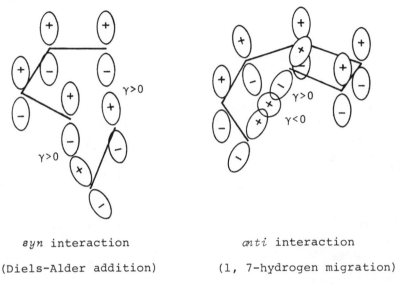

syn interaction

(Diels-Alder addition)

anti interaction

(1, 7-hydrogen migration)

Fig. 3-3. A schematic illustration of syn and anti interactions.

sign, while it is positive when the two AOs interact in opposite phase. Dewar classified the two modes of interactions as Hückel and anti-Hückel interactions in his PMO theory.[71,100] Zimmerman used the terms Hückel and Möbius in order to distinguish the two types of interactions.[72,73] Table 3-1 shows that the thermal interaction between ethylene and butadiene 1,4 prefers to take place through syn interaction while interaction between two ethylenes favours anti interaction. These examples correspond to the [4s + 2s] and [2s + 2a] cases of Woodward and Hoffmann.[8] In general, thermal syn interaction is favored when the number of electrons forming a cycle is $4n + 2$, and thermal anti interaction is facilitated when it is $4n$. The conclusion is consistent with those of Woodward and Hoffmann,[6-9] Dewar,[71,100] Zimmerman,[72,73] Salem,[35] and Fukui and Fujimoto.[70] Woodward and Hoffmann derived the selection rules, considering the correlations of the MOs by the symmetries conserved throughout reactions.[8] The reason why the symmetries of MOs of a reacting system are retained was discussed by Fukui in the formulation of reaction coordinate.[101] We can also regard a two-centric unimolecular reaction as if it were a cyclic intermolecular interaction between two parts of a molecule appropriately partitioned.[67-69,74,102]

Next let us consider single-centric reactions. When the attacking reagents is a single orbital species with the energy α, we have the reactivity index, superdelocalizability, from Eq. (1) by multiplying the energy ΔE by β/γ^2 to have dimensionless quantity. When the reagent orbital is unoccupied,

$$S_t^{(E)} = 2 \sum_i^{occ} \frac{(c_t^{(i)})^2}{\lambda_i} \quad (\varepsilon_i = \alpha + \lambda_i \beta)$$

When the reagent orbital is occupied,

$$S_t^{(N)} = 2 \sum_j^{uno} \frac{(c_t^{(j)})^2}{-\lambda_j} \quad (\varepsilon_j = \alpha + \lambda_j \beta)$$

Since the SOMO of the reagent can interact with both the occupied and unoccupied MOs of the reactant, superdelocalizability for radical attack is defined by

$$S_t^{(R)} = \sum_i^{occ} \frac{(c_t^{(i)})^2}{\lambda_i} + \sum_j^{uno} \frac{(c_t^{(j)})^2}{-\lambda_j}$$

When the one-term approximation like Eq. (2) is valid, the chemical reactivity index, frontier electron density, can be used as a measure of the relative reactivity of the various positions of the molecule

$$f_t^{(E)} = 2(c_t^{(HO)})^2$$

$$f_t^{(N)} = 2(c_t^{(LU)})^2$$

$$f_t^{(R)} = (c_t^{(HO)})^2 + (c_t^{(LU)})^2$$

These indices were first defined in the framework of the simple Hückel MO method.[48-51] When we use the MOs which are normalized including overlap integrals, the partial population of the AO t in the ith MO consists of two parts,

$$n_t^{(i)} = p_t^{(i)} + v_t^{(i)}$$

$$p_t^{(i)} = 2(c_t^{(i)})^2$$

$$v_t^{(i)} = 2 \sum_{\substack{t' \\ (\neq t)}} c_t^{(i)} c_{t'}^{(i)} s_{tt'}$$

where $p_t^{(i)}$ is the valence-inactive part and $v_t^{(i)}$ is the valence-active part.[103] Equations (12) and (18) indicate that it is appropriate to take the valence-inactive part which feels free from intramolecular bonding as a measure of chemical reactivity.[104]

The rate-determining step of the base catalyzed rearrangement of olefins has been shown to be the abstraction of hydrogen by base.[105]

$$
\begin{array}{c}
\ce{>C=C-CH<} + \text{B} \underset{}{\overset{\text{slow}}{\rightleftharpoons}} \ce{>C=C-\bar{C}<} + \text{BH}^+ \\
\Updownarrow \\
\ce{>CH-C=C<} + \text{B} \underset{}{\overset{\text{slow}}{\rightleftharpoons}} \ce{>\bar{C}-C=C<} + \text{BH}^+
\end{array}
$$

Table 3-2 gives the frontier electron densities of allylic hydrogens of olefins toward a nucleophilic attack.[106,107] The extended Hückel MO calculation gives almost the same energy value for LUMOs of a series of olefins. While the atomic population of the hydrogens fail to give the correct order, we find a parallel between the frontier electron densities and the observed reaction rates.

The nucleophilic bimolecular elimination reactions of substituted hydrocarbons are known to occur stereoselectively with the preferential abstraction

Table 3-2. Frontier Electron Densities of Hydrogens and Rate Constants of Isomerization Reaction.

R	CH_3	C_2H_5	$i\text{-}C_3H_7$	$t\text{-}C_4H_9$	$CH_2=CH$
		$CH_2=CHCH_2R$			
k (sec^{-1})[105]	8.4×10^{-4}	4.8×10^{-4}	1.5×10^{-5}	6.0×10^{-6}	3.6×10^{-3}
$p_t^{(LU)}$	0.0371	0.0365	0.0361	0.0353	0.0765
$v_t^{(LU)}$	-0.0135	-0.0133	-0.0133	-0.0130	-0.0282
N_t	0.873	0.872	0.870	0.869	0.854

R	H	CH_3	C_2H_5	$i\text{-}C_3H_7$
		$CH_2=CHCH(CH_3)R$		
k (sec^{-1})[105]	8.4×10^{-4}	2.0×10^{-4}	1.1×10^{-4}	3.0×10^{-5}
$p_t^{(LU)}$	0.0371	0.0324	0.0321	0.0312
$v_t^{(LU)}$	-0.0135	-0.0120	-0.0119	-0.0115
N_t	0.873	0.890	0.889	0.887

$CH_2=CHCH_2R$ $\qquad$ $CH_2=CHCH(CH_3)R$

of *trans-β*-hydrogen.[108] In Fig. 3-4 are given the frontier electron densities of hydrogens for nucleophilic attack.[55] We see that the β-hydrogens which are located trans to chlorine are predicted to be the most reactive in both ethyl chloride and 2-chlorobutane. In 2-*exo*-chloronorbornane, the *exo*-hydrogen attached to the carbon 3 has the largest frontier electron density value. The E2 reaction of 2-*exo*-bromonorbornane was shown to take place predominantly in cis fashion to yield norbornene.[109] In these single-centric reactions, the stereoselection or spatial direction of reagent attack is governed by the inequality of the chemical reactivities of the same sort of atoms at different

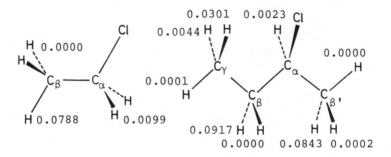

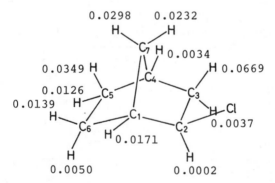

Fig. 3-4. Frontier electron densities of hydrogens toward nucleophilic reagents.

spatial positions. Another kind of inequality of spatial direction of reagent approach will play an important role in the homolytic halogenation of norbornane. The chlorination of norbornane was reported to yield 2-*exo*-chloronorbornane as the major product.[110] The first step of this reaction may be hydrogen abstraction by a radical. The extended Hückel MO calculation on 2-norbornyl radical showed that the trivalent carbon 2 did not have the sp^2 planar structure, having the hydrogen attached to the carbon 2 bent toward the endo direction.[111] Therefore, as is illustrated in Figure 3-5, the

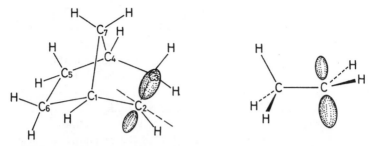

Fig. 3-5. Orbital extension in 2-norbornyl radical and in ethyl radical.

unpaired electron orbital would have larger extension to the exo direction than to the endo direction, by the mixing of the $2s$ and the $2p$ AOs of that carbon atom. Although the coefficient of this hybrid in SOMO has a definite value, the absolute value of the interaction integral γ may be larger for the exo approach of chlorine than for the endo approach. Such a deformation of the radical center of 2-norbornyl from planar structure was recently observed.[112] Inequality in the spatial direction of orbital extension may also have a significance in determining the stereoselectivity of addition of hydrogen halides, halogens, and so on, to olefins. *Ab initio* calculation on the ethyl radical suggested strongly the potentiality of such a directing effect.[113]

C. Isolated Molecule Approximation

In the above discussion, the interaction energy ΔW was discussed in the framework of no configuration change. That is, the interaction was considered, neglecting the destabilization due to the geometrical changes of both reagent and reactant from the most stable equilibrium nuclear configurations. Destabilization due to the molecular deformations of reagent and reactant as well as solvation and desolvation energy in liquid-phase reactions may unquestionably account for a part of the activation energy. Moreover, the use of the method discussed here is limited to the cases of weak interaction, owing to the approximations employed. We have assumed that the reaction path which yields the largest stabilization of the interaction in the early stage of a reaction is the path of the lowest activation energy. In other words, our discussion presented here is based on the assumption of the "noncrossing" of potential curves. Hence, it must be examined whether such an approximation is really sound or not.

In the simple Hückel MO theory, an atom is usually bonding with the neighboring atoms in occupied MOs and antibonding in unoccupied MOs.[114] That is, we have (for hydrocarbons)

$$\sum_{t'}^{\text{nei}} c_t^{(i)} c_{t'}^{(i)} = \frac{\alpha - \varepsilon_i}{-\beta}(c_t^{(i)})^2 = \lambda_i (c_t^{(i)})^2 \tag{25}$$

where $\sum_{t'}^{\text{nei}}$ means the summation over all the neighboring atoms t' of the atom t. As usual, the energies of occupied MOs are lower than α and $\sum_{t'}^{\text{nei}} c_t^{(i)} c_{t'}^{(i)}$ is positive for any t in bonding levels. On the contrary, the energies of the unoccupied MOs are higher than α, in general, making $\sum_{t'}^{\text{nei}} c_t^{(j)} c_{t'}^{(j)}$ negative for any t. Equation (25) indicates that the sum of the partial bond order(s) with the neighboring atom(s) is parallel to the partial electron density of the central atom. The following electron configurations contribute to the interactions. In electrophilic reactions, an electron is removed from occupied MOs (most dominantly from HOMO) of the reactant. In nucleophilic reactions, an electron is transferred into unoccupied MOs (most dominantly into LUMO). Both the charge-transfer from occupied MOs and that into unoccupied MOs bring about the decrease in the sum of bond order(s) of an atom with the neighboring atom(s). This will cause the isolation of the atom under the attack of the reagent from the remaining part of the reactant. The decrease in bond order reflects the stretching of the bond, which will result in the decrease in the absolute value of the resonance integral. The loosening of a bond will cause the unstabilization of MOs which are bonding in that bond region, and the stabilization of MOs which are antibonding in that bond region. When the reagent is not a single atom, the same may be said for the reagent part. The charge-transfer interaction between HOMO of the reagent and LUMO of the reactant and that between LUMO of the reagent and HOMO of the reactant will bring about the elevation of HOMOs and the lowering of LUMOs, resulting in the narrowing of the interfrontier level separations. By this, the HOMO-LUMO interactions become more and more dominant over other terms. The change in nuclear framework in the case of charge-transfer interaction which is expected from the nodal properties of the frontier orbitals usually conforms to the molecular deformation which really takes place in the reaction. For instance, we may consider the addition of a reagent to butadiene. The most reactive position of butadiene is 1 or 4, both for electrophilic and for nucleophilic reagents as shown in Fig. 3-6. HOMO of butadiene is bonding in 1,2- and 3,4-π bonds and antibonding in 2,3-π bond. LUMO is antibonding in 1,2- and 3,4-π bonds and bonding in 2,3-π bond. Charge transfer from the reagent into LUMO of butadiene and

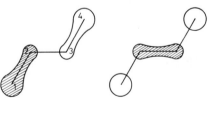

HOMO   LUMO

Fig. 3-6. Nodal properties of frontier orbitals of butadiene (hatched and nonhatched areas indicate the plus and minus parts of real MOs).

that into the reagent from HOMO of butadiene through the overlap between the reagent orbital and the π AO of a terminal carbon of butadiene, say 1, tend to isolate the carbon atom from the remainder of the molecule. With this, the HOMO is elevated and the LUMO is lowered as shown in Fig. 3-7. At the same time, the 2,3-π bond is strengthened, while the 3,4-π bond is weakened. This implies that the remainder part of butadiene by deleting the carbon 1 gets allylic properties. In many cases, the frontier electron density at the reaction center becomes larger as it is isolated from the remainder of the

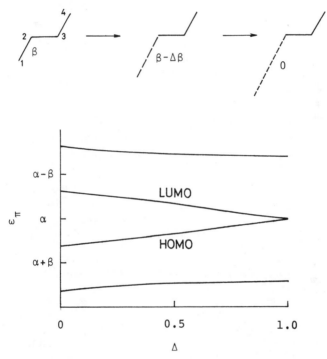

Fig. 3-7. Changes in MO energies of butadiene due to the stretching of C_1—C_2 bond.

molecule. In the example mentioned above, frontier electron density of the carbon 1 increases from 0.724 of butadiene to 1.000 of an isolated π AO, as shown in Fig. 3-8, since both the single π AO of the carbon 1 and the nonbonding MO of the allyl have the same energy $\varepsilon = \alpha$. Even in the cases in which the remainder part of a reactant by deleting an AO has no nonbonding level, the growing of the frontier electron density with the progress of the reaction seems to be a general tendency of chemical reactions. Thus, the cooperation of three principles of chemical reactions mentioned above:

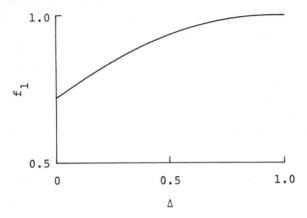

Fig. 3-8. Change in frontier electron density of the carbon 1 of butadiene due to the stretching of C_1—C_2 bond.

a positional parallelism between the charge-transfer and bond interchange, the narrowing of interfrontier level separation with bond interchange, and the growing frontier electron density along the reaction path, may not only supply the reason why HOMO and LUMO can predict the most reactive positions but also give a reasonable basis why we could extend the comparison of reaction paths in the early stage of chemical reaction to the neighborhood of the transition state. The self-acceleration of chemical interaction in this manner might be one of the basic natures of chemical reactions.

D. Charge-Transfer Interactions and Intermolecular Bond Formation

Charge-transfer interactions are responsible for the formation of new bonds and disappearance of old bonds in the reactant and the reagent. The nodal properties of HOMO and LUMO are important in predicting what change in molecular shape will take place in chemical reactions. Such a concept was also applied to the metal-catalyzed disproportionation of olefins.[79] Another problem to be considered here concerns the origin of the intermolecular bond formation.

The Coulomb interaction may be responsible for purely ionic bonds between reactant and reagent. The increase in electron densities in intermolecular region originates from the overlapping of the occupied MOs (particularly HOMO) of reactant and the unoccupied MOs (particularly LUMO) of reagent and the converse.[94,95] We consider here, as an example, the interaction between two two-electron systems A and B, A having a doubly occupied MO a_i and B having a doubly occupied MO b_k and an unoccupied

MO b_l. Neglecting the polarization interaction, the wave function is approximated by

$$\Psi = C_0\Psi_0 + C_{i\to l}\Psi_{i\to l} \tag{26}$$

The electron density is given by

$$\rho = C_0{}^2\rho_{0,0} + C_0C_{i\to l}(\rho_{0,i\to l} + \rho_{i\to l,0}) + C_{i\to l}^2\rho_{i\to l,i\to l} \tag{27}$$

where

$$\rho_{p,p'} = 4\int\int\int\int \Psi_p^*(1, 2, 3, 4)\Psi_{p'}(1, 2, 3, 4)\, d\tau_1\, d\tau_2\, d\tau_3\, d\tau_4$$

The electron density in the Ψ_0 state is given by

$$\rho_{0,0} = (\rho_{A0} + \rho_{B0}) + \frac{2(a_i{}^2 + b_k{}^2)S_{ik}{}^2 - 4a_ib_kS_{ik}}{(1 - S_{ik}{}^2)} \tag{28}$$

where ρ_{A0} and ρ_{B0} give the electron densities of A and B, respectively, in their isolated states. The second term on the right-hand side of Eq. (28) implies the distortion of the electron cloud due to the interference of the wave functions of A and B. Here we have

$$-\int a_i(1)b_k(1)S_{ik}\, dv(1) = -S_{ik}{}^2$$

This means that the adiabatic interaction between A and B cannot be the origin of intermolecular bond formation, because the changes in electron densities in intermolecular region due to electron exchange sum up to $\sim -4S_{ik}{}^2$.[94,95,115] The same can be said from MO population analysis.[116,117]

Introducing the charge-transfer state $\Psi_{i\to l}$, we have

$$\rho_{0,i\to l} \cong \sqrt{2}(\rho_{A0} + \rho_{B0})S_{il} + \sqrt{2}(a_ib_l - a_i{}^2S_{il} + b_kb_lS_{ik}) \tag{29}$$

and

$$\begin{aligned}\rho_{i\to l,i\to l} \cong\ & \rho_{A0} + \rho_{B0} - a_i{}^2 + b_l{}^2 \\ & + \{a_i{}^2(S_{ik}{}^2 - S_{il}{}^2) + b_k{}^2S_{ik}{}^2 - b_l{}^2S_{il}{}^2 \\ & - 2a_ib_kS_{ik} + 2a_ib_lS_{il} - b_kb_lS_{ik}S_{il} - b_lb_kS_{ik}S_{il}\} \end{aligned} \tag{30}$$

The integral of the second term on the right-hand side of Eq. (29) over all the space vanishes. The first term in the parentheses clearly serves as an origin of increase in electron densities in the intermolecular region between A and B, because we have

$$\int a_i(1)b_l(1)\, dv(1) = \sum_t^A \sum_u^B c_t^{(i)}c_u^{(l)}s_{tu} = S_{il}$$

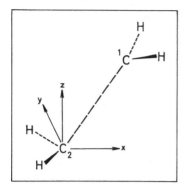 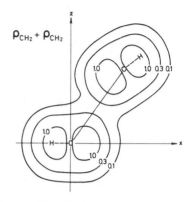

Fig. 3-9. Electron densities of two singlet methylenes without interaction in a non-least motion approach. The figures show the distribution of valence electrons in x–z cross section at $y = 0$ ($e^-/Å^3$).

and the product of the coefficients C_0 and $C_{i \to l}$ have the same sign as that of S_{il}, making $C_0 C_{i \to l} S_{il}$ positive for the lowest-energy state. The second term in the parentheses indicates that the electrons which participate in the intermolecular bond formation are supplied from the occupied MO a_i of electron donor A. The last term represents the intramolecular electron rearrangement in B due to the interaction with A.

We show the results of a semiempirical MO calculation on the changes in electron densities in the interaction of two singlet methylenes.[95] Dimerization of methylenes was first discussed by Hoffmann, Gleiter, and Mallory[21] and later by Kollmar,[118] by Sustmann and Binsch,[39] and by Basch[119] mainly from an energetic viewpoint. In Fig. 3-9 is shown the simple sum of electron densities of two singlet methylenes located 2.5 Å apart from each other in a non-least-motion approach. Figure 3-10 gives the electron densities in the Ψ'_0

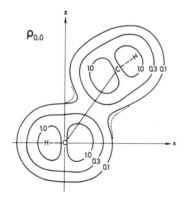

Fig. 3-10. Electron densities of two singlet methylenes without charge-transfer interactions in a non-least-motion approach.

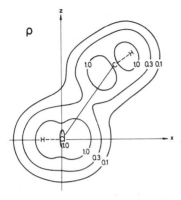

Fig. 3-11. Electron densities of two singlet methylenes with charge-transfer interactions in a non-least-motion approach.

state. We see that the electron densities in the region between the carbon atoms of two methylenes decrease by the electron exchange interaction in that state. By including charge-transferred states, we have the electron densities shown in Fig. 3-11. Considerable increase in electron densities in the region between the two carbons is observed. Figure 3-12 gives the electron densities of the points on the line connecting two carbon atoms. The electron densities of the least-motion approach of two singlet methylenes are shown in Fig. 3-13. No effective bond formation is achieved, since the HOMO-LUMO interaction is symmetry unfavorable in this process.

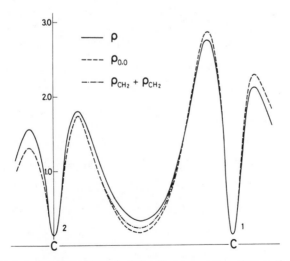

Fig. 3-12. Electron densities of two singlet methylenes at the points on the line connecting two carbons in a non-least-motion approach.

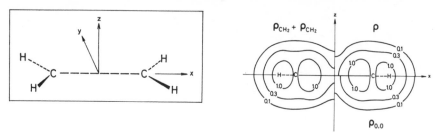

Fig. 3-13. Electron densities of two singlet methylenes in a least-motion approach.

E. Application to Other Systems

In the above discussion, we have limited ourselves to the interactions between two closed-shell systems. Some modifications are needed for the systems in which radicals and excited molecules take part. Figure 3-14 shows the mode of orbital interaction between a radical and a closed-shell molecule. SOMO of radical interacts both with HOMO and with LUMO of the closed-shell partner. The discussion mentioned in the previous sections for the interaction between two closed-shell systems almost holds for the doublet interaction between a radical and a closed-shell system.[120] In the case of photochemical reactions, there appear two SOMOs. The mode of interaction is illustrated in Fig. 3-15.[68,69,74,121] Comparing Figs. 3-1 and 3-15, we may understand why the stereoselectivity in photochemical reactions is opposite to that of thermal reactions. Photochemical additions of carbonyls to olefins were studied by Herndon and Giles, using the perturbation method in the framework of the simple Hückel MO theory.[122] Salem and Devaquet studied photochemical

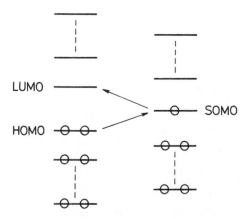

Fig. 3-14. Mode of orbital interaction between a closed-shell system and a radical.

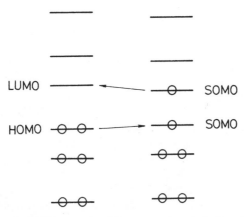

Fig. 3-15. Mode of orbital interaction between a closed-shell system and an excited open-shell system.

dimerization of dienes and keto steroids.[35,37] Detailed treatments of photo-induced reactions now seem to be somewhat difficult, because the nuclear configurations and the electronic structures of photoexcited molecules are not necessarily easy to find and the excited singlet MO wave functions cannot be obtained as simply as those of the ground state and excited triplet state. In the interaction between two radicals, the exchange interaction is responsible for the multiplet separation and may contribute to the stabilization of singlet interaction.[36] Charge-transfer between SOMOs may still be important in this case.

III. CONCLUSION

The methods of discussing chemical reactivity of molecular systems in terms of MOs in isolated states have been presented. In order to make the comparison of chemical reactivities distinct and quantitative, the interaction energy between two systems has been given by the combined sum of several energy terms. In some reactions, one of them will be important and in some other reactions, another term will dominate. In the early stage of ionic reactions, the Coulomb interaction will be the major source of stabilization. In nonionic reactions of nonpolar molecules, the reaction path may be governed by the exchange interaction energy and the delocalization interaction energy. Even in ionic cases, the role of the delocalization stabilization will be enhanced with the progress of reactions, as the net charges will be neutralized through charge-transfer interactions.[123,124] Charge-transfer interaction is also important as the major origin of bond interchange accompanying chemical reactions. Orientation in single-centric reactions and stereoselection in multicentric reactions have been discussed in a unified manner, using the second-order perturbation equation of the delocalization interaction energy.

A chemical reactivity index is a theoretical measure for chemical reactivities of molecules which is obtained by the employment of the most simple quantum chemical tool applicable for chemical reactions, namely, MO methods. The chemical reactivity index, calculated from the MOs of isolated reactants, is correlated with reaction path, under the assumption of the non-crossing of potential energy curves for more than two different modes of interactions. Although it is not easy to verify the correctness of such an assumption, we have discussed about one possible reason why we could recognize the pertinency of the isolated molecule approximation.

Before finishing our discussion, we may point out the following rule for the "frontier controlled" reactions:

The majority of chemical reactions are liable to take place at the position and in the direction where the overlapping of HOMO and LUMO of respective reactants is at its maximum; in an electron-donating species, HOMO predominates in the overlapping interaction, while LUMO does so in an electron-accepting reactant; in the reacting species which possess SOMOs, these play the part of HOMO or of LUMO or of both.

These particular MOs may be termed the generalized frontier orbitals.

References

1. A. Streitwieser, Jr., *Molecular Orbital Theory for Organic Chemists*, Wiley, New York, 1961, pp. 307–356.
2. K. Fukui, in *Molecular Orbitals in Chemistry, Physics, and Biology*, P.-O. Löwdin and B. Pullman, Eds., Academic, New York, 1964.
3. K. Fukui, in *Modern Quantum Chemistry, Istanbul Lectures, Part I*, O. Sinanoğlu, Ed., Academic, New York, 1965.
4. K. Fukui, *Sigma Molecular Orbital Theory*, O. Sinanoğlu and K. B. Wiberg, Eds., Yale Univ. Press, New Haven, Conn., 1970.
5. K. Fukui, Fortschr. *Chem. Forsch.* **15**, 1 (1970).
6. R. B. Woodward and R. Hoffmann, *J. Am. Chem. Soc.* **87**, 395, 2511 (1965).
7. R. Hoffmann and R. B. Woodward, *J. Am. Chem. Soc.* **87**, 2046, 4388, 4389 (1965).
8. R. B. Woodward and R. Hoffmann, *Angew. Chem. Int. Ed. Engl.* **8**, 781 (1969); *The Conservation of Orbital Symmetry*, Academic, New York, 1969.
9. R. Hoffmann and R. B. Woodward, *Acct. Chem. Res.* **1**, 17 (1969).
10. R. Hoffmann, *J. Chem. Phys.* **39**, 1397 (1963); **40**, 2047, 2474, 2480 (1964).
11. J. A. Pople, D. P. Santry, and G. A. Segal, *J. Chem. Phys.* **43**, s129 (1965).

12. J. A. Pople and D. L. Beveridge, *Approximate Molecular Orbital Theory*, McGraw-Hill, New York, 1970.
13. J. A. Pople, D. L. Beveridge, and P. A. Dobosh, *J. Chem. Phys.* **47**, 2026 (1967).
14. M. J. S. Dewar and G. Klopman, *J. Am. Chem. Soc.* **89**, 3089 (1967).
15. N. C. Baird and M. J. S. Dewar, *J. Chem. Phys.* **50**, 1262 (1969).
16. N. C. Baird, M. J. S. Dewar, and R. Sustmann, *J. Chem. Phys.* **50**, 1275 (1969).
17. T. Yonezawa, K. Yamaguchi, and H. Kato, *Bull. Chem. Soc. Jap.* **40**, 536 (1967).
18. H. Kato, H. Konishi, H. Yamabe, and T. Yonezawa, *Bull. Chem. Soc. Jap.* **40**, 2761 (1967).
19. H. Kato, K. Morokuma, T. Yonezawa, and K. Fukui, *Bull. Chem. Soc. Jap.* **38**, 1749 (1965).
20. R. Hoffmann, *J. Am. Chem. Soc.* **90**, 1475 (1968).
21. R. Hoffmann, R. Gleiter, and F. B. Mallory, *J. Am. Chem. Soc.* **92**, 1460 (1970).
22. R. Hoffmann, S. Swaminathan, B. G. Odell, and R. Gleiter, *J. Am. Chem. Soc.* **92**, 7091 (1970).
23. R. Hoffmann, C. C. Wan, and V. Neagu, *Mol. Phys.* **19**, 113 (1970).
24. M. J. S. Dewar, E. Haselbach, and M. Shanshal, *J. Am. Chem. Soc.* **92**, 3505 (1970).
25. J. R. Hoyland, *Theor. Chim. Acta* **22**, 229 (1971).
26. M. J. S. Dewar and W. W. Schoeller, *J. Am. Chem. Soc.* **93**, 1481 (1971).
27. E. Clementi, *J. Chem. Phys.* **46**, 3851 (1967).
28. W. Th. A. M. van der Lugt and P. Ros, *Chem. Phys. Lett.* **4**, 389 (1969).
29. G. Berthier, D.-J. David, and A. Veillard, *Theor. Chim. Acta* **14**, 329 (1969).
30. A. Dedieu and A. Veillard, *Chem. Phys. Lett.* **5**, 328 (1970).
31. A. J. Duke and R. F. W. Bader, *Chem. Phys. Lett.* **10**, 631 (1971).
32. J. N. Murrell, M. Randić, and D. R. Williams, *Proc. Roy. Soc. Lond.* Ser. A **284**, 566 (1965).
33. G. Klopman and R. F. Hudson, *Theor. Chim. Acta* **8**, 165 (1967).
34. G. Klopman, *J. Am. Chem. Soc.* **90**, 223 (1968).
35. L. Salem, *J. Am. Chem. Soc.* **90**, 543, 553 (1968).
36. K. Fukui and H. Fujimoto, *Bull. Chem. Soc. Jap.* **41**, 1989 (1968).
37. A. Devaquet and L. Salem, *J. Am. Chem. Soc.* **91**, 3793 (1969).
38. A. Devaquet, *Mol. Phys.* **18**, 233 (1970).
39. R. Sustmann and G. Binsch, *Mol. Phys.* **20**, 1, 9 (1971).
40. T. Ri and H. Eyring, *J. Chem. Phys.* **8**, 433 (1940).
41. A. Pullman and B. Pullman, *Experientia* **2**, 364 (1946).
42. M. J. S. Dewar, *Trans. Faraday Soc.* **42**, 764 (1946).
43. E. Hückel, *Z. Physik* **70**, 204 (1931).
44. C. A. Coulson and H. C. Longuet-Higgins, *Proc. Roy. Soc. Lond.*, Ser. A **191**, 39 (1947); **192**, 16 (1947).
45. C. A. Coulson, *Discussions Faraday Soc.* **2**, 9 (1947).

46. G. W. Wheland, *J. Am. Chem. Soc.* **64**, 900 (1942).
47. M. J. S. Dewar, *J. Am. Chem. Soc.* **74**, 3357 (1952).
48. K. Fukui, T. Yonezawa, and H. Shingu, *J. Chem. Phys.* **20**, 722 (1952).
49. K. Fukui, T. Yonezawa, C. Nagata, and H. Shingu, *J. Chem. Phys.* **22**, 1433 (1954).
50. K. Fukui, T. Yonezawa, and C. Nagata, *Bull. Chem. Soc. Jap.* **27**, 423 (1954).
51. K. Fukui, T. Yonezawa, and C. Nagata, *J. Chem. Phys.* **26**, 831 (1957).
52. R. D. Brown, *J. Chem. Soc.* 2232 (1959).
53. S. Nagakura and J. Tanaka, *J. Chem. Soc. Jap., Pure Chem. Sec.* **75**, 993 (1954).
54. K. Fukui, H. Kato, and T. Yonezawa, *Bull. Chem. Soc. Jap.*, **34**, 1111 (1961).
55. K. Fukui and H. Fujimoto, *Tetrahedron Lett.* 4303 (1965).
56. K. Fukui, H. Hao, and H. Fujimoto, *Bull. Chem. Soc. Jap.* **42**, 348 (1969).
57. H. Fujimoto, S. Yambe, and K. Fukui, *Bull. Chem. Soc. Jap.* **44**, 971 (1971).
58. K. Fueki and K. Hirota, *J. Chem. Soc. Jap., Pure Chem. Sec.* **81**, 209 (1960).
59. H. C. Longuet-Higgins and E. W. Abrahamson, *J. Am. Chem. Soc.* **87**, 2045 (1965).
60. S. I. Miller, in *Advances in Physical Organic Chemistry*, V. Gold, Ed., Academic, New York, 1968, Vol. 6, pp. 185–332.
61. N. T. Anh, *Les Regles de Woodward-Hoffmann*, Ediscience, Paris, 1970.
62. G. B. Gill, *Quart. Rev.* **22**, 338 (1968).
63. R. D. Brown, *J. Chem. Soc.* 691, 2730 (1950).
64. R. D. Brown, *J. Chem. Soc.* 691, 2730 (1950).
65. K. Fukui, C. Nagata, T. Yonezawa, H. Kato, and K. Morokuma, *J. Chem. Phys.* **31**, 287 (1959).
66. K. Fukui, K. Morokuma, T. Yonezawa, and C. Nagata, *Bull. Chem. Soc. Jap.* **33**, 963 (1960).
67. K. Fukui and H. Fujimoto, *Bull. Chem. Soc. Jap.* **39**, 2116 (1966).
68. K. Fukui, *Tetrahedron Lett.* 2009 (1965).
69. K. Fukui, *Bull. Chem. Soc. Jap.* **39**, 498 (1966).
70. K. Fukui and H. Fujimoto, *Bull. Chem. Soc. Jap.* **40**, 2018 (1967).
71. M. J. S. Dewar, *Tetrahedron Suppl.* **8**, Part I, 75 (1966).
72. H. E. Zimmerman, *J. Am. Chem. Soc.* **88**, 1564, 1566 (1966).
73. H. E. Zimmerman, *Acct. Chem. Res.* **4**, 272 (1971).
74. K. Fukui and H. Fujimoto, *Mechanisms of Molecular Migrations*, B. S. Thyagarajan, Ed., Interscience, New York, 1969, Vol. 2, pp. 117–190.
75. W. C. Herndon and L. H. Hall, *Tetrahedron Lett.* 3095 (1967).
76. M. J. Goldstein, *J. Am. Chem. Soc.* **89**, 6357 (1967).
77. H. E. Simmons and T. Fukunaga, *J. Am. Chem. Soc.* **89**, 5208 (1967).
78. R. Hoffmann, A. Imamura, and G. D. Zeiss, *J. Am. Chem. Soc.* **89**, 5215 (1967).
79. F. D. Mango and J. H. Schachtschneider, *J. Am. Chem. Soc.* **89**, 2484 (1967); **91**, 1030 (1969); **93**, 1123 (1971).
80. F. D. Mango, *Tetrahedron Lett.* 505 (1971).

81. W. Th. A. M. van der Lugt, *Tetrahedron Lett.* 2281 (1970).
82. G. S. Lewandos and R. Pettit, *Tetrahedron Lett.* 789 (1971).
83. W. Kutzelnigg, *Tetrahedron Lett.* 4965 (1967).
84. D. T. Clark and G. Smale, *Tetrahedron Lett.* 3673 (1968).
85. D. T. Clark and G. Smale, *Tetrahedron* 25, 13 (1969).
86. D. T. Clark and D. R. Armstrong, *Theor. Chim. Acta* 13, 365 (1969); 14, 370 (1969).
87. K. Hsu, R. J. Buenker, and S. D. Peyerimhoff, *J. Am. Chem. Soc.* 93, 2117 (1971).
88. M. J. S. Dewar and S. Kirschner, *J. Am. Chem. Soc.* 93, 4290, 4291, 4292 (1971).
89. W. Th. A. M. van der Lugt and L. J. Oosterhoff, *J. Am. Chem. Soc.* 91, 6042 (1969).
90. C. C. J. Roothaan, *Rev. Mod. Phys.* 23, 69 (1951).
91. F. B. van Duijneveldt and J. N. Murrell, *J. Chem. Phys.* 46, 1759 (1967).
92. J. L. Lippert, M. W. Hanna, and P. J. Trotter, *J. Am. Chem. Soc.* 91, 4035 (1969).
93. E. G. Cook, Jr. and J. C. Schug, *J. Chem. Phys.* 53, 723 (1970).
94. H. Fujimoto, S. Yamabe, and K. Fukui, *Tetrahedron Lett.* 443 (1971).
95. H. Fujimoto, S. Yamabe, and K. Fukui, *Bull. Chem. Soc. Jap.* 44, 2936 (1971).
96. M. Born and J. R. Oppenheimer, *Ann. Physik* 84, 457 (1927).
97. R. S. Mulliken, *J. Chim. Phys.* 46, 497 (1949).
98. R. S. Mulliken, *J. Chem. Phys.* 23, 1833 (1955).
99. T. Koopmans, *Physica* 1, 104 (1933).
100. M. J. S. Dewar, *The Molecular Orbital Theory of Organic Chemistry*, McGraw-Hill, New York, 1969.
101. K. Fukui, *J. Phys. Chem.* 74, 4161 (1970).
102. K. Fukui, *Acct. Chem. Res.* 4, 57 (1971).
103. K. Ruedenberg, *Rev. Mod. Phys.* 34, 326 (1962).
104. K. Fukui and H. Fujimoto, *Bull. Chem. Soc. Jap.* 40, 2787 (1967).
105. A. Schriesheim, J. E. Hofmann, and C. A. Rowe, Jr., *J. Am. Chem. Soc.* 83, 3731 (1961), and their subsequent papers.
106. H. Fujimoto, H. Oba, and K. Fukui, *J. Chem. Soc. Jap., Pure Chem. Sec.* 90, 1005 (1969).
107. H. Fujimoto, S. Sugihara, S. Yamabe, and K. Fukui, *Bull. Chem. Soc. Jap.* 44, 2565 (1971).
108. E. S. Gould, *Mechanism and Structure in Organic Chemistry*, Holt and Rinehart, New York, 1959, pp. 472–513.
109. H. Kwart, T. Takeshita, and J. L. Nyce, *J. Am. Chem. Soc.* 86, 2606 (1964).
110. E. C. Kooyman and G. C. Vegter, *Tetrahedron* 4, 382 (1958).
111. H. Fujimoto and K. Fukui, *Tetrahedron Lett.* 5551 (1966).
112. J. Gloux, M. Guglielmi, and H. Lemaire, *Mol. Phys.* 19, 833 (1970).
113. W. A. Lathan, W. J. Hehre, and J. A. Pople, *J. Am. Chem. Soc.* 93, 808 (1971).
114. K. Fukui and H. Fujimoto, *Bull. Chem. Soc. Jap.* 42, 3399 (1969).

115. V. Magnasco, *Theor. Chim. Acta* **21**, 267 (1971).

116. R. S. Mulliken, *J. Chem. Phys.* **23**, 1841 (1955).

117. L. Salem, *Proc. Roy. Soc. Lond.*, Ser. A **264**, 379 (1961).

118. H. Kollmar, *Tetrahedron Lett.* 3337 (1970).

119. H. Basch, *J. Chem. Phys.* **55**, 1700 (1971).

120. H. Fujimoto, S. Yamabe, and K. Fukui, unpublished results.

121. K. Fukui, K. Morokuma, and T. Yonezawa, *Bull. Chem. Soc. Jap.* **34**, 1178 (1961).

122. W. C. Herndon and W. B. Giles, *Mol. Photochem.* **2**, 277 (1970).

123. H. Fujimoto, S. Yamabe, and K. Fukui, *Tetrahedron Lett.* 439 (1971).

124. K. Fukui, H. Fujimoto, and S. Yamabe, *J. Phys. Chem.*, in press.

CHAPTER 4

The Generalized Perturbation Theory of Chemical Reactivity and its Applications

G. Klopman
Case Western Reserve University
Cleveland, Ohio

I. INTRODUCTION

The field of applied theoretical organic chemistry came out in the last couple of years from a long period of stagnation and is presently undergoing a real revolution. Two breakthroughs in related areas have prompted this renaissance. Sophisticated self-consistent quantum-mechanical methods became available that allow the calculation of the stability of large organic molecules[1,2] to be made. These methods have recently started to be used for the determination of the stability of possible reaction intermediates[3] and have already proven to be of invaluable importance in such areas as carbonium ion chemistry[4] and gas-phase[5] reactions. The second one is the real understanding on quantum-mechanical grounds of the intimate process of bond formation that finally emerged from the numerous studies of the factors responsible for the characteristic properties of specific bonds.

Up to the middle of the sixties, only reactivity indices determined from the Hückel theory[6] were available for correlating reactivity data. The whole scope of applied theoretical organic chemistry was then restricted to properties of aromatic derivatives only. However, most of the needed data and all the ingredients necessary to produce a generalized theory of chemical reactivity were available. Orbital energies and orbital symmetry properties were known but not exploited to their full potential for the purpose of gaining knowledge of reactivity constraints. The perturbation theory was developed in the fifties[7] but used only for such trivial matters as determining the Hückel resonance energy of large molecules from smaller components, or approximating reactivity indices of aromatic derivatives.

In the mid-sixties, two developments prompted a revival of interest in the application of theoretical models to chemical reactivity and proved since to be of wide importance. The first one is the discovery by Woodward and Hofmann[8] and Longuet-Higgins and Abrahamson[9] of the laws of conservation of symmetry that govern concerted reactions. These laws have since been extensively tested[10] and rank now among the most useful theoretical tools of the organic chemist. The second one results from the development of generalized perturbation theories[11,12] that attracted attention to the special role played by some specific orbitals[13] of the molecules engaged in a chemical reaction.

The subject of this paper is to present a resume of the applications of the polyelectronic perturbation treatment of chemical reactivity that we started developing several years ago and that now allows practically every aspect of chemical reactivity to be covered.

The basic problems of chemical reactivity are the determination of the nature of the products formed in a reaction and the rate at which the processes occur. Both of these properties are, however, intimately related and depend on internal (structure of the reagents) as well as external (conditions under which the experiment is performed) factors.

The generalized perturbation theory addresses itself to the problem of what happens to the energy when two reagents interact, and deals with the process of electron transfer during bond formation. It is convenient to classify reactions into two general categories:

1. Donor-acceptor interactions encompass all reactions in which the rate determining step involves a formal transfer of electrons from one of the reagents to the other (Fig. 4-1), for example, nucleophilic and electrophilic substitution and vinylogous substitution. These reactions can either be

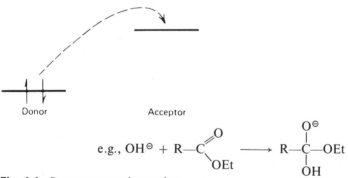

Donor   Acceptor

e.g., $OH^{\ominus}$ + R—C$\underset{OEt}{\overset{O}{\lessgtr}}$ $\longrightarrow$ R—$\overset{O^{\ominus}}{\underset{OH}{\overset{|}{\underset{|}{C}}}}$—OEt

Fig. 4-1. Donor-acceptor interaction.

single-site or *double-site* depending on whether the rate-determining step involves the interaction of only one or several atoms of each of the reagents, that is, the reaction proceeds via a linear or cyclic intermediate. For example,

$$H^{\oplus} + CH_2{=}CH_2 \xrightarrow[\text{d.a. process}]{\text{single-site}} [H_2C{-}\overset{H}{\overset{\displaystyle\diagdown}{C}}\overset{\oplus}{H_2}] \xrightarrow{X^{\ominus}} \text{product}$$

$$H^{\oplus} + CH_2{=}CH_2 \xrightarrow[\text{d.a. process}]{\text{double-site}} [H_2\overset{H}{\overset{\displaystyle\diagup\ \diagdown}{C}}\overset{\oplus}{{-}CH_2}] \xrightarrow{X^{\ominus}} \text{product}$$

2. *Exchange interactions* comprise reactions involving no formal transfer of electrons in the rate-determining step, but rather two transfers in opposite directions (Fig. 4-2). These reactions involve the more or less synchronous interaction of two different sites of each of the undissociated reagents, for example, the Diels-Alder reaction.

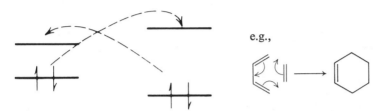

Fig. 4-2. Exchange interaction.

Although exchange interactions are basically double sited (i.e., occur via a cyclic intermediate), one may envisage the possibility that the formation of one of the bonds runs so much ahead of the other that the transition state is reached before the second site starts interacting. In these stepwise processes, a single-site interaction path describes more appropriately the reaction.

$$
\begin{array}{c}
H_2C \\ \| \\ H_2C
\end{array} +
\begin{array}{c}
CH_2 \\ \| \\ CH_2
\end{array}
\xrightarrow[\text{e. interaction}]{\text{double-site}}
\left[
\begin{array}{cc}
H_2C & CH_2 \\ \| & \| \\ H_2C & CH_2
\end{array}
\right]
\longleftrightarrow
\left[
\begin{array}{cc}
H_2C & CH_2 \\ \| & \| \\ H_2C & CH_2
\end{array}
\right]
\equiv
\left[
\begin{array}{cc}
H_2C & CH_2 \\ & \\ H_2C & CH_2
\end{array}
\right]
\longrightarrow
$$

$$
\begin{array}{c}
H_2C \\ \| \\ H_2C
\end{array} +
\begin{array}{c}
CH_2 \\ \| \\ CH_2
\end{array}
\xrightarrow[\text{e. interaction}]{\text{single-site}}
\left[
\begin{array}{cc}
H_2C & CH_2 \\ \| & \\ H_2C & CH_2
\end{array}
\right]
\longleftrightarrow
\left[
\begin{array}{cc}
H_2C & CH_2 \\ & \| \\ H_2C & CH_2
\end{array}
\right]
\equiv
\left[
\begin{array}{cc}
H_2C{-}CH_2 \\ | \quad | \\ H_2C \cdot \cdot CH
\end{array}
\right]
\longrightarrow
$$

II. SINGLE-SITE DONOR-ACCEPTOR INTERACTIONS

A. Charge and Orbital Control

The result of the application of the generalized polyelectronic perturbation theory (GP) to the donor-acceptor (Fig. 4-3) interaction case yields the

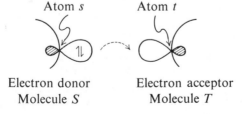

Electron donor   Electron acceptor
Molecule *S*   Molecule *T*

Fig. 4-3. Orbitals in single-site donor-acceptor interactions.

following fundamental equation,[11,14] consisting basically of two terms, an electrostatic term and a covalent term:

$$\Delta E_{\text{total}} = -\underbrace{\frac{q_s q_t}{R_{st}\varepsilon}}_{\substack{\text{electro-}\\\text{static}\\\text{term}}} + 2 \sum_{\substack{\text{occupied orbitals } m \\ \text{of molecule } S}} \sum_{\substack{\text{unoccupied orbitals} \\ n \text{ of molecule } T}} \underbrace{\frac{(c_s^m c_t^n \Delta\beta_{st})^2}{E_m^* - E_n^*}}_{\text{covalent term}} \quad (1)$$

where ΔE_{total} is the total change in energy due to the partial formation of a bond between an atom s of the electron donor molecule S and an atom t of the electron acceptor molecule T, in a solvent of effective dielectric constant ε. q_s and q_t are the total charges of atoms s and t in the isolated molecules. R_{st} is the distance between the two atoms s and t for which the interaction energy is calculated; c_s^m are the coefficients of the atomic orbitals of atom s in the various molecular orbitals m of molecule S; c_t^n are the coefficients of the atomic orbitals of atom t in the various molecular orbitals n of molecule T; $\Delta\beta_{st}$ is the change in the resonance integral between the interacting orbitals of atoms s and t at the distance R_{st}; E_m^* and E_n^* are quantities related to the energy of the various molecular orbitals m and n of the isolated S and T molecules in the same medium as that being used to perform the reaction.

Before proceeding further, let us first try to familiarize ourselves with the meaning of this equation and understand qualitatively its implications. Let us consider the interaction of two chemical species and assume that the particular reaction under consideration may proceed along several distinct pathways, yielding in each case a different product. Thus the donor molecule S (base or nucleophile) may react either via its atom s or s' (Fig. 4-4).

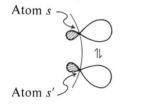

e.g., $SCN^{\ominus}$ may react either through the sulfur or the nitrogen end

Molecule S

Fig. 4-4. Ambident electrophile.

Being an electron donor, the molecule S must have occupied orbitals from which electrons will be extracted during the reaction. Let us assume two such molecular orbitals characterized by their energy, E_m^* and $E_{m'}^*$. Let us similarly assume that the acceptor molecule T may react with the donor via two alternative centers t and t' (Fig. 4-5). Being an acceptor, the molecule T must have at least one unoccupied orbital E_n^* that may accommodate the electrons it is receiving from the donor.

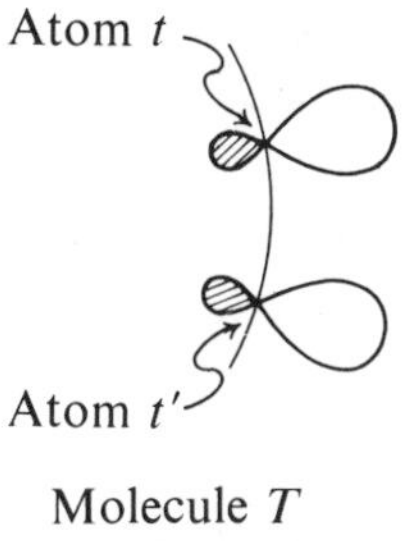

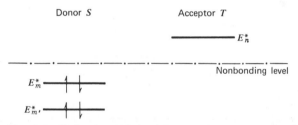

e.g., 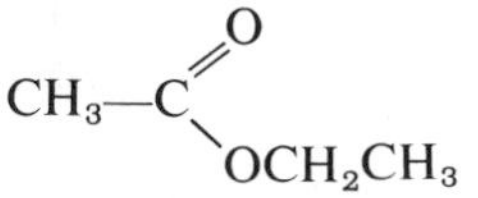 CH_3—C (with =O and OCH$_2$CH$_3$ substituents) is attacked by a nucleophile either at the carbonyl carbon (acylation) or at the saturated carbon atom (alkylation)

Molecule T

Fig. 4-5. Ambident nucleophile.

The orbital picture of interest for the chemical process can then be schematically represented as in Fig. 4-6. The total energy change produced by the

Donor S Acceptor T

E_n^*

Nonbonding level

E_m^*

$E_{m'}^*$

Fig. 4-6. Molecular orbitals of donor and acceptor molecules.

interaction of S and T can be evaluated from the GP equation [Eq. (1)] for each of the possible reaction intermediates,

$$\Delta E_{total} = -\frac{q_s q_t}{R_{st}\varepsilon} + \frac{2(c_s{}^m c_t{}^n \Delta\beta_{st})^2}{E_m^* - E_n^*} + \frac{2(c_s{}^{m'} c_t{}^n \Delta\beta_{st})^2}{E_{m'}^* - E_n^*} \qquad (2)$$

electro-   covalent terms
static
term

for the interaction of s with t (similarly for the other possibilities $s't$, st', and $s't'$).

The first term merely represents the electrostatic interaction between two atoms carrying a formal charge q_s and q_t. The second and third terms introduce the partial transfer of electrons respectively from m to n and m' to n, and measure the stabilization that accompanies the resulting covalent bond formation. For the covalent terms to be significant, an important requirement is that a channel be provided for the electrons of S to migrate, at least partially, to T. This is achieved if the corresponding β term is nonzero, which occurs only if adequate overlap between the atomic orbitals of the atoms s or s' and t or t' is provided and will obviously depend on the ability of the reaction intermediate to attain the proper geometry. If the molecules can achieve such required positioning, then a splitting will occur between the interacting orbitals and stabilization of the intermediate will usually occur (Fig. 4-7).

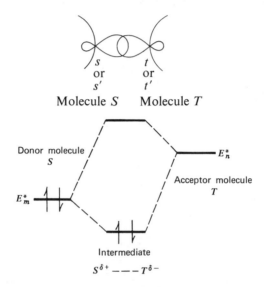

Fig. 4-7. The splitting resulting from the interaction of two overlapping orbitals.

The magnitude of the stabilization depends, among other things, on the corresponding resonance integral β, function itself of the overlap between the interacting orbital lobes.

One may consider two important general cases. The first case occurs if we assume that, for some reason that will be analyzed later, the two sets of molecular orbitals are separated by a large *energy gap*, such as is represented in Fig. 4-8. In such a case, the covalent terms of the GP equation are relatively small since the denominators $E_m^* - E_n^*$ and $E_{m'}^* - E_n^*$ are large. Furthermore, if the separation between the orbitals m and m' is small compared to that between m and n, we may replace in the above expressions, E_m^* and $E_{m'}^*$, by their average, E_{av}^*, without creating a serious error.

Under these conditions, the covalent terms can be combined into

$$\text{covalent terms} = 2\,\frac{(c_s{}^m c_t{}^n \beta_{st})^2 + (c_s{}^{m'} c_t{}^n \Delta\beta_{st})^2}{E_{av}^* - E_n}$$

$$= 2(c_s{}^{m2} + c_s{}^{m'2})c_t{}^{n2}\,\frac{(\Delta\beta_{st})^2}{E_{av}^* - E_n} \tag{3}$$

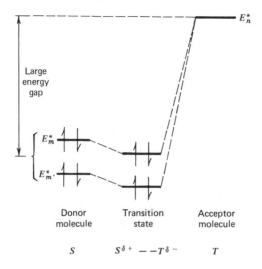

Fig. 4-8. Charge controlled reaction.

In this final expression, the sum $c_s^{m^2} + c_s^{m'^2}$ measures the total electronic occupancy of the orbital s in the molecule S, and $c_t^{n^2}$ the availability (or positive hole) left on orbital t of molecule T. The covalent term will therefore be optimized when the interaction involves the donor center carrying the highest electronic density and the acceptor center with the largest positive hole.

On the other hand, the electrostatic term, which is the first term of the GP equation, always favors the interaction of the atoms carrying the highest opposite charges. As a result, we find that when a large energy gap separates the occupied orbitals of the donor and the unoccupied orbitals of the acceptor, both the covalent and the electrostatic terms favor the interaction between the atoms carrying the highest opposite charge densities. For this reason, we will refer to such a situation as a *charge controlled reaction*. Thus, if in our example of the interaction of the donor S with the acceptor T, the molecular orbitals are such as described here, then the reaction will occur predominantly between the atoms s or s' and t or t' depending on which pair provides the best *charge* combination.

Before proceeding to the second case, let us first recognize that the occurrence of a charge controlled reaction does not necessarily imply charged reagents. The existence of a large energy gap will often be sufficient to force the reaction to proceed along such a path even though the reagents do not contain highly charged atoms. The nature of the product, in such a case, is determined by the *covalent* term which simply requires interaction between the donor center carrying the *highest "electronic" density* and the acceptor center with the *lowest "electronic" density*. Obviously, however, the reaction will not be very facile as the covalent stabilization of the intermediate would remain small.

The second case we wish to consider is that which occurs when the highest occupied orbital of the donor and the lowest unoccupied orbital of the acceptor are nearly degenerate. This is represented schematically in Fig. 4-9 together with the anticipated splitting produced by the interaction. The stabilization of the intermediate is here also given by the GP equation. However, due to the near degeneracy between the m and n orbitals, the denominator $E_m^* - E_n^*$ becomes very small and as a result the covalent term for the specific interaction of the orbitals m and n [second term of Eq. (2)] becomes extremely large.

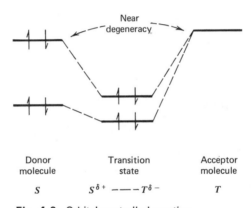

Fig. 4-9. Orbital controlled reaction.

The covalent interaction between the other orbital of the donor (m') and that of the acceptor (n) as well as the electrostatic interaction term, remain relatively small and can, to a first approximation, be neglected. It thus appears that the major term contributing towards the stability of the intermediate is one proportional to the numerator of the covalent term between the two *"vicinal"* orbitals m and n, that is,

$$\Delta E_{\text{total}} \approx 2c_s{}^m c_t{}^n \Delta\beta_{st} \qquad (4)$$

The reaction between a donor and an acceptor whose orbitals are as characterized above will thus occur between the two centers possessing the highest density of charge in the frontier orbitals, that is, the highest occupied orbital of the donor and lowest unoccupied orbital of the acceptor. We will refer to such a situation as a *frontier* or more generally as an *orbital-controlled reaction*. Thus if the donor S and the acceptor T are such that there is no "energy gap" between their frontier orbitals, the reaction, rather than yielding the product formed by the best charge combination, will proceed to give the product formed by the interaction of the two atoms s or s' and t or t' carrying the highest absolute charge (coefficient) in the "vicinal" orbitals.

As an example illustrating such a dual behavior, let us consider the nucleophilic attack by a thiocyanate ion of an electrophilic agent such as a metal cation characterized by an undefined empty orbital n. The molecular orbitals corresponding to the thiocyanate ion can be calculated approximately by the Hückel method. The specific parameters used to characterize the heteroatoms are practically irrelevant to the problem. The result of such calculation provides three molecular orbitals of which the two lowest are occupied. For example,

$$\psi_3 = 0.59\phi_S - 0.74\phi_C + 0.33\phi_N \quad \text{(empty)}$$

$$\psi_2 = 0.74\phi_S + 0.33\phi_C - 0.59\phi_N \quad \text{(occupied)}$$

$$\psi_1 = 0.33\phi_S + 0.59\phi_C + 0.74\phi_N \quad \text{(occupied)}$$

Schematically the relevant molecular orbitals can be represented as in Fig. 4-10. The size of the circles represents the contribution of each of the

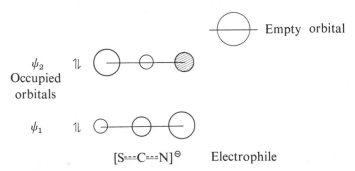

Fig. 4-10. Molecular orbitals in the reaction of SCN⁻ with an electrophile.

atomic $p\pi$ orbitals (i.e., coefficient) of the corresponding atoms in each molecular orbital (and not the actual size of the orbitals). Negative coefficients are represented by dashed circles.

The total charge on the sulfur atom is obtained as the core charge minus the double sum, over all occupied molecular orbitals, of the square of the coefficients of the atomic orbital of sulfur, that is,

$$q_S = 1 - 2 \times (0.33^2 + 0.74^2) = -0.31$$

Similarly for nitrogen

$$q_N = 1 - 2 \times (0.74^2 + 0.59^2) = -0.79$$

The charge on carbon, $+0.09$, and that on the cation are irrelevant here.

If a large gap exists between the set of orbitals of the two reagents, then the reaction is charge controlled and thiocyanate will "tend" to coordinate the cation through its nitrogen end which carries the highest negative charge.

If, on the other hand, there is no big gap between the two sets of orbitals, the reaction is orbital controlled and the cation will seek the largest electronic density pertaining to the highest occupied orbital of SCN$^-$. Assuming all other things to be equal, the metal cation will thus "tend" to coordinate the sulfur atom, which carries the highest electronic charge in this particular orbital.

$$c_S^{m^2} = 0.74^2 = 0.55 \qquad c_N^m = (-0.59)^2 = 0.35$$

Thus we have established two possible pathways and energetic controls for any given reaction. What remains to be discussed is when does one kind or the other occur, that is, how to determine whether an energy gap exists or not. The answer to that question is trivial if we can attach a specific value to the quantities E^* that characterize the interacting orbitals. This can easily be done theoretically and it is found[14] that the value E^* characterizing an occupied orbital is equal to the energy necessary to extract half of its pair or electrons. Similarly, the value E_n^* characterizing an empty orbital n is the energy gained by filling it by half a pair of electrons.

Specifically these quantities refer to the derivative of the respective orbital energy with respect to their occupancy, that is, the tendency to donate or receive half a pair of electrons,

$$-E_x^* = \frac{\partial E_x}{\partial v_x} \tag{5}$$

where E_x is the orbital energy and v_x its occupancy ($v_x = \frac{1}{4}$ for an acceptor's empty orbital and $\frac{3}{4}$ for a donor's occupied orbital). This definition has a striking resemblance with that of "atomic electronegativity"[15] which was defined[16] as being the tendency for a "neutral" atom to accept additional electrons in a molecule.

$$-\chi = \left(\frac{\partial E_x}{\partial v_x}\right) \qquad \text{for } v_x = \frac{1}{2} \tag{6}$$

We will thus refer to the quantities $-E^*$ as the *orbital electronegativities*.

The orbital electronegativity is a function of the occupancy of the orbital and reduces to the classical (neutral) electronegativity if the total occupancy of the orbital is one electron.

Let us now go back to the two kinds of reaction controls and try to define the conditions under which they will occur from a qualitative analysis of the electronegativities of the orbitals of interest.

For a charge controlled reaction to occur, an energy gap must exist between the highest occupied molecular orbital (HOMO) of the donor and the lowest unoccupied molecular orbital (LUMO) of the acceptor. This is achieved if the donor has low-lying occupied orbitals, that is, no tendency to donate its electrons, or high electronegativity. For reactions occurring in solution, the process of removing electrons from the donor, which usually is a negatively charged species (base or nucleophile), involves a loss of solvation energy; this contributes to the difficulty of performing the process. The smaller the ionic radius of the atom from which the electrons are removed, the larger the solvation energy and the lower the orbital will be. Consequently, charge controlled reactions involve atoms of high electronegativity and small ionic radii, that is, high orbital electronegativity. Examples of such charged species are F^- and OH^-; uncharged species favoring charge controlled reactions are H_2O and amines.

For an acceptor to be suitable for a charge controlled process requires that it have no low-lying unoccupied orbitals. This requirement is met when we deal with a poor acceptor, one that has a small affinity for electrons and, if positively charged, one that is strongly solvated. Examples of such acceptors are Al^{3+}, Mg^{2+}, and protons.

The reverse requirement applies to reagents, candidates for orbital controlled reactions. The donors must have high-lying occupied orbitals, relatively low electronegativity and a large ionic radius. Examples of such donors are I^-, RS^-, and H^-. The acceptors must be good acceptors; that is they must be easily reduced, have low-lying unoccupied orbitals, relatively high affinity for electrons, and large ionic radii. Examples are Hg^{2+}, Ag^+, and Pt^+.

It can be shown that only interactions involving acceptors and donors seeking to react via the same route, charge control or orbital control, are energetically favorable. This is easily understandable as, for example, if a donor is seeking an acceptor to donate its electrons, the process will be easy only if the acceptor is capable of accommodating them. Thus we may anticipate favorable coordination in solution between a proton or a cation like aluminum(III) and fluoride or hydroxyl ions, and between Hg^{2+} or Ag^+ and iodide or sulfhydride ions. On the other hand, the combination of H^+ or Al^{3+} with iodide or sulfhydride ions or that between Hg^{2+} or Ag^+ and fluoride or hydroxyl ions is less favorable. As a consequence, we expect, for example, HI and H_2S to be stronger acids (more dissociated), respectively, than HF or H_2O.

B. Hard and Soft Acids and Bases

The above results of our perturbation theory bear a striking similarity to those derived from Pearson's treatment of hard and soft acids and bases.[17] Pearson proposed a general principle according to which hard acids coordinate best with hard bases, and soft acids with soft bases. This principle was based on the observation that some acceptors (acids) such as Mg^{2+} and

Al^{3+} tend to coordinate best with ligands (donors) in the order

$$F > Cl > Br > I$$

$$N > P > As > Sb > Bi$$

$$O > S > Se > Te$$

whereas others such as Ag^+ and Hg^{2+} tend to coordinate with these ligands in the opposite order. It was noticed that the metals and the ligands behaving in the first fashion were small unpolarizable ions; hence the name hard, whereas those behaving in the second fashion bear the opposite properties and were called soft. The treatment was further extended to include a large number of acceptors and donors, including organic reagents. The results are included in Table 4-1 for donors (bases) and in Table 4-2 for acceptors (acids).

However, a precise measure of the hardness of a reagent could not be defined except from its chemical behavior. Furthermore, the reasons for such behavior were difficult to assess, and the whole concept therefore lacked a theoretical basis.

The treatment we have developed here provides such a theoretical basis. Thus a hard-base–hard-acid interaction is typically a charge controlled reaction. It mostly results from a favorable *electrostatic* interaction between a donor of high orbital electronegativity and an acceptor of low orbital electronegativity. On the other hand, the interaction between a soft acid and a soft base is that resulting from the *covalent* coordination of a donor of low orbital electronegativity and an acceptor of high orbital electronegativity.

In the following pages we will adopt the hard and soft nomenclature to characterize reagents that tend to react respectively via charge controlled and orbital controlled paths. Since the behavior of reagents depends primarily on the difference between their orbital electronegativities in solution, a reasonable classification of hard and soft reagents should be obtained by a tabulation of this value. This can easily be done from a knowledge of the ionization

Table 4-1. Empirical Classification of Hard and Soft Donors (Bases)[17]

Hard	Soft	Borderline
H_2O, OH^-, F^-	R_2S, RSH, RS^-	$C_6H_5NH_2$
$CH_3CO_2^-$, PO_4^{3-}, SO_4^{2-}	I^-, SCN^-, $S_2O_3^{2-}$	C_5H_5N, N_3^-
Cl^-, CO_3^{2-}, ClO_4^-, NO_3^-	R_3P, R_3As, $(RO)_3P$	Br^-, NO_2^-, SO_3^{2-}
$ROH,^a$ RO^-, R_2O	CN^-, RNC, CO	N_2
NH_3, RNH_2, N_2H_4	C_2H_4, C_6H_6	
	H^-, R^-	

^{a} The symbol R stands for alkyl group such as CH_3 or C_2H_5.

Table 4-2. Empirical Classification of Hard and Soft Acceptors (Acids)[17]

Hard	Soft	Borderline
H^+, Li^+, Na^+, K^+	Cu^+, Ag^+, Au^+, Tl^+, Hg^+	Fe^{2+}, Co^{2+}, Ni^{2+}
Be^{2+}, Mg^{2+}, Ca^{2+}, Sr^{2+}, Mn^{2+}	Pd^{2+}, Cd^{2+}, Pt^{2+}, Hg^{2+},	Cu^{2+}, Zn^{2+}, Pb^{2+},
Al^{3+}, Sc^{3+}, Ga^{3+}, In^{3+}, La^{3+},	CH_3Hg^+, $Co(CN)_5{}^{2-}$,	Sn^{2+}, Sb^{3+},
N^{3+}, Cl^{3+}, Gd^{3+}, Lu^{3+}	Pt^{4+}, Te^{4+}	Bi^{3+}, Rh^{3+},
Cr^{3+}, Co^{3+}, Fe^{3+}, As^{3+}, CH_3Sn^{3+}	Ti^{3+}, $Tl(CH_3)_3$, BH_3,	Ir^{3+}, $B(CH_3)_3$,
Si^{4+}, Ti^{4+}, Zn^{4+}, Th^{4+}, U^{4+},	$Ga(CH_3)_3$, $GaCl_3$, GaI_3,	SO_2, No^+, Ru^{2+},
Pu^{4+}, Ce^{8+}, Hf^{4+}, WO^{4+}	$InCl_3$	Os^{2+}, R_3C^+,
$UO_2{}^{2-}$, $(CH_3)_2Sn^{2+}$, VO^{2+},	RS^+, RSe^+, RTe^+	$C_6H_5{}^+$, GaH_3
MoO^{3+}	I^+, Br^+, HO^+, RO^+	
$BeMe_2$, BF_3, $B(OR)_3$	I_2, Br_2, ICN, etc.	
$Al(CH_3)_3$, $AlCl_3$, AlH_3	Trinitrobenzene, etc.	
$RPO_2{}^+$, $ROPO_2{}^+$	Chloranil, quinones, etc.	
$RSO_2{}^+$, $ROSO_2{}^+$, SO_3	Tetracyanoethylene, etc.	
I^{7+}, I^{5+}, Cl^{7+}, Cr^{6+}	O, Cl, Br, I, N, RO·, RO_2·	
RCO·, CO_2 NC^+	M^0 (metal atoms)	
HX (hydrogen bonding molecules)	Bulk metals	
	CH_2, carbenes	

potential IP, electron affinity EA, and the effective ionic radii R_{ion} (in Å) of the reacting atoms by using the following equations[14]:

$$\text{For acceptors: } E_n^* = \left(\frac{\partial E}{\partial \nu_n}\right)_{\nu_n = 1/4} = -\frac{3IP + EA}{4}$$

$$+ \frac{14.388(q - 0.5x)x}{R_{ion} + 0.82}\left(1 - \frac{1}{\varepsilon}\right) \quad \text{(in eV)}$$

$$\text{For donors: } E_m^* = \left(\frac{\partial E}{\partial \nu_m}\right)_{\nu_m = 3/4} = -\frac{IP + 3EA}{4} \tag{7}$$

$$+ \frac{14.388(q + 0.5)}{R_{ion}}\left(1 - \frac{1}{\varepsilon}\right) \quad \text{(in eV)}$$

where q is the initial charge of the ion (negative for bases, positive for acids), ε is the dielectric constant of the solvent (assumed to be 80 for water!) and x is a parameter used to take into consideration the variation of size of the ions upon oxidation or reduction and defined arbitrarily as

$$x = q - (q - 1)\sqrt{0.75} \quad \text{for } q \geqslant 1 \tag{8}$$

The first term in each of these equations refers to the gas-phase orbital electronegativity and the second term is the correction due to the influence of the solvent.

The results which are given in Table 4-3 for a series of acceptors and in Table 4-4 for a few donors agree quite well with the classification in terms of hardness and softness as determined empirically by Pearson from their experimental behavior. As a result the expression for E^* can be used *a priori* to determine the kind of behavior to be expected from a yet unclassified reagent.

Table 4-3. Calculated Orbital Electronegativity of Cations in Water

X^a	IP,^{a} (eV)	EA,^{a} (eV)	Orbital Energy (eV)	$r + 0.82$ (Å)	Desolvation	E_n (eV)	
Al^{3+}	28.44	18.82	26.04	1.33	32.05	6.01	↑
La^{3+}	19.17	11.43	17.24	1.96	21.75	4.51	Hard
Ti^{4+}	43.24	28.14	39.46	1.50	43.81	4.35	
Be^{2+}	18.21	9.32	15.98	1.17	19.73	3.75	
Mg^{2+}	15.03	7.64	13.18	1.48	15.60	2.42	
Ca^{2+}	11.87	6.11	10.43	1.81	12.76	2.33	
Fe^{3+}	30.64	15.96 (16.18)	26.97	1.46	29.19	2.22	
Sr^{2+}	11.03	5.69	9.69	1.94	11.90	2.21	
Cr^{3+}	30.95	16.49	27.33	1.45	29.39	2.06	
Ba^{2+}	10.00	5.21	8.80	2.16	10.69	1.89	
Ga^{3+}	30.70	20.51	28.15	1.44	29.60	1.45	
Cr^{2+}	15.01 (16.49)	7.28 (6.76)	13.08	1.65	13.99	0.91	
Fe^{2+}	16.18	7.90	14.11	1.56	14.80	0.69	Borderline
Mn^{2+}	15.64	7.43	13.59	1.62	14.25	0.66	
Co^{2+}	16.49 (17.05)	8.42 (7.86)	14.47	1.54	14.99	0.52	
Li^+	5.39	0.82	4.25	1.50	4.74	0.49	
H^+	13.60	0.75	10.38	—	10.8	0.42	
Ni^{2+}	17.11 (18.15)	8.67 (7.63)	15.00	1.51	15.29	0.29	
Na^+	5.14	0.47	3.97	1.79	3.97	0	
Cu^{2+}	17.57 (20.29)	9.05 (7.72)	15.44	1.54	14.99	−0.55	
Zn^{2+}	17.96	9.39	15.82	1.56	14.80	−1.02	
Tl^+	6.10	(2.0)	5.08	2.22	3.20	−1.88	
Cd^{2+}	16.9	8.99	14.93	1.79	12.89	−2.04	
Cu^+	7.72	2.0	6.29	1.78	3.99	−2.30	
Ag^+	7.57	2.2	6.23	2.08	3.41	−2.82	
Tl^{3+}	29.30	20.42	27.45	1.77	24.08	−3.37	Soft
Au^+	9.22	2.7	7.59	2.19	3.24	−4.35	
Hg^{2+}	18.75	10.43	16.67	1.92	12.03	−4.64	↓

^{a} C. E. Moore, "Atomic Energy Levels," National Bureau of Standards Circular 467, U.S. Government Printing Office, Washington, D.C., 1949.

Table 4-4. Calculated Orbital Electronegativity of Anions in Water

X	IP (eV)	EA (eV)	Orbital Energy (eV)	r (Å)	Desolvation (eV)	E_m (eV)	
F⁻	17.42	3.48	6.96	1.36	5.22	−12.18	↑
H₂O	25.4	12.6	15.8	(1.40)	(−5.07)	−(10.73)	Hard
OH⁻	13.10	2.8	5.38	1.40	5.07	−10.45	
Cl⁻	13.01	3.69	6.02	1.81	3.92	−9.94	
Br⁻	11.84	3.49	5.58	1.95	3.64	−9.22	
CN⁻	14.6	3.2	6.05	2.60	2.73	−8.78	
SH⁻	11.1	2.6	4.73	1.84	3.86	−8.59	Soft
I⁻	10.45	3.21	5.02	2.16	3.29	−8.31	
H⁻	13.6	0.75	3.96	2.08	3.41	−7.37	↓

C. Nucleophilic Orders

Providing that some reasonable approximations can be made concerning the nature of the transition states and the value of the quantum-mechanical terms appearing in the GP equation, reactivity orders toward various centers can be estimated.

The results of Table 4-5 illustrate the semiquantitative agreement between the calculated and observed behavior of various bases towards a set of electrophiles treated as single orbitals characterized by decreasing orbital electronegativities.[14]

It is found, for example, that toward a very soft center ($E_n^* = -7$, first line) sulfur and iodide ions are very reactive, whereas hydroxyl ions are relatively inert. These results compare satisfactorily well with the observed sequence of reactivity toward the very soft peroxide oxygen.[18]

Table 4-5. Nucleophilic Order[a]

E_n^*		Nucleophilic Order						
−7	Calcd	HS⁻	> I⁻	> NC⁻	> Br⁻	> Cl⁻	> HO⁻	> F⁻
	$k \times 10^4$, reaction with peroxide oxygen	Too fast	6900	10	0.23	0.0011	≈ 0	
−5	Calcd	HS⁻	> NC⁻	> I⁻	> HO⁻	> Br⁻	> Cl⁻	> F⁻
	$k \times 10^6$, attack on saturated carbon	(25)	10	12	1.2	0.5	0.11	
	E (Edwards)		2.79	2.06	1.65	1.51	1.24	1.0
+1	Calcd	HO⁻	> NC⁻	> HS⁻	> F⁻	> Cl⁻	> Br⁻	> I⁻
	k for attack on carbonyl carbon (acylation)	890	10.8	—	0.001	—	Unreactive —	
	pK_a	15.7	9.1	7.1	3.2	—	(−4.3)	(−7.3)

[a] For more details on the calculation procedure see ref. 14.

The second sequence is obtained for the reaction with a slightly harder reagent ($E_n^* = -5$) and may apparently be compared with the observed reactivity toward saturated carbon atoms. It is found that the sulfur atom is still the most nucleophilic toward such centers, but the hydroxyl ion has increased its relative nucleophilicity. The cyanide and iodide ions have exchanged their places and, even though this is not the observed behavior, the trend is undoubtedly correct.[19] Finally, toward a hard center ($E_n^* = 1$) the reactivity sequence drastically changed. The hydroxyl ion continued to improve and became the most nucleophilic of all. Furthermore, the sequence of halogen reactivity is now running in the opposite direction than that of the previous cases. Such behavior correlates well with that observed for the reactions toward a carbonyl carbon,[20] where a substantial positive charge must exist due to the polarization of the C=O bond, and also toward a proton, as shown by the satisfactory correlation obtained with the pK_a's.

Before proceeding further, it might also be interesting here to emphasize the parallelism that exists between the GP equation and yet another empirical theory of chemical reactivity, the so-called free-energy relationships. In particular, the Edwards equation,[21] sometimes also called the oxybase equation,[22] presents a definite similarity to the equation resulting from our present treatment.[23]

In the oxybase equation, the relative rate of reaction of molecules with a specific electrophile are set equal to the sum of two terms,

$$\log \frac{k}{k_{\text{ref}}} = \alpha E + \beta H$$

where α and β are parameters characteristic of the type of reaction, the nature of the electrophile and the conditions under which the reaction is performed, but not of the properties of the nucleophile. On the other hand, H and E are properties of the nucleophile, H is related to its pK_a and E to its redox potential.

It was shown previously that the relative reactivity of bases toward a proton (see Table 4-4) is illustrative of a charge controlled behavior. On the other hand, the redox potential is a measure of the ease of removal of electrons from the base, and can thus be associated with its orbital electronegativity.

We thus have in the oxybase equation the two elements responsible for the charge and orbital control of reactions, and it is not surprising therefore that its results parallel those of the generalized perturbation theory.

D. Nucleophilic Attack on Ambient Electrophiles

In addition to the semiquantitative correlations with nucleophilic orders presented in the previous section, the theory also provides a rationale for the qualitative prediction of the nature of the product to be expected in the reaction between a specific nucleophile and an ambident electrophile.

Thus when 2-bromo-propane is treated with a base, either nucleophilic substitution or elimination of HBr may occur. The first case is initiated by attack by the nucleophile on the carbon atom bearing the halogen, whereas the second process requires the removal of one of the protons in α to carbon 2. The problem thus consists in determining the relative reactivity of the specific nucleophile toward a saturated carbon atom and a proton.

From various previous studies, it appears that the orbital electronegativity of a saturated carbon is smaller (soft) than that of the proton (hard). Based on the previously discussed rules, we may therefore anticipate that, in the absence of dominant steric effects, a soft nucleophile (good electron donor) will favor attack on the carbon, whereas a hard nucleophile (strong base) will react with the proton. This conclusion certainly provides us with a guideline to select the right nucleophile to produce either product; however, the theory is not quantitative enough to allow us to estimate whether with our arsenal of reagents we will be able to find one that is soft or hard enough to perform the desired reaction.

In the present case, both products can be obtained; a hydroxylic base is hard enough to remove one of the protons and yields the elimination product, whereas the enolate ion of ethyl malonate is a sufficiently good electron donor to perform the substitution reaction.[24]

A similar case is provided by the competition between substitution and elimination in the nucleophilic attack on 1,2-dichloro-ethane.[25]

Here again the hard hydroxylic base produces elimination via attack on the proton, whereas the softer sulfhydric nucleophile prefers to react with the softer carbon center to yield the substitution product.

A systematic study of the influence of the basicity (hardness) of a set of phenolic nucleophiles on the ratio of elimination versus substitution obtained upon reaction with phenethyl bromide[26] reveals the generality of the correlation.

Table 4-6 shows that, as predicted, the percentage of elimination increases steadily with the basicity of the phenol used as nucleophile.

A slightly more complex situation arises if the electrophile is a conjugated species, which has several sites suitable for attack by a nucleophile. Among

Table 4-6. Competition between Elimination and Substitution for the Reaction of Phenethyl Bromide and Substituted Phenol.[26]

Phenol	$pK_a{}^a$	Ratio $\dfrac{\text{Elimination}}{\text{Substitution}}$
p-OCH$_3$	11.52	0.94
p-CH$_3$	11.67	0.93
p-H	11.28	0.86
p-Br	10.50	0.52
p-COCH$_3$	9.21	0.30
p-NO$_2$	7.63	0.20

^{a} pK_a determined in 45.9% alcohol by G. Schwarzenbach and E. Ruchin *Helv. Chim. Acta* **22**, 360 (1939).

such species, heterocyclic derivatives have led to the breakdown of practically every theory that was set to correlate their reactivity. Pyridinium ions react with various electrophiles to yield either the 2- or 4-addition product. Its molecular orbitals can be calculated by quantum-mechanical methods, and for our purpose, satisfactorily enough with the Hückel method. The only orbitals of interest for our purpose are the empty orbitals that may accept electrons from a donor. These are represented schematically in Fig. 4-11

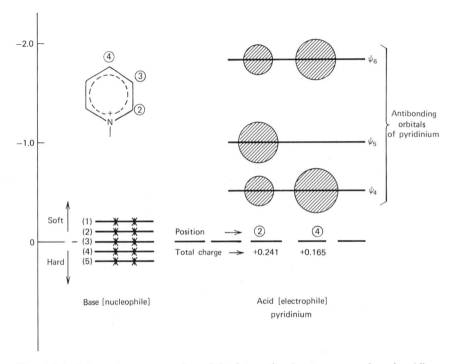

Fig. 4-11. Schematic representation of the interaction between several nucleophiles and the antibonding orbitals of pyridinium salts.

together with a set of various possible donors (1 to 5) characterized by decreasing orbital electronegativities. The circles represent, as in the previously described thiocyanate ion, the coefficient of the atomic $p\pi$ orbital of atoms 2 and 4 in the various empty molecular orbitals of the pyridinium ion. In the Hückel theory, the electron repulsion terms are not specifically taken into consideration and therefore the orbital electronegativities are simply approximated here by the molecular orbital energies.

The GP equation can now be used to calculate the perturbation between each of the five nucleophiles and positions 2 and 4. The results are illustrated

Table 4-7. Perturbation Calculation on Pyridinium Ion
             (in β Units)

Nucleophile	Orbital Electronegativity of Perturbing Acceptor	Position 2	Position 4
n^{r}5	−0.4	5.5474	6.3381 ↑
n^{r}4	−0.2	2.3613	2.4016
n^{r}3	0	1.5958	1.5458
n^{r}2	0.2	1.2281	1.1613
n^{r}1	0.4	1.0061	0.9391 ↓

in Table 4-7, where the perturbation energies (in β units) are listed for the various interactions.

It can be seen that for nucleophile n^{r}1, characterized by a large orbital electronegativity, the value for the interaction with position 2 is larger than that for position 4. This is still true for nucleophile n^{r}2 and 3, but as the orbital electronegativity decreases, corresponding to a softer reagent, that is, a frontier controlling nucleophile, the perturbation with position 4 increases and nucleophiles n^{r}4 and 5 show preferential affinity for attack at this position 4. As a result, we may anticipate that a hard nucleophile will add at position 2, whereas a soft nucleophile will add at position 4. This is borne out by experiment, as it has been shown that charge controlling bases such as hydroxylic bases, amines, and BH_4^- react along the first suggested path, whereas CN^- and $S_2O_4^{2-}$ react along the second one.[27] (Fig. 4-12).

Orbital control, CN $^\ominus$

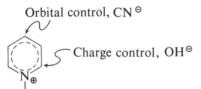

Charge control, OH$^\ominus$

Fig. 4-12. Alternative sites of nucleophilic attack on pyridinium ion.

Although the agreement between the predicted and observed behavior is quite satisfactory, it is important to realize that, at this stage, the results given by the theory only point toward trends rather than identify specific reagents. For example, there is no available method that would allow us to identify each of the five nucleophiles chosen to illustrate the ambident reactivity of the pyridinium ion with actual reagents. But the results obtained by calculating the perturbation energy produced by each of these reagents allowed us to identify the controlling factors that yield the alternative products. Thus we may only say that if addition at position 4 is desired, our best chance to obtain the highest yield of this product is to use the softest possible reagent at our disposal, and vice versa for addition at position 2.

E. Electrophilic Attack on Ambident Nucleophiles

The study of the behavior of ambident nucleophiles follows along the same lines as those concerning the study of the ambident electrophiles. The major difference is that we are concerned here with the occupied orbitals of the donor rather than the unoccupied orbitals of the acceptor in the previous case.

We have already discussed briefly the behavior of the thiocyanate ion and found that charge controlled reactions proceed favorably with the nitrogen end, whereas orbital controlled reactions prefer the sulfur end. This is experimentally observed, as hard acceptors tend to form the isothiocyano derivatives whereas the soft acceptors usually bind as thiocyano compounds.[28]

$$CH_3SCN \xleftarrow{CH_3I} [S \text{-----} C \text{=====} N]^\ominus \xrightarrow{RC(O)Cl} R-C \overset{O}{\underset{NCS}{<}}$$

Orbital control   Charge control

A similar situation is encountered with enolate ions where the soft center is the carbon atom and the hard center is the oxygen atom.

$$CH_3-\overset{|}{\underset{|}{C}}-C=O \xleftarrow[\text{C alkylation}]{CH_3Br} \left[\overset{}{>}C \text{-----} C \text{-----} O \right]^\ominus \xrightarrow[\text{O alkylation}]{(CH_3CH_2)_3O^\oplus} \overset{}{>}C=C-O-CH_2CH_3$$

Orbital control   Charge control

An interesting case is provided by the nitrous anion where the lone pair of the nitrogen is easily available to a soft electrophile but the highest concentration of negative charge resides on the oxygen atoms. Depending on the counter ion, the reaction of nitrous salts with alkyl halides may lead either to the

Charge controlled reaction

$$O \text{---} \overset{..}{N} \text{---} O$$

Orbital controlled reaction

nitro derivatives or to nitrites.[29] In the presence of sodium ions, the nitrogen pair is mostly free and is easily transferred to an acceptor's empty orbital. On the other hand, a very soft counterion such as a silver ion may compete effectively with the acceptor for the lone pair of nitrogen, leaving the oxygens available for charge controlled coordination.

Another interesting example is provided by the reaction between triphenylphosphine and α-halogenated ketones.[30]

The reaction was shown to proceed via initial attack of the halogen to form a quaternary phosphonium enolate ion pair. The two ions then recombine, either through the carbon or the oxygen atom of the enolate (paths 1 and 2). The rate of the reaction is apparently determined by the first step, but the nature of the product is determined by the ease of recombination in the second step (Fig. 4-13). Both of these steps can be correlated theoretically, even with results obtained from the Hückel theory. Thus the rate of the reaction should be proportional to the ease of formation of the enolate ion and will depend on the change in resonance energy produced by enolization. These values are in column 2 of Table 4-8 and reproduce the qualitative pattern of the experimentally observed rate values.

Table 4-8. Rate and Product Correlations for the Reaction of
α-Haloketones with Triphenylphosphine[30]

Reactant	ΔW (β Units)	Reactivity	Frontier Electron Density (f)		f_0/f_c	Position of Reaction
			Oxygen	Carbon		
Reference[a]	1.186	Fast (0°)	0.187	0.650	0.288	Oxygen
Bromomalonate	1.911	Very fast (−40°)	0.143	0.559	0.259	Oxygen
Phenacyl bromide	1.1265	Slow (80°)	0.123	0.606	0.203	Carbon
Bromolactone	1.046	Very slow (80°)	0.116	0.680	0.170	Carbon

[a] For example, α-bromocyclohexanone and α-bromocyclopentanone.

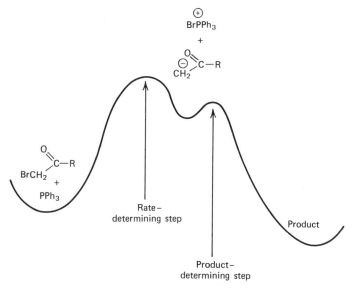

Fig. 4-13. Reaction path for the reaction of α-haloketones with triphenylphosphine.

In order to determine the controlling factors in the recombination step, it must first be recognized that triphenylphosphinium is a soft reagent, that is, one that accepts electrons readily. The ratio of the electronic charges on the oxygen and carbon atoms in the highest occupied orbital of the enolate ion should therefore provide a measure of their ability to combine with triphenylphosphine. This is given in column 4 of Table 3-8, where it is shown that, as anticipated, the smaller this ratio, the more favorable the attack on carbon.

The electrophilic substitution on pyridine N-oxide provides an example of application of the concept along the same lines as those used in the context of nucleophilic addition to pyridinium ions. Here again, we deal with a heterocycle that may react via several different centers depending on the nature of the electrophile. A similar calculation can be performed here by perturbing the occupied orbitals of pyridine N-oxide (Table 4-9) by a set of external reagents represented by empty orbitals of decreasing orbital electronegativity.

The calculated perturbation for positions 1, 2, and 3 of pyridinium oxide is reported in Table 4-10. It can be seen from these results that, with an orbital controlling reagent (orbital electronegativity -0.4 to -1.1β) pyridine N-oxide is predicted to react via atom 1. With harder reagents, the most favorable position is 3 and with very hard ones, the preferred position is 2. This is observed experimentally[31] as pyridinium oxide can be mercurated at position 1, nitrated at position 3, and sulfonated at position 2.

Let us reiterate here again that there was no way for us to predict this behavior exactly as it occurs. The only result the theory provided us with was

Table 4-9. Occupied Molecular Orbitals of Pyridine N-Oxide

Orbital				Atom Number			
Energy	1	2	3	4	5	6	7
			Coefficients (eigenvectors)				
0.0723 β	−0.292	0.010	0.293	0.010	−0.292	−0.031	0.861
1.0000 β	−0.500	−0.500	0	0.500	0.500	0	0
1.5473 β	0.123	0.484	0.626	0.484	0.123	−0.293	−0.152
2.9804 β	0.326	0.141	0.095	0.141	0.326	0.831	0.216

guidelines, telling us that our best chance to attack position 1 is to use an orbital controlling reagent, that is, a soft reagent, and the best chance to reach position 2 or 3 is to use a hard or very hard reagent. The theory does not tell us whether there will be a reagent soft enough or hard enough to achieve our goal.

Table 4-10. Perturbation Calculation on Pyridine N-oxide (in β Units)

Orbital Electronegativity of Perturbing Donor	Position		
	1	2	3
−0.40	0.7973	0.6102	0.7714
−0.60	0.6405	0.5423	0.6254
−0.80	0.5431	0.4883	0.5355
−1.00	0.4749	0.4443	0.4723
−1.20	0.4237	0.4077	0.4245
−1.40	0.3834	0.3787	0.3860
−1.60	0.3507	0.3501	0.3556
−1.80	0.3235	0.3271	0.3296
−2.00	0.3005	0.3070	0.3073
−2.20	0.2807	0.2892	0.2881
−2.40	0.2635	0.2733	0.2713
−2.60	0.2484	0.2592	0.2584
−2.80	0.2350	0.2464	0.2461

F. Electrophilic Aromatic Substitution

The last important example that we wish to illustrate in this section is that concerning electrophilic substitution on aromatic derivatives. This was, for many years, the only kind of reaction that could be handled by semiempirical quantum-mechanical methods, and for this reason became one of the most studied areas of theoretical organic reactivity. Numerous reactivity indices were suggested to correlate the reactivity of the various aromatic derivatives. Most of these indices are intercorrelated and thus yield similar results. Most of the correlations are satisfactory for regular alternant hydrocarbons and most fail when used for nonalternant or heterocyclic derivatives. One of the indices that was used was the frontier orbital density,* introduced by Fukui, and consisting of the square of the coefficient of the atomic orbital in the highest occupied molecular orbital of the aromatic derivative. The use of this reactivity index was criticized[32] on the grounds that it does not make sense, and fails badly for nonalternant, substituted, and heterocyclic derivatives.

However, with the development of the generalized perturbation treatment of chemical reactivity, new light is shed on the problem. By definition, all carbon atoms in an alternant aromatic hydrocarbon carry the same charge. As a result, the charge controlling factors are constant for all positions of all aromatic hydrocarbons and *only orbital (or frontier) controlling factors affect their reactivity*. We thus explain the observed correlation between frontier orbital density and reactivity of alternant aromatic hydrocarbons.

Our explanation, however, does much more than that; for example, a direct consequence of the established fact that only orbital controlling factors are relevant to aromatic reactivity is that the order of reactivity of various aromatic centers cannot be changed by changing the electrophile with which it reacts. The only change that can be made is that related to the ability of the electrophile to combine in an orbital controlling fashion, which will result in the contraction or expansion of the reactivity range, that is, determine the selectivity of the reagent.

Thus a charge controlling reagent will not be able to distinguish between various aromatic sites and be unselective. On the other hand, a soft reagent will react quickly with those aromatic sites that may easily transfer electrons to it, for example, 9-anthracene, and slowly with the others.

The prediction of reactivity based on the frontier electron density index failed badly for heterocycles, nonalternant, and substituted aromatic derivatives.[32] This can now easily be understood, since all those are cases in which substantial charge differences exist between the possible reaction sites. As a result, the frontier electron density indices fail to produce the correct answers.

---

* We adopted the name frontier controlled reactions by analogy with the frontier orbitals defined in this fashion by Fukui.[13]

We have already shown how this problem is handled for heterocycles in the present theory (see pyridium ion and pyridine N-oxide).

The treatment can easily be extended for nonalternant and substituted aromatic hydrocarbons. For example, for toluene, both PNDO and CNDO theories[33] show carbon 2 as carrying the highest negative charge and carbon 4 the highest electronic charge in the highest occupied molecular orbital. Therefore, in the absence of dominant steric effects, carbon 2 is expected to be the preferred site for electrophilic attack by a charge controlling reagent, for example, chlorination or nitration, and carbon 4 is the preferred site for orbital controlling reagents,[14] for example, bromination or mercuration. Furthermore, we may say that, in accordance with the previous conclusions, reagents that prefer to react with the ortho position of toluene will show small selectivity between various sites of alternant hydrocarbons, whereas those yielding mostly para toluene derivatives will be more selective (Fig. 4-14). The experimental values needed to prove the point are not as yet available, but the argument can to a certain extent be substantiated by the results of Table 4-11,[34] where the relative rate ratio k_{Tol}/k_{benz} is shown to increase regularly with the ratio of the para to ortho product formed in the reaction with toluene.

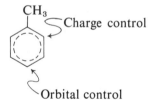

CH_3

Charge control

Orbital control

Fig. 4-14. Alternative sites for electrophilic attack on toluene.

In these results, one may also notice that, as predicted, the largest ortho isomer is found to form with electrophiles substituted by electron-withdrawing groups. These lead to a larger positive charge on the alkylating carbon, hence charge control. The reverse effect is true with electron-donating substituents, hence orbital control.

Table 4-11. Electrophilic Aromatic Substitution on Toluene and Benzene

X	TiCl$_4$ Catalyzed Benzylation by $XC_6H_4CH_2Cl$		AlCl$_3$ Catalyzed Phenylsulfonylation by $XC_6H_4SO_2Cl$	
	k_{Tol}/k_{Benz}	2 para/ortho	k_{Tol}/k_{Benz}	2 para/ortho
p-NO$_2$	2.5	1.15	2.8	1.61
p-Cl	6.2	2.70	7.5	2.94
p-H	6.3	2.70	9.0	4.35
p-CH$_3$	29	4.17	17	12.5
p-CH$_3$O	97	5.0	83	33.3

III. DOUBLE-SITE VERSUS SINGLE-SITE DONOR-ACCEPTOR INTERACTIONS

A. Intermolecular Donor-Acceptor Interactions

One of the simplest and most interesting results of this treatment is to provide an explanation for the well-known back-side attack in SN_2 reactions to yield inversion of configuration of the electrophilic carbon. In addition, the theory strongly supports the postulate of front-side attack and retention of configuration in SE_2 reactions. Let us first examine the case of the bimolecular nucleophilic substitution.

1. The Mechanism of Nucleophilic Attack.   The reaction in point is known to proceed along a path such as illustrated below:

$$NaI + \underset{a}{\overset{b}{\diagdown}}C-Cl \longrightarrow [I \cdots \overset{b}{\underset{a}{\overset{c}{C}}} \cdots Cl]^{\ominus} \; Na^{\oplus} \longrightarrow I-\overset{c}{\underset{a}{\overset{b}{C}}} + NaCl$$

From the GP theory point of view, what has to be considered is the *transfer of electrons from the occupied orbital of the donor to the unoccupied orbital of the acceptor*. Such transfer weakens both the bond from which the electrons are removed and that in which they are transferred. The former because the bonding orbitals are depleted of their electrons and the latter because the antibonding orbitals are being populated.

In each case, the total bond index decreases and both bonds are being broken, for example,

$$NaBr + -\overset{|}{\underset{|}{C}}-I \longrightarrow + Na^{\oplus} + Br - \overset{|}{\underset{|}{C}}- + I^{\ominus}$$

The process involves the formations of one bond (CBr) and the breaking of two (NaBr and CI), and therefore, in general, is unfavorable. Most reactions will occur only if one of the bonds has been predissociated, usually by the action of the solvent.

$$NaBr \xrightarrow{\text{solvent}} Na^{\oplus} + Br^{\ominus}$$

$$Br^{\ominus} + -\overset{|}{\underset{|}{C}}-I \longrightarrow Br - \overset{|}{\underset{|}{C}}- + I^{\ominus}$$

The situation is illustrated in Fig. 4-15 where the donor, nucleophile, is represented by a single bioccupied orbital, and the acceptor, electrophile, by a biatomic antibonding molecular orbital characterized by a central node. In Fig. 4-15 and following diagrams the dashed parts of each orbital represent a negative sign of the wave function that characterizes it. Within each independent molecule the choice of the positive and negative lobes is purely

Donor's occupied orbital Acceptor's empty orbital

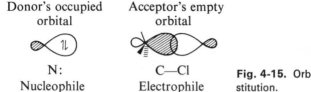

N: C—Cl   **Fig. 4-15.** Orbitals in nucleophilic sub-
Nucleophile   Electrophile   stitution.

arbitrary, but as soon as one of the lobes has been assigned, all the others are determined by symmetry considerations. Thus, if we choose the back lobe of the sp^3 orbital of carbon to be positive, the other lobes of the orbitals pertaining to the molecule are determined by the antibonding character of the orbital, and this requires here a central node. Since the molecular orbitals of the donor are originally determined independently from those of the acceptor, we may again choose arbitrarily one of the lobes to be positive. For convenience, however, we usually choose the pair of lobes that will be interacting during the course of the reaction to be of the same sign. However, a different choice does not affect the results if the GP equation is properly applied.

Two possible mechanisms of transfer may *a priori* be imagined:

1. A back-side attack whereby the transfer of electrons occurs via the back lobe of the carbon sp^3 orbital (Fig. 4-16a).

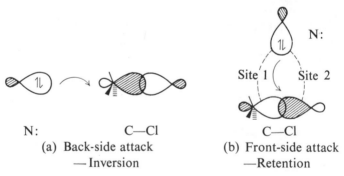

N: C—Cl   C—Cl
(a) Back-side attack   (b) Front-side attack
— Inversion   —Retention

Fig. 4-16. Possible mechanisms of nucleophilic attack.

2. A front-side attack involving simultaneous interaction between the donor's occupied orbital and the front lobe of both orbitals constituting the molecular empty bond orbital of the acceptor, thus leading to a cyclic intermediate (Fig. 4-16b).

The former case is a simple "single-site interaction" but the latter case is a "double-site" interaction since it involves a three-membered ring produced by the interaction of the donor's atomic orbital with the *two* adjoining atomic orbitals constituting the molecular orbital of the acceptor. In order to differentiate between these two possible processes, we may again make use of the

GP equation. Thus, for the back-side attack, the change in energy produced by the perturbation is given by

$$\Delta E = - \frac{q_{Nu}q_C}{R_{Nu C}\varepsilon} + 2 \frac{(c_C c_{Nu}\Delta\beta_{Nu C})^2}{E_{Nu} - E_{CCl}} \tag{9}$$

For a double-site reaction involving concerted interactions, the appropriate GP equation is identical to that given previously except that each term is summed over all interacting sites. Thus the general GP equation for multiple-sited interactions is

$$\Delta E = - \sum_{st} \frac{q_s q_t}{R_{st}\varepsilon} + 2 \sum_{\substack{m \\ occ}} \sum_{\substack{n \\ unocc}} \frac{\overline{(\sum_{st} c_s{}^m c_t{}^n \Delta\beta_{st})^2}}{E_m^* - E_n^*} \tag{10}$$

where the sum over st is that over the atomic orbitals involved in each site's interaction. As a result, the change in energy produced by the front-side attack (attack on the bond) is

$$\Delta E = - \frac{q_{Nu}q_C}{R_{Nu C}\varepsilon} - \frac{q_{Nu}q_{Cl}}{R_{Nu Cl}\varepsilon} + 2 \sum_{\substack{m \\ occ}} \sum_{\substack{n \\ unocc}} \frac{(c_C c_{Nu}\Delta\beta_{Nu C} + c_{Cl} c_{Nu}\Delta\beta_{Cl Nu})^2}{E_{Nu}^* - E_{CCl}^*} \tag{11}$$

Since there is a node between the atomic orbital of the carbon and that of the leaving group, that is, c_C and c_{Cl} have opposite signs, the last term of Eq. (11) is very small. Similarly, the two polar contributions have opposite signs due to the dipole of the bond and cancel each other to a large extent. As a result, the stabilization of the double-site intermediate is negligible compared to that of the single site and the reaction does not occur along such a path.

This conclusion can very simply be reached from an examination of the schematic representation of the orbitals, Fig. 4-16, and by applying the following "matching" rule:

For a reaction to be allowed along a certain path, all interacting orbital pairs must match. A concerted multisite reaction is forbidden unless all sites orbitals are matched with their counterpart of the other molecule.

Thus, in the case of the back-side attack, the two lobes that will be interacting in the process can be chosen to match each other, that is, to have the same sign. This leads to a reasonably favorable interaction. On the other hand, for a front-side attack, the match of the donor's orbital with one of the acceptor's orbitals is canceled by its mismatch with the other, and as a result, the reaction is not favorable along such a path.

Before proceeding further, let us qualify the applicability of the above matching rule by the following remark. If a concerted reaction is allowed, this does not mean that the reaction is going to proceed along this path, since another one may be even more facile. However, if a concerted reaction is forbidden, it will not usually occur, and a stepwise alternative will be preferred.

The situation occurring in the nucleophilic attack on a double bond can be described along similar lines. Here, however, no back lobe is available for the transfer of electrons (Fig. 4-17).

C=C

Fig. 4-17. Antibonding π orbital of ethylene.

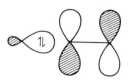

Fig. 4-18. Nucleophilic attack of a double bond along a σ path.

An approach along the axis of the bond will be forbidden for attack on the π system as no overlap results from it (Fig. 4-18).

On the other hand, the overlap with the σ framework is favorable, but requires the breaking of a C—C σ bond, which needs too much energy. The only alternative is along a path such that the reagent interacts with one of the lobes of the π orbital in the plane of the π system (Fig. 4-19).

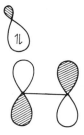

Fig. 4-19. Preferred path for the nucleophilic attack of a double bond.

By doing this, however, the reagent interacts at least partially with the lobe of the other π orbital also. This interaction is unfavorable and as a result, nucleophilic attack on a double bond cannot usually be performed.

The situation is quite different, however, if the π bond is strongly polarized, as is the case when the double bond is substituted by a strong electron withdrawing group, such as CF_3, or in a carbonyl group. Here the contribution of the π atomic orbital of the carbon to the antibonding molecular orbital is far greater than that of the π orbital of oxygen (Fig. 4-20) and as a result, the interaction remains favorable enough for the reaction to proceed:

$$N^{\ominus} + \text{C=O} \longrightarrow \underset{}{\overset{N}{\diagdown}}C-O^{\ominus}$$

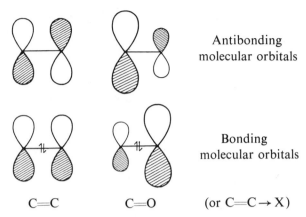

Antibonding
molecular orbitals

Bonding
molecular orbitals

C=C   C=O   (or C=C → X)

Fig. 4-20. π molecular orbitals of C=C and C=O.

2. The Mechanism of Electrophilic Attack. Let us now turn to the study of electrophilic attack and apply the concepts developed above to determine its most favorable path. As usual, the reaction involves an occupied orbital of the donor, represented in Fig. 4-21 by a biatomic bonding orbital, and an empty orbital of the acceptor represented as a single atomic orbital.

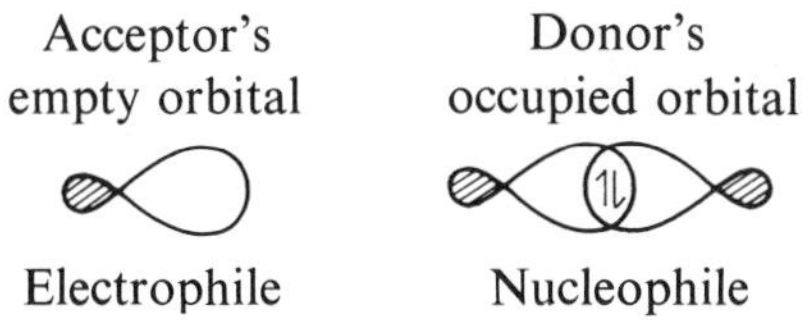

Acceptor's   Donor's
empty orbital   occupied orbital

Electrophile   Nucleophile

Fig. 4-21. Orbitals for electrophilic attack on σ bonds.

An examination of the schematic representation of back-side and front-side attacks reveals that here the sign of the lobes can be chosen to provide adequate matching in both situations (Fig. 4-22). Thus an allowed path is

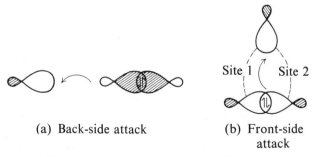

Site 1   Site 2

(a) Back-side attack   (b) Front-side attack

Fig. 4-22. Possible mechanisms of electrophilic attack.

provided for both mechanisms, and, in principle, both may occur. However, a far better overlap situation is provided by the attack on the bond and therefore strongly favors the mechanism by which it may occur, that is, front-side attack. A further incentive for such a mechanism is provided by the fact that the electrons mostly reside between the atoms and can therefore most efficiently be extracted from this region. This conclusion is borne out both by experiment and sophisticated quantum-mechanical calculations.[35] It was shown, for example, that protonation of methane occurs in strong acid, via front-side attack:[4a]

$$\underset{/}{\overset{\backslash}{\text{C}}}\text{--H} \xrightarrow{\text{H}^{\oplus}} \underset{/}{\overset{\backslash}{\text{C}}} \overset{\text{H}}{\underset{\text{H}}{\ominus}}$$

Depending on the heterogeneity of the molecular orbital, however, one may envisage various possible cases, from a central attack on a covalent bond all the way to a back-side attack on a very electronegative atom, that is, one that has a large atomic contribution to the bonding orbital (hydrogen bond!):

$$\left[\begin{array}{c} \text{H} \\ \text{H------H} \end{array} \right]^{\oplus} \qquad [\text{H---F---H---}]^{\oplus}$$

The electrophilic attack on a double bond follows a similar pattern, as shown by the schematic representation of Fig. 4-23. Here also the concerted

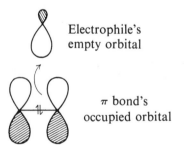

Electrophile's
empty orbital

π bond's
occupied orbital

Fig. 4-23. Electrophilic attack on a double bond.

attack on the two lobes of the π bond is allowed and the electrophile may, in the intermediate state, occupy any position from a central one to a more directional one, depending on the symmetry of the occupied π orbital that controls the reaction.

The Markownikoff rule can easily be deduced from such a treatment. Thus if electron-donating substituents, such as methyl groups, are attached to an olefinic carbon, they transfer some electrons into the inner σ orbitals of the carbon and therefore weaken its ability to accept additional electrons in its π orbital, that is, decrease its orbital electronegativity. As a result, the π

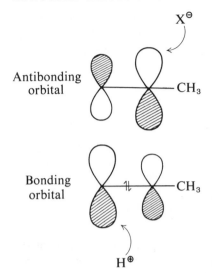

Antibonding
orbital

Bonding
orbital

Fig. 4-24. Inductive effect on position of electrophilic attack.

molecular orbitals are distorted in such a fashion as to reflect the inductive effect of the substituents, yielding a lower contribution to the bonding π orbital and a symmetrical increased contribution to the antibonding π orbital (Fig. 4-24). The electrophilic reagent will tend to occupy a position as close as possible to the largest concentration of charge in the bonding orbital and therefore settle for the other carbon atom. The process is further helped by the fact that the geigen nucleophile, seeking to interact with the antibonding orbital, will be best accommodated by the methyl substituted carbon atom.

An interesting case is provided by aromatic electrophilic substitution, where it can be shown that only a few sites are suitable for such front-side attack. Thus the HOMO of toluene has the symmetry shown in Fig. 4-25 and may accommodate an electrophile only on bonds formed by orbitals bearing the same sign, that is, 1 and 2, 1 and 6, 3 and 4, or 4 and 5.

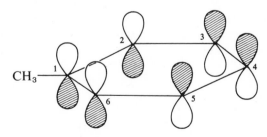

Fig. 4-25. Highest occupied molecular orbital of toluene.

The regions between atoms 2 and 3 or atoms 5 and 6 are not suited for attack by an electrophile since the molecular orbital is antibonding in this region. We may thus anticipate electrophilic attack on toluene to proceed only along one of the paths illustrated in Fig. 4-26.

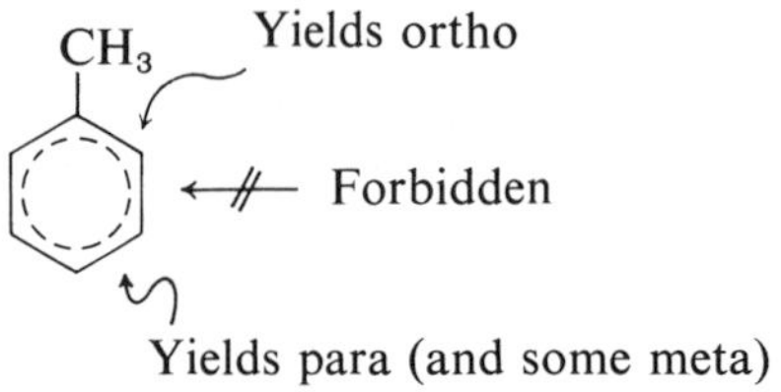

Fig. 4-26. Possible sites for electrophilic attack on toluene.

B. Intramolecular Donor-Acceptor Interactions

In some instances, a donor-acceptor interaction may occur within a molecule. The process consists of a partial transfer of electrons from one part of the molecule to another part of the same molecule. In most cases, the results are obtained as trivial extensions of the intermolecular donor-acceptor interaction case. However, when the two reaction sites involved in the process are either adjacent or conjugated, specific constrains apply and the GP treatment provides further insight into the mechanism of the reaction.

Intramolecular donor-acceptor interaction steps are usually initiated by an external reagent, nucleophilic and/or electrophilic, for example,

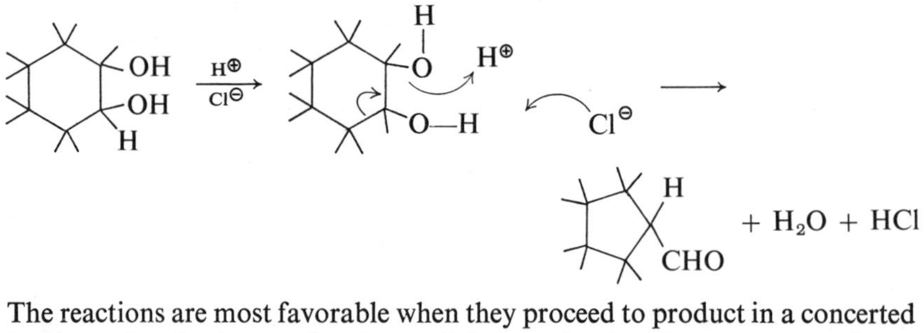

The reactions are most favorable when they proceed to product in a concerted fashion, that is, intermolecular and intramolecular donor-acceptor interaction steps occur simultaneously. In these cases, the position of attack by the external reagents is often determined by the relative geometry of the reactive sites, for example,

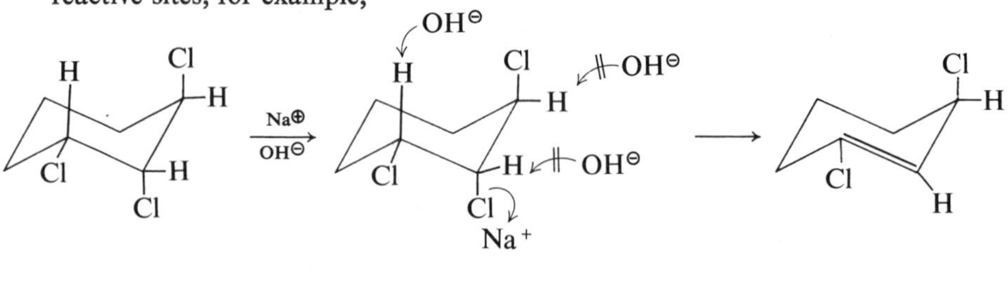

This is due to the fact that electrons proceed along the favored path from the electrophilic site (that received the electrons from the external nucleophile) to the nucleophilic site (that releases it to the external electrophile). These electronic transfers occur via intramolecular donor-acceptor interactions and their most favorable path can be discussed in terms of the GP equation. This is illustrated in the following examples.

1. The β Elimination Reaction. In the β elimination reaction two groups X and Y are lost from adjacent saturated carbon atoms that simultaneously combine to form a double bond.

$$\begin{array}{cc} X & Y \\ | & | \\ -C-C- \\ | & | \end{array} \longrightarrow \begin{array}{c} \diagdown \\ \diagup \end{array}C=C\begin{array}{c} \diagup \\ \diagdown \end{array} + X-Y$$
$$(1) \ (2)$$

Let us define the reactive *"fragments"* as those molecular entities that are separated by the bond(s), or partial bonds to be formed during the rate determining step, that is, the C—X and the C—Y bonds separated by the CC double bond to be formed. Let us postulate that each of the fragments can be defined independently and calculate the perturbation generated by their interaction. In this way, the process can be visualized as a donor-acceptor reaction in which the C—X bond acts as the donor and the C—Y bond as the acceptor via a single-site carbon(1)-carbon(2) interaction.

From the GP point of view, what has to be considered is, as before, the transfer of electrons, which occurs from the occupied molecular orbital of the C—X bond (fragment 1) to the unoccupied molecular orbital of the C—Y bond (fragment 2) (Fig. 4-27).

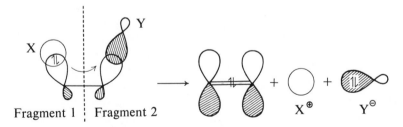

Fig. 4-27. Orbitals in the β elimination reaction.

This transfer results in the breaking of both the bond from which the electrons are removed, that is, the C—X bond, and that on which they are transferred, that is, the C—Y bond. In contrast, the overlapping region between the two carbon atoms has now become populated by the two remaining electrons and the double bond forms. Possibly also the X and Y atoms combine to form the XY molecule.

One of the important questions raised by this mechanism is whether the bond formation between atoms X and Y occurs in a concerted fashion with their release of the molecule. If this was the case, one would anticipate that this recombination participates to the lowering of the energy of the transition state and hence favors cis elimination.

The resulting reaction would then be a double-site reaction since we would have both the C—C and the XY interactions involved in the rate determining step. A simple examination of Fig. 4-27 shows, however, that this mechanism is not appropriate.

In setting up the orbital picture for the reaction, one arbitrarily chooses one of the sites orbital pair to match and determine whether the other sites orbital pair does so too. For a concerted mechanism to be allowed, all sites orbital pairs must match. In Fig. 4-27 the two carbon orbitals were chosen to match and it is seen that as a result, the orbital lobes of the X and Y group have opposite sign. Consequently, the XY recombination does not participate in the rate-determining step and cis elimination is not favored over trans elimination. The reaction is therefore a single-site reaction and must be initiated by an appropriate external reagent:

$$\mathrm{Nu}^{\ominus}$$
$$\mathrm{X}$$
$$\mathrm{C{-}C{-}Y} \longrightarrow \mathrm{NuX} + \mathrm{C{=}C} + \mathrm{Y}^{\ominus}$$

The ease with which this is done depends on the magnitude of the perturbational energy as given by the GP equation. The reaction is obviously charge controlled since its rate determining step involves proton abstraction. Consequently, strong bases are the most effective reagents (see Section II.B). Of special interest is the study of the influence of the relative position of the two bonds C—X and C—Y upon the rate of the reaction. Considerable insight in this problem is obtained when we notice that the relative rotation of the two fragments only affects the overlap between the two interacting lobes of the carbon atoms and leaves all other factors unchanged.

The resonance integral which appears in the GP equation is the only term that depends on this overlap and its variation will thus determine the magnitude of the perturbation. The larger the overlap, the larger β and the more favorable the reaction.

Three possible cases may normally occur: trans elimination, cis elimination, and gauche elimination. These are illustrated in Fig. 4-28 where for convenience only the sp^3 orbitals of the two carbon atoms are represented. It can be seen that only trans and cis elimination involve significant overlap between

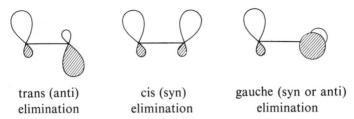

trans (anti)   cis (syn)   gauche (syn or anti)
elimination   elimination   elimination

Fig. 4-28. Carbon sp^3 orbitals as precursors of double bond in the β elimination reaction.

the relevant orbitals. In the gauche elimination, the two orbitals are in different planes and therefore cannot overlap appreciably. As a result, we find that both cis and trans elimination are more favorable than gauche elimination.

In open-chain hydrocarbons and in most cyclic hydrocarbons the two adjacent carbon atoms are staggered with respect to each other and as a result the X and Y group cannot occupy a cis position (Fig. 4-29). In these

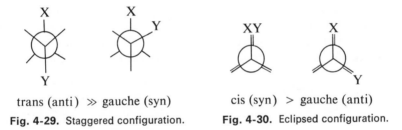

trans (anti) $\gg$ gauche (syn)   cis (syn) > gauche (anti)

Fig. 4-29. Staggered configuration.   **Fig. 4-30.** Eclipsed configuration.

cases, it is thus predicted and amply confirmed that trans (anti) elimination occurs much more readily than gauche (syn) elimination. However, in the special case of molecules forced into an eclipsed situation, we predict that cis elimination should occur more efficiently than gauche elimination (Fig. 4-30). This conclusion is supported by experimental evidence[36,37] as cis elimination was shown to occur predominantly in the compounds shown in Fig. 4-31.

Finally, one may notice that trans elimination occurs via an intramolecular transition state that resembles the intermolecular back-side attack whereas

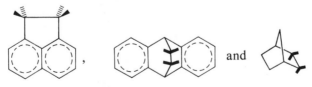

Fig. 4-31. Candidates for cis elimination.

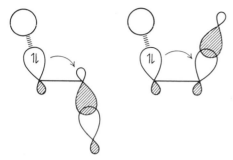

trans elimination　　　cis elimination

Fig. 4-32. Orbital picture in trans and cis elimination.

cis elimination requires front-side attack (Fig. 4-32). As a result, trans elimination is more favorable than cis elimination and the general order of reactivity is

trans elimination > cis elimination > gauche elimination

2. Vinylogous Nucleophilic Substitution. The general observation that for a donor-acceptor reaction to be favorable, the orbitals that combine must overlap appreciably, leads to interesting conclusions concerning the stereochemistry of molecules undergoing vinylogous substitution.

The reaction in point proceeds as illustrated below, for example,

$$N^{\ominus} + \overset{1}{C}H_2 = \overset{2}{C}H - \overset{3}{C}H_2Cl \longrightarrow$$

$$N - CH_2 \cdots CH \cdots CH_2 \cdots Cl^{\ominus} \longrightarrow NCH_2 - CH = CH_2 + Cl^{\ominus}$$

As in the case of the β elimination, the initial step of the reaction involves attack by an external nucleophile. A pair of electrons liberated by this initial attack can migrate towards a remote leaving group and, in the process, perform an intramolecular donor-acceptor substitution.

The position of attack is determined, as in ordinary nucleophilic substitution, by applying the GP equation to the various candidate atoms (i.e., carbon 1 and 3). However, in the event that vinylogous substitution is found to be the case, an interesting problem is generated by the possibility for the nucleophile to attack either from the top (same side as leaving group) or the bottom (opposite side from the leaving group) of the plane of the double bond (Fig. 4-33).

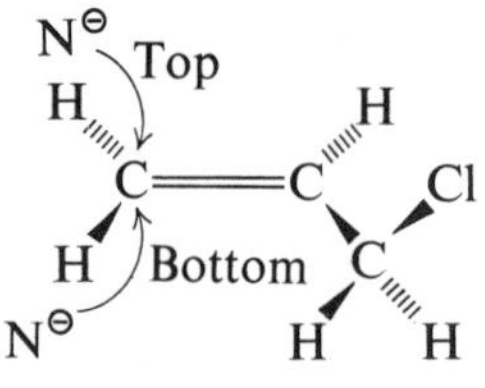

$$\begin{array}{c} N^{\ominus}\diagdown \text{Top} \\ H_{\prime\prime\prime\prime} \diagdown \downarrow \qquad H \\ \diagup C = C \diagdown \diagup Cl \\ H \diagup \text{Bottom} \quad C_{\prime\prime\prime\prime} \\ N^{\ominus} \qquad H \qquad H \end{array}$$

Fig. 4-33. Alternative path for nucleophilic vinylogous substitution.

These two possibilities, however, lead to different isomers only if the rotation around the C—C single bond is restricted. Such a case is illustrated in the following example where rotation around the single C—C bond is prevented by the ring structure (Fig. 4-34).

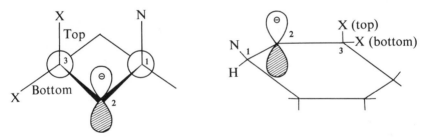

Fig. 4-34. Alternative path for vinylogous substitution of ring systems.

The mode of attack here will be determined by the ability of the molecule to process the electrons from the region of attack to the leaving group, and, as shown below, is different for attack from the top or the bottom (Fig. 4-35).

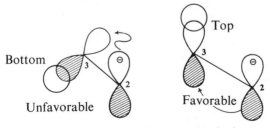

X = possible leaving group

Fig. 4-35. Orbital diagram in nucleophilic vinylogous substitution.

Let us assume that initial attack has occurred and examine the Newman projection of the intermediate cyclohexyl anion. It can easily be seen that the donor (π orbital on carbon 2) and the acceptor (antibonding orbital of the C—X bond) overlap appreciably only if they are in the same plane, that is, the leaving group is at the top position (Fig. 4-36). The reaction is thus

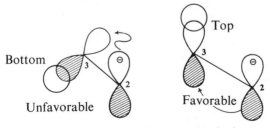

Fig. 4-36. Overlap in top and bottom substitution.

favorable only if the nucleophiles attack from the same side (top) as that from where the leaving group departs.

This prediction is borne out by the following experiment[38] where only the indicated product was found to form:

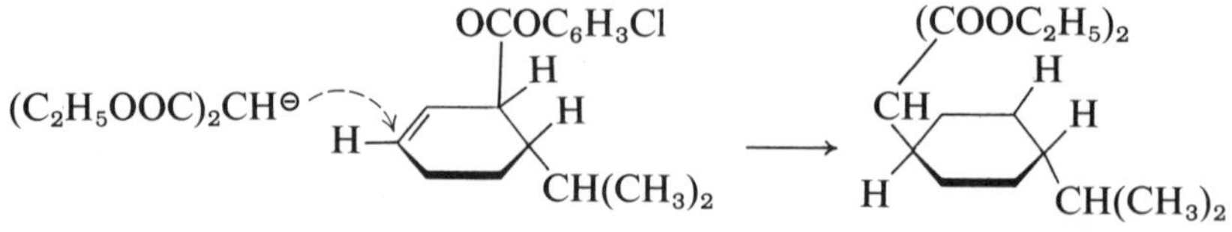

3. 1–2 Shifts in Organic Cations—Wagner–Meerwein and Other Rearrangements. Another type of reaction that can be described as an intramolecular donor-acceptor interaction is the migration of an alkyl group X to an electron deficient α position.

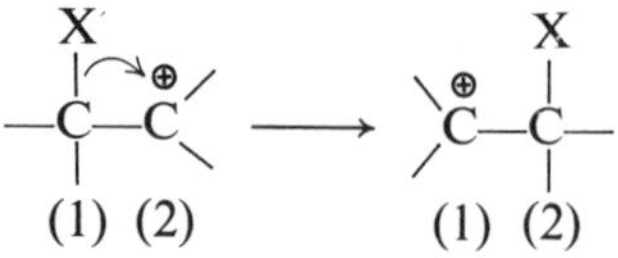

Many such rearrangements are known to occur in carbocations (e.g., Wagner-Meerwein rearrangement, Pinacol rearrangement, and Wolff rearrangement), as well as in heteroatomic electron deficient species (e.g., Hofman rearrangement).

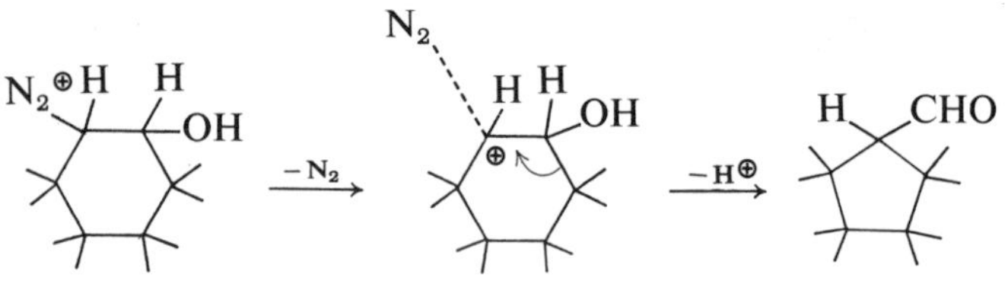

Wagner-Merwein
rearrangement

Two possible methods can be considered as reasonable to deal with such reactions, and both yield the same results.

In the first method, we consider two fragments, one consisting of the C—X bond and the other of the electron-deficient α atom. In this way, the reaction may be considered as an intramolecular electrophilic attack, where the acceptor is the empty p orbital of the electron deficient atom and the donor is the bonding orbital of the C—X bond.

The molecular orbital picture for the reaction is thus analogous to that involved in the double-site intermolecular electrophilic reaction (see Section III.B.3) (Fig. 4-37). The two fragments being independent, we may choose

Site 2

Site 1

Fragment 1   Fragment 2

Fig. 4-37. Molecular orbital diagram for 1,2 shifts in organic cations.

the sign of the empty orbital to match (give positive overlap) with the orbital of the neighboring carbon atom (site 1). By doing so, we find that it also matches the orbital of the migrating group (site 2), thus allowing the reaction to proceed intramolecularly via a three center bond (Fig. 4-38).

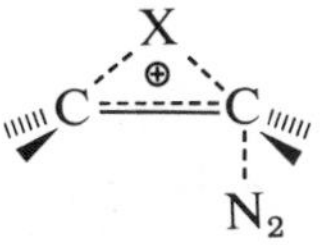

$$X$$

$$\oplus$$

C ====== C

$$N_2$$

Fig. 4-38. Intermediate in the Wagner-Meerwein rearrangement.

An interesting question also arises here concerning the possibility that the migration of the X group occurs concurrently with the departure of the leaving group (e.g., N_2 in the Wagner-Meerwein reaction).

Obviously, as soon as the nitrogen has left the molecule, there is no other alternative for the system than to proceed to product. However, if the migration may occur simultaneously, one may expect an enhancement of the reaction rate. The concerted reaction may occur only if the orbital vacated by the departure of the leaving group can appreciably overlap with the orbitals of the C—X bond. This is obtained, in usual staggered systems, only if the group that migrates is trans to the group that leaves (Fig. 4-39).

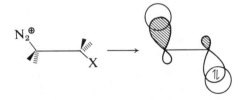

$$N_2^{\oplus}$$

X

Fig. 4-39. Orbital diagram in the Wagner-Meerwein rearrangement.

If the two groups are gauche with respect to each other, a situation similar to that encountered in the β elimination arises and the two processes will not occur concurrently. Thus, we expect that in a cyclohexane derivative forced into a specific configuration, the Wagner-Meerwein reduction yielding ring contraction will occur most favorably when the leaving N_2 group is in an equatorial position (Fig. 4-40).

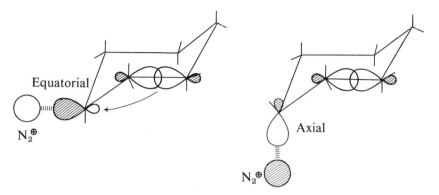

(a) Trans staggered (Same plane) (b) Gauche staggered (Different plane)
Good overlap Bad overlap

Fig. 4-40. Molecular orbitals involved in the Wagner-Meerwein elimination reaction of cylohexane derivatives with the leaving N_2 group in (a) equatorial and (b) axial positions.

In the second method available to us to deal with this reaction, we choose the fragments as separated by the *overall* bonds that are formed in the reaction rather than by the bonds formed in the rate determining step (Fig. 4-41).

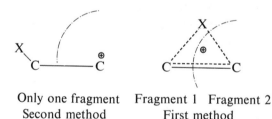

Only one fragment Fragment 1 Fragment 2
Second method First method

Fig. 4-41. Alternative representation of reaction fragments.

In this way, one obtains only one fragment, consisting of all orbitals involved in the reaction. In such a case, the sign of the atomic components of the fragments molecular orbitals must be such as to satisfy the symmetry of the system. A simple method that will be used throughout the rest of this paper is to use the "equivalent π" molecular orbitals to determine the symmetry of the orbitals of a fragment. The procedure consists in reproducing the nodes of the π orbitals in the fragments orbitals. This is illustrated in Fig. 4-42 for fragments of size 2 and 4 correlated with the bonding molecular orbital of ethane and the LUMO of butadiene.

For the reaction in point, the fragment involves three atomic orbitals of which the "equivalent π" is determined by analogy with the π molecular

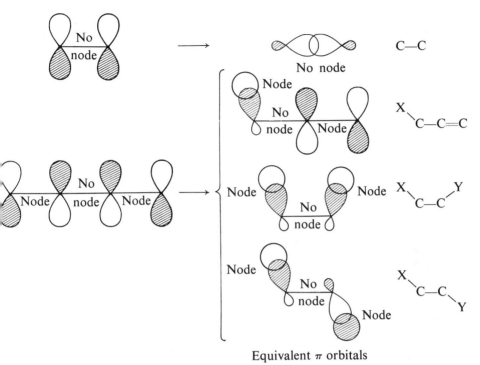

Equivalent π orbitals

Fig. 4-42. Correlation between π molecular orbitals and "equivalent π orbitals" of reactive fragments.

orbitals of the allyl ion as shown in Fig. 4-43. During the course of the reaction, two or more atomic orbital components of the fragment start interacting between themselves via formation of a new intramolecular bond, and as a result the energy of the molecular orbitals changes. The GP equation remains valid but involves the transfer of electrons from one region to another region of the *same* molecular orbital. In other terms, both the donor and the acceptor sites are linked by a common molecular orbital, that is, $E_m^* = E_n^*$ and the perturbation deals only with the change in energy experienced by the electrons of each of the occupied molecular orbitals upon "self-perturbation."

The GP equation for such a "degenerate" case is simply

$$E = -\frac{q_s q_t}{R_{st}\varepsilon} + 2\sum_{\substack{m \\ occ}} (c_s{}^m c_t{}^m \Delta\beta_{st}) \tag{12}$$

Of the two terms composing this equation only the second one, orbital control, is dependent on the symmetry of the orbitals and therefore bears an overwhelming importance as to the geometry of the transition state. Of the

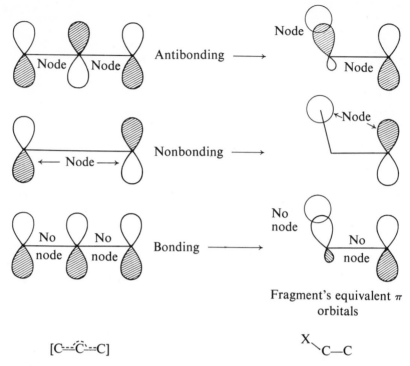

Fig. 4-43. Correlation between the equivalent π orbitals of a three-center fragment and the π molecular orbitals of the allyl molecule.

three equivalent π orbitals of the fragment obtained for the electron deficient X migration, only the lowest one is occupied, since the system involves only two electrons (2 for the C—X bond and 0 on the α atom). As a result, the perturbation, and the stereochemistry of the reaction is controlled by the symmetry of the bonding fragment orbital only. Thus, the intramolecular migration is possible here if the lobes of the X group and of the α atom match. This is seen to be the case (Fig. 4-43) and the reaction is thus allowed. From this point on all conclusions are the same as those derived from the first approach.

4. Electrocyclic Reactions of Ionic Species. Although most electrocyclic reactions are exchange reactions and thus most appropriately handled in the next chapter, those involving ions are true donor-acceptor interactions and can be studied at this stage.

a. Ring Closure of Conjugated Alkenium Ions. As a central case in point let us consider the ring closure of the allyl cation to form the cyclopropyl

Fig. 4-44. Ring closure of the allyl cation.

cation (Fig. 4-44). The procedure can be interpreted along each of the methods described in the previous section, that is, double- or single-site donor-acceptor interaction. In the first method, we consider as before two fragments, one consisting of the double bond and the other of the α carbon atom. The orbital picture is then set so as to allow positive overlap to occur between the adjacent carbon atomic orbitals of the two fragments (site 1). The ring closure occurs due to the transfer of electrons from the occupied orbital of the π bond (donor) to the p empty orbital of the α carbon atom (acceptor) (Fig. 4-45).

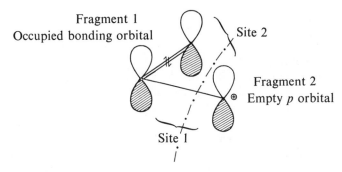

Fig. 4-45. Molecular orbitals in the allyl cation.

The stereochemistry of the ring closure is determined by the ability of the fragments to achieve such conformation as to produce positive overlap [matching between the interacting atomic components of the fragments (site 2)]. In the present case this is seen to occur upon rotation in opposite directions of the atomic orbitals 1 and 3, that is in a disrotatory mode (Fig. 4-46).

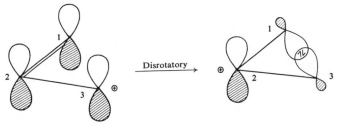

Fig. 4-46. Disrotatory ring closure of the allyl cation.

The ring closure of the allyl anion can be handled in a similar fashion. The process here involves as the donor, the occupied p orbital and as the acceptor the antibonding orbital of the double bond (Fig. 4-47).

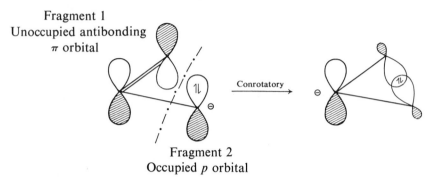

Fragment 1
Unoccupied antibonding
π orbital

Conrotatory

Fragment 2
Occupied p orbital

Fig. 4-47. Conrotatory ring closure of the allyl anion.

In contrast to the previous case, however, it is seen that the process now requires conrotatory motion of the interacting orbitals.

The interpretation of the ring closure of larger conjugated alkenium ions follows along the same line. For example, the ring closure mode of the pentadienyl anion (Fig. 4-48) is found by examining the interaction between

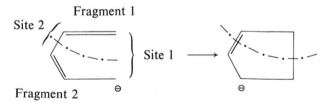

Fragment 1

Site 2

Site 1

Fragment 2

Fig. 4-48. Ring closure of the pentadienyl anion.

fragment 1 consisting of the unoccupied orbital of a double bond (acceptor) and fragment 2 composed of the *highest* occupied orbital of an allyl anion (donor) (Fig. 4-49). Here again, the molecular orbitals of the two fragments

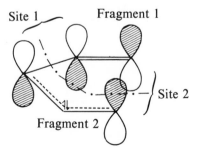

Site 1   Fragment 1

Site 2

Fragment 2

Fig. 4-49. π molecular orbitals involved in the ring closure of the pentadienyl anion.

are chosen so as to match at site 1. Positive overlap is then obtained at site 2 by disrotatory motion of the interacting orbitals. In this example we have used the highest occupied molecular (HOMO) of the donor, and we should use the lowest unoccupied molecular orbital (LUMO) of the acceptor whenever several empty orbitals are available.

A very important question is why is it that those "frontier orbitals" are so important in determining the stereochemistry of the product. The question has been very extensively debated and may perhaps best be answered on the basis of the results obtained by applying the second method, that is, single-site self-perturbation, to the problem.

It was shown previously that the total perturbation is obtained as the sum of the self-perturbation experienced by each occupied molecular orbital. However, it must be realized that when the formation of a new bond is accompanied by a rotation of the interacting atomic orbitals out of the plane of overlap with the other atomic orbitals of the system, then only two electrons may be involved in the process in order to occupy only the bonding orbital of the new bond.

This means that one of the covalent terms of the GP equation [Eq. (11)] contributed overwhelmingly to the energy change. Consequently, during the transfer of the two electrons from an initial molecular orbital to the bonding orbital of the new bond, the remaining part of the molecular orbital must fade away while it becomes more important in those molecular orbitals that remain occupied in the region not affected by the electron transfer.

It is therefore normal that the energy is maximized when the least bonding molecular orbital fades away; hence the stereochemical importance of the highest occupied orbital.

This "stereodirecting" orbital in the allyl cation has the symmetry represented in Fig. 4-50, and leads to disrotatory ring closure.

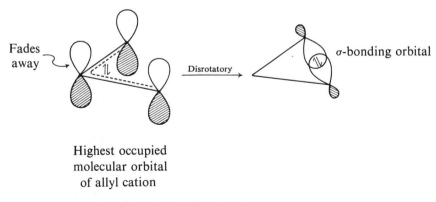

Fig. 4-50. Orbital transformations during ring closure.

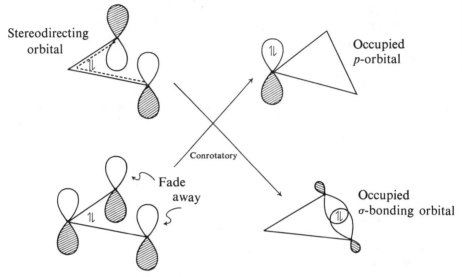

Fig. 4-51. Orbital transformations in the conrotatory ring closure of the allyl anion.

In the allyl anion the stereodirecting orbital, that is, the molecular orbital that transforms into the new σ bonding orbital, is the highest occupied molecular orbital. The lowest molecular orbital does not contribute to the stereochemical direction of the reaction since its atomic components in the region of the new bond fades away during the process (Fig. 4-51).

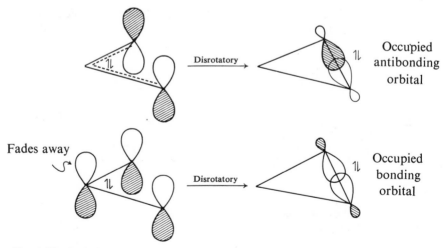

Fig. 4-52. Orbital transformations in the disrotatory ring closure of the allyl anion.

The ring closure therefore occurs in a conrotatory mode. One may note that if ring closure occurred in a disrotatory mode, the stereodirecting orbital would stransform into a σ antibonding orbital, thus creating a very unstable situation (Fig. 4-52).

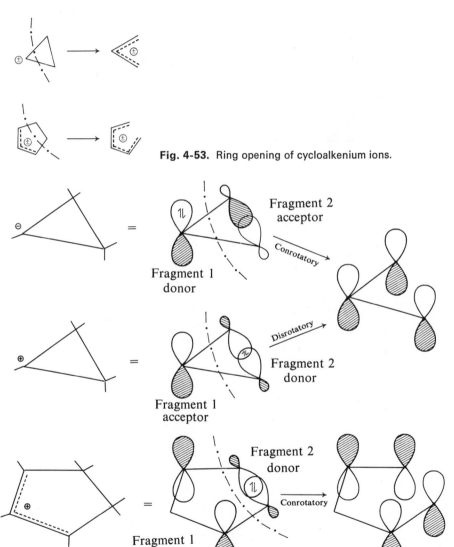

Fig. 4-53. Ring opening of cycloalkenium ions.

Fig. 4-54. "π equivalent" molecular orbitals and mechanism of ring opening of cyclopropyl and cyclopentenyl ions.

b. Ring Opening of Conjugated Cycloalkenium Ions. The ring opening of cycloalkenium ions is the reverse of the previous ring-closure reaction (Fig. 4-53). As usual the two fragments are determined as those molecular entities separated by the bonds (or partial bonds) which form during the reaction.

The disrotatory or conrotatory ring opening is thus determined from the symmetry of the HOMO and LUMO of the fragments orbitals by the matching orbital pairs of both sites. The process is illustrated in Fig. 4-54 for the ring opening of the cyclopropyl anion, the cyclopropyl cation, and the cyclopentenyl cation.

5. *1,2-Anionic Rearrangements.* 1,2-Anionic rearrangements include such reactions as the Stevens rearrangement, the Wittig rearrangement, and the Sommelet rearrangement.[39]

$$R\text{—}\overset{\oplus}{N}\text{—}\overset{\ominus}{C}\text{—} \longrightarrow N\text{—}C\text{—}R \qquad \text{Stevens (Sommelet)}$$

$$R\text{—}O\text{—}\overset{\ominus}{C}\text{—} \longrightarrow {}^{\ominus}O\text{—}C\text{—}R \qquad \text{(Wittig)}$$

The reaction is believed to proceed via the intermediacy of a carbanion generated by the action of a strong base and yields a product in which the migrating group has retained its configuration; that is,

(Wittig rearrangement)

The fragments involved in this reaction are, on the one hand, a *p* orbital, and on the other, a σ bond (Fig. 4-55). The shift of electrons occurs from the

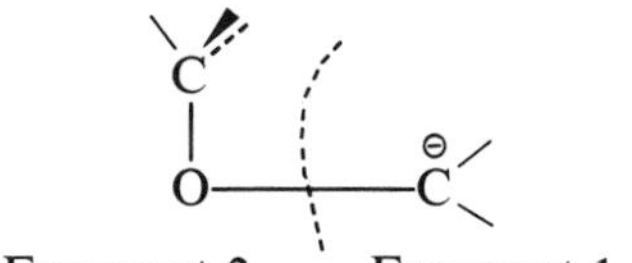

Fig. 4-55. Fragments in the 1,2-anionic rearrange-
Fragment 2   Fragment 1   ment.

bioccupied p (or sp^3) orbital to the antibonding σ orbital and results in the transfer of the $CR_1R_2R_3$ group. The orbitals involved in the reaction may be written in the two ways shown in Fig. 4-56. Both are representative of a *front-*

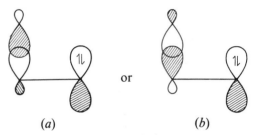

(a)   (b)

Fig. 4-56. Alternative representations of the orbitals involved in the 1,2-anionic shifts.

side nucleophilic attack! Since there is no possibility of matching the p (sp^3) orbital lobe with both atomic orbitals of the σ bond, we may rule out the concerted migration involving a three-centered bond of the kind that was found in the 1-2 rearrangement of carbocations (Fig. 4-57).

Allowed   Forbidden
Carbocation   Carbanion   **Fig. 4-57.** Three-membered ring intermediates.

The two alternative representations of the molecular orbitals yield to two different conclusions.

In the first case (Fig. 4-56a), positive overlap is provided between the occupied p orbital and the atomic orbital of the heteroatom involved in the σ bond. As a result, it is seen that negative overlap is generated by the interaction between the carbanion and the migrating group, and consequently the reaction cannot proceed in a concerted fashion. The mechanism resulting from such a description is analogous to that suggested originally by Stevens[40] and involves a dissociation step followed by recombinations (Fig. 4-58). This

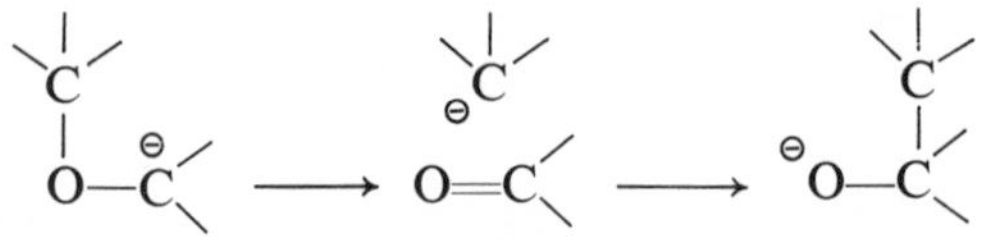

Fig. 4-58. Possible mechanism of the Stevens rearrangement.

mechanism, however, is improbable since it involves a very unstable carbanion. Furthermore, it does not explain the stereospecificity observed for the rearrangement.

In the second case, positive overlap is provided between the migrating group and the p_Π orbital. (Fig. 4-56b), and as a result negative overlap is found between the p orbital and the neighboring heteroatomic σ orbital component. Thus, no possibility of forming a partial double bond between the carbanion and the heteroatom exists in this case. The proper structure of the transition state is then most appropriately represented as in Fig. 4-59 where

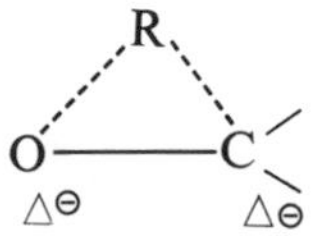

Fig. 4-59. Possible transition state in the Stevens rearrangement.

the R group, while migrating, provides a channel for the negative charge to go from the least favorable carbon site to the most favorable oxygen site. In this mechanism, the R group migrates as a neutral (free radical!)[41] entity and, in contrast to the carbocation case, at no time is there a partial double bond character between the carbon and the oxygen atom. The driving force is provided here by the ability of the heteroatom to absorb the negative charge liberated by the process.

This mechanism may possibly be made even more appealing if, rather than assuming a planar intermediate, one chooses to rotate the occupied p (sp^3) orbital of the carbanion out of the plane (Fig. 4-60). This relieves its repulsion

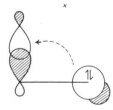

Fig. 4-60. Possible position of the interacting orbitals in the 1,2-anionic shifts.

with the heteroatomic component of the σ bond without destroying the overlap channel with the migrating group.

Finally, one should mention a third possibility, which, however, should occur with inversion of configuration. In this, the back lobe of the migrating group may interact with the carbanion orbital to yield the product in a concerted fashion (Fig. 4-61).

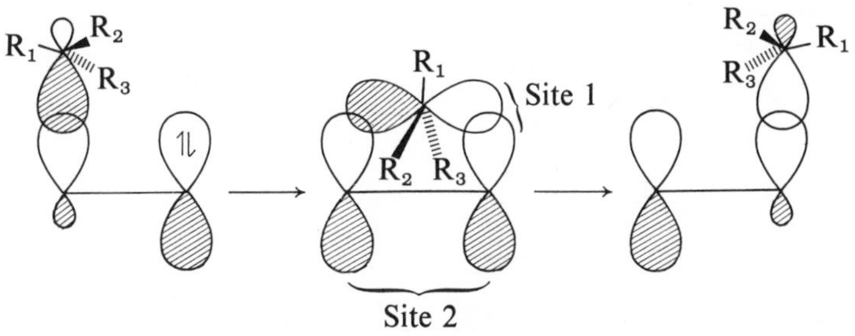

Fig. 4-61. Inversion of configuration in 1,2-anionic shifts.

However, the steric hindrance in this case is probably prohibitive.

Further experiment is needed before we may assess whether only one or several of these possible reaction paths really occurs.

IV. EXCHANGE REACTIONS

A. Theory

Exchange reactions were defined as single- or double-site reactions in which no formal charge is transferred from one of the reagents to the other. Actually, electronic transfers do occur, but in opposite directions, and probably at different sites, for example, addition of diazomethane to olefins:

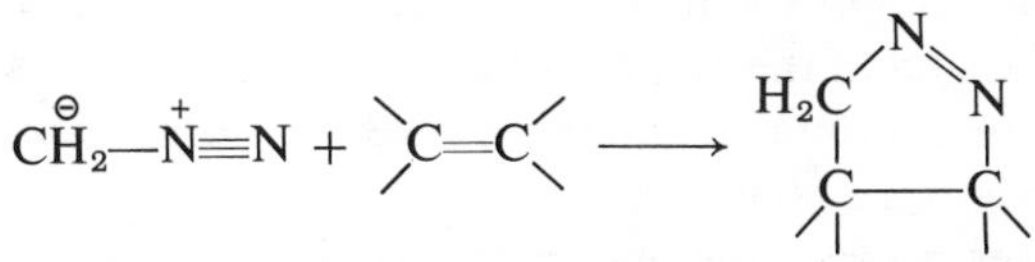

However, the specific direction of the electronic process at each of the sites remains often undefined because of the characteristic feature that the two reaction sites are linked by a continuous string of overlapping orbitals, as for example in the Diels-Alder Reaction (Fig. 4-62).

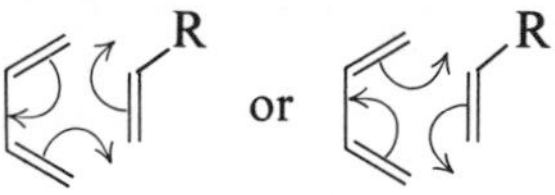

Fig. 4-62. Electron migration path in the Diels-Alder reaction.

The perturbation energy for the concerted exchange interaction of two species through two sites s-t and s'-t' is given by the sum of the two simultaneous charge-transfer interactions occurring in opposite directions (Fig. 4-63).

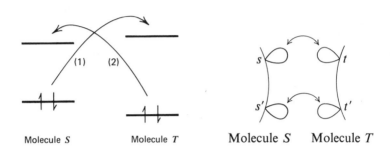

Molecule S Molecule T Molecule S Molecule T

(1) + (2) is an Exchange Reaction
(1) Transfer of electrons from S to T (Donor-Acceptor Interaction)
(2) Transfer of electrons from T to S (Donor-Acceptor Interaction)

Fig. 4-63. Exchange Reaction.

The result is given by the most general form of the GP equation,

$$\Delta E = -\frac{q_s q_t}{R_{st}\varepsilon} - \frac{q_{s'} q_{t'}}{R_{s't'}\varepsilon} + \sum_m \sum_n \nu_{mn} \frac{(c_s{}^m c_t{}^n \Delta\beta_{st} + c_{s'}{}^m c_{t'}{}^n \Delta\beta_{s't'})^2}{E_m^* - E_n^*} \qquad (13)$$

where ν_{mn} equals 0 if the sum of the number of electrons in m and n is 0 or 4, ν_{mn} equals 1 if the sum of the number of electrons in m and n is 1 or 3, and ν_{mn} equals 2 if the sum of the number of electrons in m and n is 2. Thus, only interactions between orbitals containing together at least one and at most three electrons are to be considered.

The interactions between the unoccupied orbitals of both reagents do not have to be considered as they do not involve electrons and thus do not lead to energy changes. Similarly, the interaction between the doubly occupied orbitals of both reagents may to a first approximation be neglected as the result of their interaction is essentially symmetrical* (Fig. 4-64).

As a result whatever energy is gained by the two electrons of the most binding orbital is lost by those in the less bonding orbital and no change in

---

* This is only true as long as one neglects the overlap. When the overlap is included, this term may lead to a small repulsion.[12]

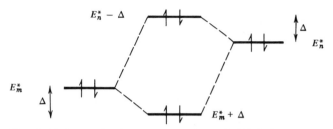

Fig. 4-64. Interaction between two bioccupied orbitals.

energy is being produced by the interactions:

$$E_{\text{initial}} = 2E_m^* + 2E_n^*$$

$$E_{\text{final}} = 2(E_m^* + \Delta) + 2(E_n^* - \Delta)$$

$$\Delta E = 0$$

In reactions involving reagents possessing singly occupied orbitals, for example, radicals or excited states, the calculations must include the interaction of the singly occupied orbitals with both occupied and unoccupied orbitals since they both lead to energy changes. If the separation between the singly occupied orbital with both an empty and bioccupied orbital is the same, both interactions will carry the same weight (Fig. 4-65). If, however, the

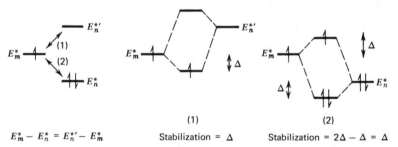

Fig. 4-65. Symmetrical interaction of a singly occupied orbital with equidistant LUMO and HOMO.

separation is different, then the "vicinal" orbital interaction will be overwhelming (Fig. 4-66). This is often the case, and reagents, even in exchange reactions can usually be classified as electrophilic or nucleophilic.

The GP equation [Eq. (13)] can be used to obtain a semiquantitative value of the stabilization of the transition state, providing that the necessary quantum-mechanical data are accessible. The most important information can, however, be gathered even without such extensive calculations. Exchange

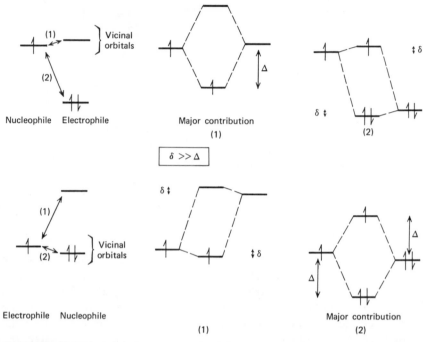

Fig. 4-66. Vicinal orbitals in the interaction of a singly occupied orbital with a closed-shell molecule.

reactions may occur either in a concerted or in a stepwise fashion, leading to stereochemically different products. The determination of the nature and stereochemistry of the product is in these complex reactions often more important than the rate at which they are formed. Among the factors responsible for the occurrence of a given concerted reaction in a definite stereochemical fashion, only the covalent term leads to strict geometrical constraints. As a result, when the stereochemistry of a concerted reaction is being investigated, it is often sufficient to examine the covalent term in order to predict the nature of the product..

Thus a simple examination of the symmetry of the controlling orbitals of the reagents, as was already carried out in the double-site donor-acceptor interactions, will often prove sufficient.[42,43] Such an approach is obviously very similar, in spirit, to that suggested by Woodward and Hofmann,[8] and applies among many others, to all the cases treated by these authors.[10] Its advantages are numerous: first of all it is a simple extension of the donor-acceptor generalized perturbation and therefore provides a consistent approach to both single-site and double-site reactions, concerted or not. In addition, the treatment lends itself to semiquantitative applications including solvent effects and finally, as will be shown in the following pages, provides the rational for a wide range of complex reactions.

According to the GP equation [Eq. (13)], only those interactions between an occupied orbital of one of the reagents and an unoccupied orbital of the other are to be considered, unless one of the reagents is a radical or an excited species. The largest contribution to the change in the covalent term of the energy is usually provided by the specific interaction of a given pair of molecular orbitals. In accord with our previous discussion of this matter (see p. 103) we will refer to these orbitals as the "stereodirecting orbitals."

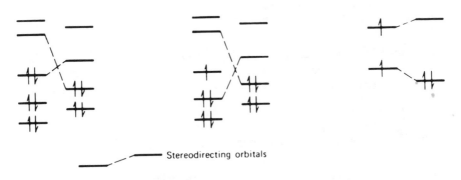

Fig. 4-67. Examples of stereodirecting orbitals.

For closed-shell reagents, the stereodirecting orbitals are simply the frontier orbitals, HOMO and LUMO, but for radicals, excited or ionized species, this is not necessarily so. The largest perturbation produced by the interaction of two molecular orbitals is produced when they are vicinal, that is, the closest. Their interaction, however, carries a weight that depends on the total occupancy of the pair of orbitals under consideration. It was shown previously that if the total occupancy is equal to 0 or 4, the perturbation is negligible, $\nu_{mn} = 0$. When one or three electrons are involved in the interaction, $\nu_{mn} = 1$ but for a total occupancy of two, ν_{mn} has the largest value and is equal to 2. As a result, the interaction of two vicinal orbitals, involving one or three electrons, may carry less weight than the interaction of two orbitals separated by a larger gap, but involving two electrons. The various possibilities are illustrated in Fig. 4-67. Figure 4-68 exemplifies possible discrepancies between frontier and stereodirecting orbitals.

$[CH_2{=}CH{-}CH{=}CH_2]^+$
Stereodirecting orbitals
Frontier orbital
Stereodirecting orbital

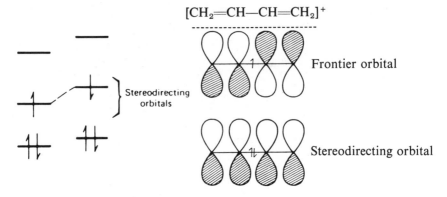

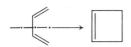

Fig. 4-68. Examples of discrepancy between stereodirecting orbitals and frontier orbitals.

Having identified the stereodirecting orbitals, we can now apply to them the matching rule (see p. 85) and predict whether a given concerted reaction is allowed or not. The full procedures thus consist of the following steps.

1. Determine the stereodirecting orbitals of the reagents.
2. Calculate the perturbation between them or simply determine if the interacting orbital pairs match at each of the reaction sites.

(*a*) Ring closure of butadiene
Butadiene treated as two ethylene fragments
(*b*) Ring opening of cyclobutene
Cyclobutene treated as one ethylene fragment and one ethane fragment
(*c*) Ring closure of hexatriene
Hexatriene treated as one butadiene fragment and one ethylene fragment

Fig. 4-69. Determination of the reactive fragments in intramolecular reactions.

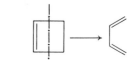

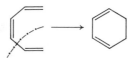

In predicting the most favorable path, the following remarks have to be kept in mind.

1. The molecular orbitals for an intramolecular reaction are determined, as described already on p. 91, by treating the molecule as two independent fragments separated by the bonds to be formed. In this, a σ bond takes preference over a π bond, and, if several possibilities still exist, any of them should be satisfactory (Fig. 4-69).

2. Even if no matching exists between the orbital pairs of the reactive fragments, a double-site exchange reaction may still be possible due to the dynamics of the system. Thus, if a slight distortion of one of the bonds is sufficient to initiate substantial overlap (matching), the reaction may proceed further and yield the desired product. However, such reaction is expected to be slow unless the two fragments are already linked together (Fig. 4-70).

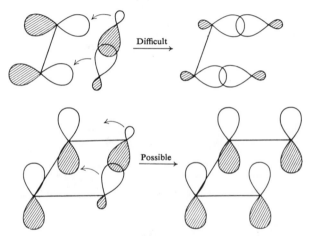

Fig. 4-70. Intermolecular and intramolecular reaction between a π and a σ bond.

3. A double-site interaction does not necessarily imply that the two bonds are to be formed to the same extent at all times; that is [see Eq. (13)]

$$c_s c_t \neq c_{s'} c_{t'} \quad \text{and} \quad \Delta \beta_{st} \neq \Delta \beta_{s't'}$$

As a matter of fact, it was found that the regiochemistry of exchange reactions is best explained when it is assumed that the formation of one of the bonds runs at least slightly ahead of the other.

When the concerted double-site reaction is allowed, this differential is too small to allow the reagents to rearrange, but large enough to favor the path that would have proceeded to the formation of the most stable single-site

interaction intermediate, for example,

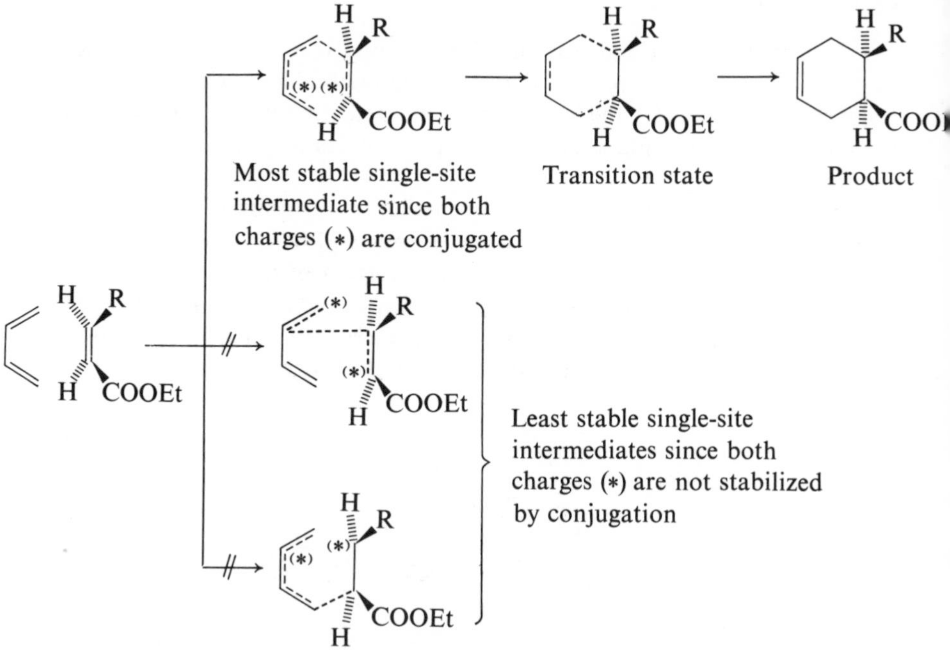

Most stable single-site
intermediate since both
charges (∗) are conjugated

Transition state

Product

Least stable single-site
intermediates since both
charges (∗) are not stabilized
by conjugation

On the other hand, when the straightforward double-site process is difficult, such as if the cyclic intermediate is strained, then the formation of the
two bonds proceed at a substantially different pace.

In the resulting single site exchange interaction, the reagents have in addition the possibility to reorganize before the cyclic intermediate is formed:

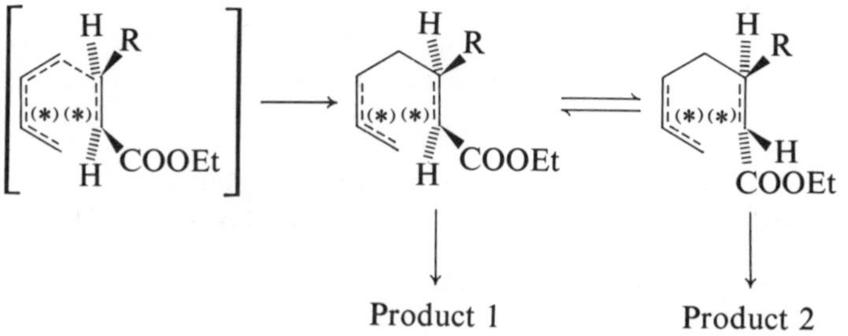

Product 1

Product 2

4. Finally, one should remember that when the conclusion of the examination is that an allowed double-site path exists, this does not mean that the
reaction is going to proceed along such a path, but merely that it is a facile
route. On the other hand, if a path is forbidden, the coresponding reaction
cannot occur.

Before proceeding to the illustration of these rules, it is useful to classify the reactions in terms of the size of the fragments that are interacting. We shall define the "fragment size" as the smallest number of orbitals that have to be counted to go from one of the sites to the other in the "π equivalent" picture of the fragments.

Thus a 1 + 1 interaction is a single-site reaction, a 2 + 3 interaction is a double-site interaction between orbitals 1 and 2 of one of the reagents with orbitals 1 and 3 of the other, and so on (Fig. 4-71).

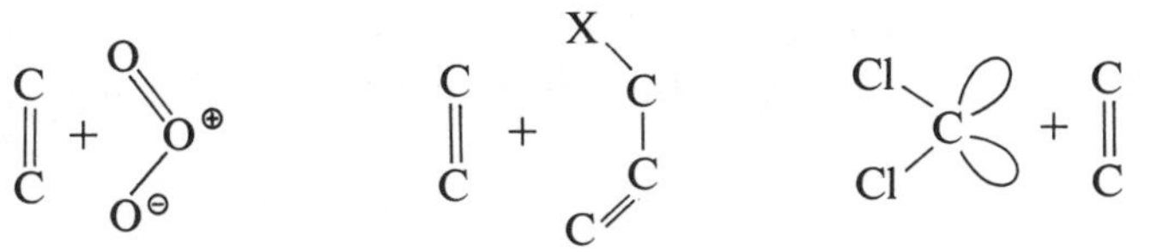

| (2 + 3) Interaction | (2 + 4) Interaction | (2 + 2) Interaction |

Fig. 4-71. Classification of reactions according to the size of the fragments.

A further classification is obtained when, within each category, reactions are classified with respect to the type of bonds that link the sites of each of the reagents, that is, π or σ bonds. The letter n is used when two orbitals of a single atom form a fragment. Thus the interactions of Fig. 4-71 are, respectively $(2 + 3)(\pi + \pi)$, $(2 + 4)(\pi + \pi\sigma)$, and $(2 + 2)(n + \pi)$. Further examples are shown in Fig. 4-72.

We have already discussed the 1 + 1 and 1 + 2 reactions and seen that both are donor-acceptor interactions. Let us now consider the higher-order reactions such as 2 + 2, 2 + 3, 2 + 4,... all of which can be exchange reactions.

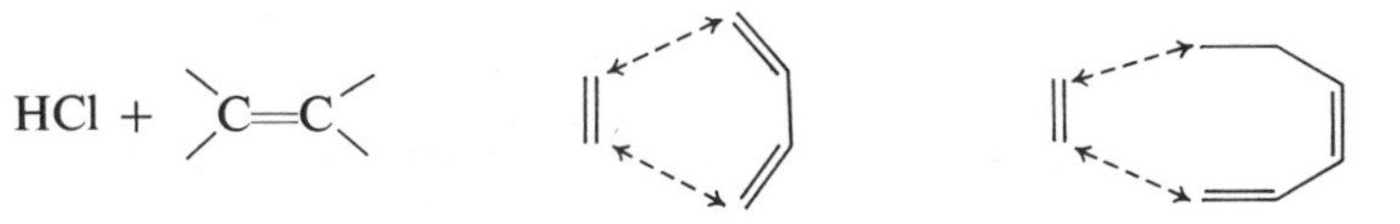

HCl + $\diagdown$C=C$\diagup$

| $(2 + 2)\ (\sigma + \pi)$ | $(2 + 4)\ (\pi + \pi\pi)$ | $(2 + 6)\ (\pi + \pi\pi\sigma)$ |

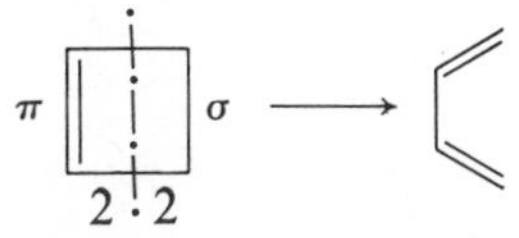

The ring opening of cyclobutene
is a $(2 + 2)\ (\pi + \sigma)$ reaction

Fig. 4-72. Classification of reactions according to the size and nature of the fragments.

B. 2 + 2 Exchange Reactions

The 2 + 2 exchange reactions may be of several different kinds depending on the nature of the bonds separating the interacting orbitals in each of the fragments. Among these, the $\pi + \pi$, $\pi + \sigma$, $\sigma + \sigma$, and $\pi + n$ are the most frequently observed combination.

1. (2 + 2) (π + π) Exchange Reactions

a. Case I. Dimerization of Ethylenic Derivatives. The $\pi + \pi$ addition which would result from the interaction of two ethylene molecules to yield cyclobutane is normally forbidden as a concerted reaction. This is seen from the fact that no matching exists between the stereodirecting orbitals in Fig. 4-73.

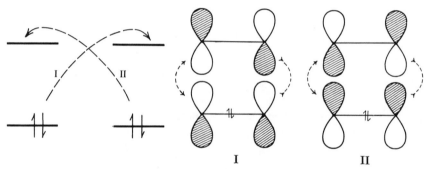

Fig. 4-73. (2 + 2)($\pi + \pi$) addition.

In the excited state, however, the process is permissible since matching does occur between the vicinal orbitals (Fig. 4-74).

$$C_2H_4 \xrightarrow{h\nu} C_2H_4^* \xrightarrow{C_2H_4} C_4H_8$$

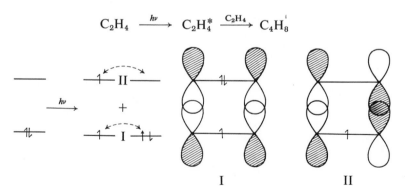

Fig. 4-74. Dimerization of ethylene under irradiation.

This is observed experimentally, as shown by the fact that butenes dimerize under irradiation, to yield exclusively the indicated products.[44]

Actually, it is found that activated double bonds such as $CF_2{=}CF_2$, $CH_2{=}CH{-}CN$, $CH_2{=}C{=}CH_2$, and $CH_2{=}C{=}O$ sometimes add to each other even in the absence of irradiation[45]; for example,

$$2\ CF_2{=}CF_2 \longrightarrow \begin{array}{c} F_2C{-}CF_2 \\ |\ \ \ \ | \\ F_2C{-}CF_2 \end{array}$$

Since the straightforward addition is forbidden, we must envision an alternative mechanism, possibly along a path such as illustrated in Fig. 4-75.

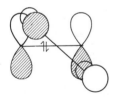

Fig. 4-75. Possible intermediate in the dimerization of ethylenic derivatives.

In this case, matching of the two orbital pairs does occur, but the very strict geometrical constraints will tend to render the concerted reaction very difficult. Thus, although the reaction must *necessarily* proceed via a cyclic intermediate such as described in Fig. 4-75, it is usually admitted that a diradical (or zwitterion) produced along a *single-site* interaction path constitutes the most plausible transition state. Thus, only after one of the bonds is formed, can the two parts of the molecule bend over each other to reach the orbital lobes of the correct symmetry and close the ring. The ring closure can, as illustrated in Fig. 4-76, proceed along two alternative routes, yielding two

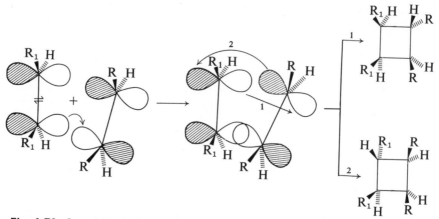

Fig. 4-76. Stereochemical consequences of the $(2 + 2)(\pi + \pi)$ exchange reaction.

different isomers. This mechanism is supported by the fact that, whenever possible, the reaction actually leads to several isomers indicating that rotation does occur at some stage of the reaction path, for example,

$$\underset{\substack{H_3C \\ H}}{\overset{}{\diagdown}} C{=}C \underset{H}{\overset{HC{=}C\overset{H}{\diagup}}{\diagdown}} CH_3 \xrightarrow{CF_2CCl_2}$$

$$\underset{F_2C\text{————}CCl_2}{\overset{H_3C \quad H \quad H \quad CH{=}CHCH_3}{\underset{C\text{————}C}{}}} + \underset{F_2C\text{————}CCl_2}{\overset{H_3C \quad H \quad \overset{CH{=}CHCH_3}{\underset{C\text{————}C}{}} H}}$$

Another interesting problem associated with these reactions concerns the rather unexpected orientations exhibited by the products. Most of these reactions may, in principle, yield at least two different positional isomers, for example,

(a) $2\ CH_2{=}CH{-}CN \longrightarrow$ [cyclobutane with CN at 1,2-positions] $+$ [cyclobutane with CN at 1,3-positions]

(b) $2\ CH_2{=}C{=}CH_2 \longrightarrow$ [cyclobutane with CH_2 at 1,2] $+$ [cyclobutane with CH_2 at 1,3]

(c) $2 CH_2=C=O \longrightarrow$

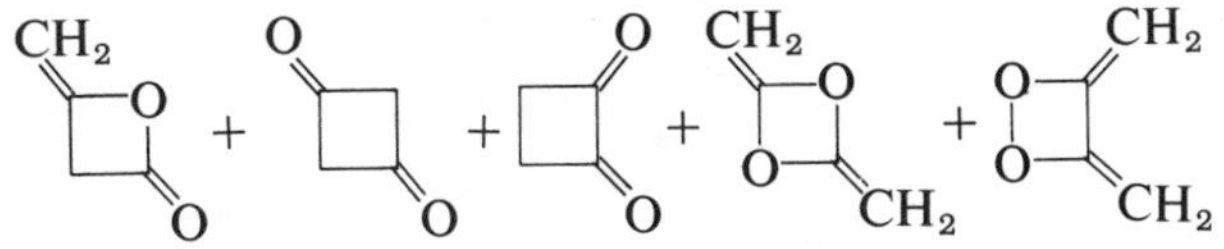

However, it is found that one of them is obtained much more abundantly in each case. In the examples above, the first of each of the products is the one experimentally found to predominate.

This can easily be understood on the basis of the suggested mechanism and the hypothesis, made on p. 115, concerning the stability of the single-site intermediates. Thus it was suggested that, if the formation of one of the bonds runs ahead of the other, the most favorable reaction path will be that initiated by the most favorable single-site interaction.

Let us consider first the case of acrylonitrile. The largest coefficients for both the HOMO and the LUMO are found at the carbon end positions. In accordance with the theory, we thus expect the bond between these two positions to form first. The resulting intermediate is also the precursor of the most stable single-site interaction intermediate, and upon ring closure must give a mixture of *cis*- and *trans*-1,2-dicyanocyclobutane but no 1,3 derivative:

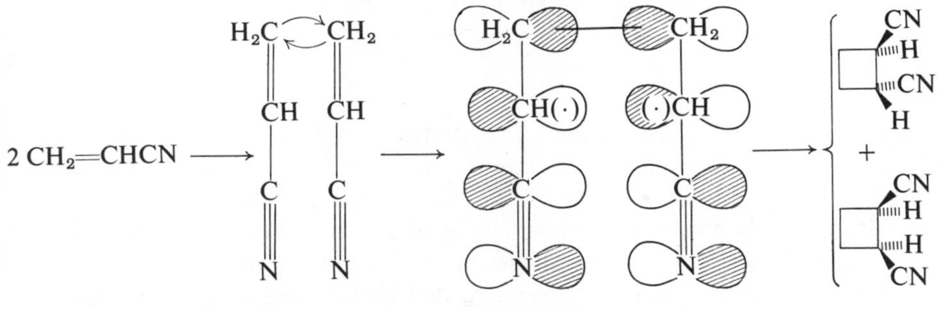

This result parallels the experimentally observed one.

Allene and ketenes present particularly interesting cases as they both react easily in 2 + 2 additions but in rather significantly different fashion.

For the dimerisation of allene, the GP equation predicts the following single-site interaction intermediates to be the most favorable:

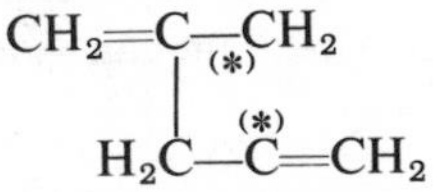

However, due to the particular ability of allene to extend its conjugation by rotating one of the end orbitals by 90°, the most favorable single-site interaction intermediate will rather be one in which the two central carbon atoms are bonded. This permits the electrons to shift into the end carbon atom orbitals where they now can participate in the extended allylic conjugation

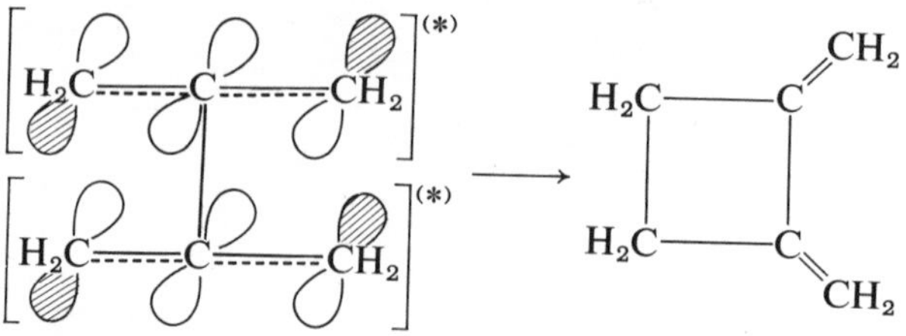

Fig. 4-77. Dimerization of allene.

(Fig. 4-77). Accordingly, the major product is expected to be the 1,2-di-methylenecyclobutane, with some 1,3 derivative produced along the former pathway.

Experimentally, one obtains 85% of the former product and 15% of the latter, in reasonable agreement with our expectations. This result however could not have been anticipated from the GP theory alone.

Ketenes have the interesting property of having two *different* π systems orthogonal with each other:

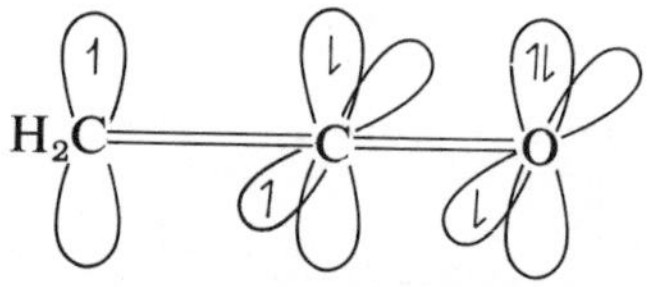

that is, a C=O bond involving two electrons and, perpendicular to it, an allylic system containing four electrons. As a result, ketenes, which are mostly excellent acceptors, via attack on the carbon orbital of the C=O bond, are also relatively good donors via attack by either the end carbon (soft) or the oxygen (hard) orbitals of the allylic system. The dimerization can be visualized as an initial donor-acceptor interaction between one π system of one of the reagents and the other π system of the other reagent, yielding the following single-site interaction intermediate:

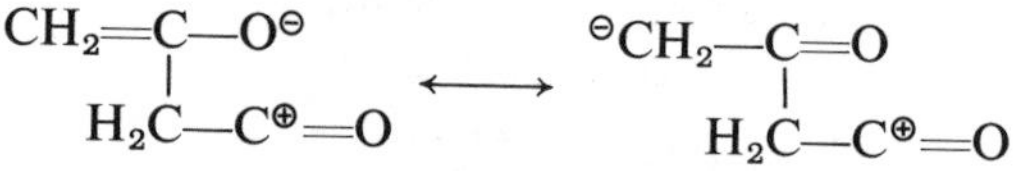

Upon ring closure, such an intermediate can give either one of the following products:

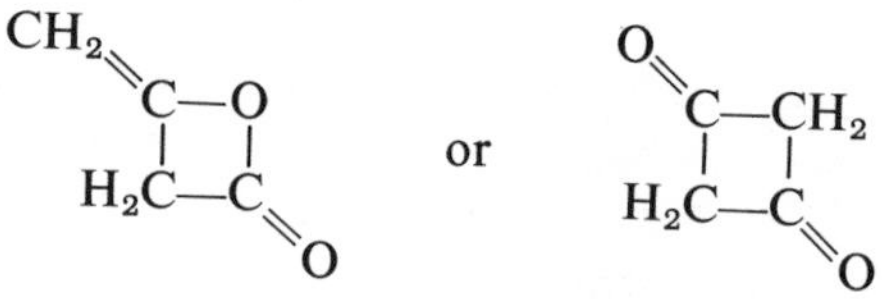

It is found experimentally that in the absence of strong steric repulsions, the former one is formed,* for example, for ketene itself, whereas with substituted ketenes, the product progressively becomes such as in the latter structure [46]:

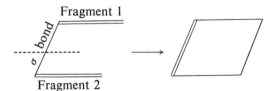

50%   +   50%

In the addition to cyclopentadiene, ketene again reacts via its central carbon with one of the ends of cyclopentadiene, and, upon ring closure yields the following product [47]:

The last question that remains to be answered is why are these derivatives so apt to undergo 2 + 2 addition? The answer is probably that each has the ability to strongly stabilize an intermediate where partial rotation has already occurred, thus preparing for an easier ring closure.

$$\overset{|}{C}F_2—\overset{(\cdot)}{C}Cl_2, \ \overset{|}{C}H_2—\overset{(\cdot)}{C}H—CN, \ \overset{(+)}{C}H_2—\overset{|}{C}=CH_2, \ CH_2=\overset{|}{C}—\overset{(-)}{O}, \ [\overset{|}{C}H_2—\overset{(+)}{C}=O]$$

b. Case II. Electrocyclic Ring Closure of Butadiene. In the ring closure of butadiene to form cyclobutene, butadiene may be considered as formed of two ethylene fragments attached together by a σ bond (Fig. 4-78). The major

Fragment 1

σ bond

Fragment 2

Fig. 4-78. Electrocyclic ring closure of butadiene.

perturbation term is that between the biooccupied (bonding) orbital of one of the fragments by the unoccupied (antibonding) orbital of the other (Fig. 4-79; see also Section III.B.4).

* This is to be expected as attack on a C=O bond is a charge controlled process most efficiently executed by an oxygen atom.

No matching normally exists between the orbital pairs but a slight distortion of the molecule such as a rotation of the end orbitals creates enough incentive for the reaction to go further along the path that was initiated (see remark 2, p. 115). However, it can be seen that only a concerted rotation of the two end orbitals in the same direction leads to a match of the pair (Fig. 4-79).

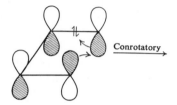

Conrotatory

Fig. 4-79. Ring closure of butadiene.

As a result, the reaction may proceed along a concerted path, but only in a *conrotatory* fashion. This conclusion is in accord with that derived from the rules of conservation of orbital symmetry and was amply proven experimentally[48] (Fig. 4-80).

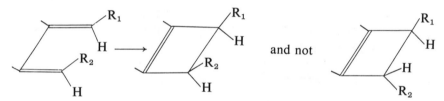

and not

Fig. 4-80. Stereochemical consequences of the conrotatory ring closure of butadiene.

The system can equally well be treated as a self-perturbation of butadiene (see Section III.B.3) in which case the highest biocuppied orbital (HOMO) is the stereodirecting orbital. The conclusions concerning the mode of ring closure are the same as those derived above, namely, conrotatory ring closure (Fig. 4-81). Under irradiation, butadiene closes in a *disrotatory* fashion, that is, the orbitals rotate in opposite directions. This is found simply by consider-

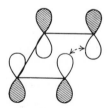

Fig. 4-81. Stereodirecting orbital of butadiene.

ing the interactions of the vicinal orbitals of the two ethylene fragments where one of the fragments is in an excited state (Fig. 4-82).

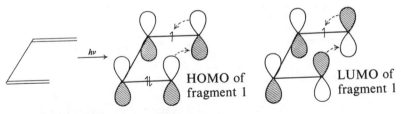

HOMO of fragment 1

LUMO of fragment 1

Fig. 4-82. Ring closure of excited butadiene.

2. *(2 + 2)(π + σ) Exchange Reactions.* The reaction between a π bond and a σ bond is normally a difficult concerted reaction. The reason is that proper overlap is initiated only by a partial breaking of the σ bond, that is, rotation of the σ orbitals, and usually requires an external reagent to promote it. However, in cases where the two fragments are already attached together, the vibrational energy generated by thermal or light excitation may be sufficient to initiate the reaction. Several such cases are known to occur and are studied below.

a. Case III. Electrocyclic Ring Opening of Cyclobutene. This reaction is the reverse of the ring closure of butadiene and can be treated in a similar fashion. The two fragments that have to be considered are a π and a σ bond (Fig. 4-83). For the thermal reaction, we consider again the interaction between the bonding orbital of one of the fragments and the antibonding

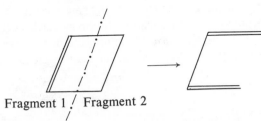

Fragment 1 / Fragment 2

Fig. 4-83. Fragments in the electrocyclic ring opening of cyclobutene.

orbital of the other. The results obtained by considering either alternative π bonding + σ antibonding or π antibonding + σ bonding are normally consistent and it is usually sufficient to examine only one of the charge transfer processes (Fig. 4-84).

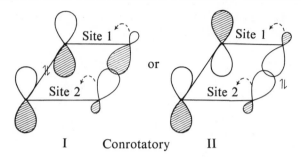

I   Conrotatory   II

(a) Bonding + Antibonding   (b) Antibonding + Bonding

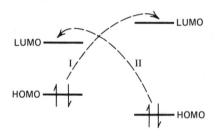

Fig. 4-84. Electrocyclic ring opening of cyclobutene (thermal).

Both processes indicate that only the conrotatory ring opening provides matching of the orbitals at both sites. The result is compatible with that obtained in the study of the ring closure of butadiene. Both processes are conrotatory allowed and the reaction therefore is reversible.

When cyclobutene is excited by UV irradiation, one of the electrons of the π fragment is transferred into the antibonding π orbital. The major contribution to the change in energy occurring during ring opening is now produced by the vicinal orbital interactions leading to the molecular orbital picture of Fig. 4-85. Here, both perturbation terms indicate disrotatory ring opening, again in accord with the results obtained for the ring closure of butadiene.

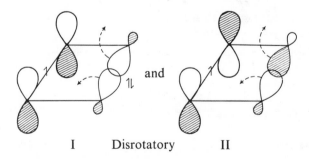

I   Disrotatory   II

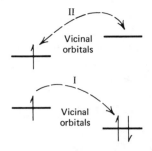

Fig. 4-85. Electrocyclic ring opening of cyclobutene (light induced).

b. *Case IV. 1,3-Sigmatropic Rearrangement.* Another $(2 + 2)(\pi + \sigma)$ reaction where the two fragments are already linked in the starting compounds is the sigmatropic rearrangement. In this case, a group R could migrate to a remote position while the double bond shifts in the opposite direction:

$$\underset{}{\sout{}}C=C-C- \longrightarrow -C-C=C\sout{}$$

Let us first consider the case where the migrating group R is a hydrogen atom. The resulting reaction can be treated as a $2 + 2$ exchange reaction or as the self-perturbation of a butadiene "π equivalent" fragment (Fig. 4-86 a and b).

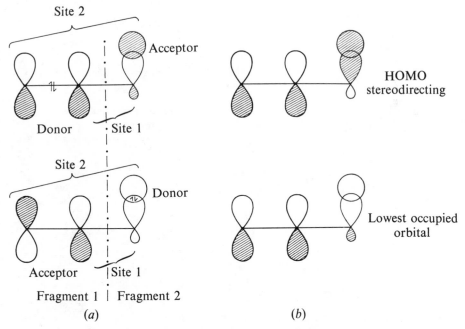

Fig. 4-86. Alternative representation of the 1-3 sigmatropic shift.

It can easily be seen that if the orbitals at site 1 are set so as to match, those at site 2 are of opposite sign. As a result the concerted "suprafacial" reaction is forbidden. A similar conclusion is reached from the examination of the stereodirecting orbital of Fig. 4-86b. However, it is conceivable that the group R can migrate to the bottom rather than to the top of the π orbital (Fig. 4-87).

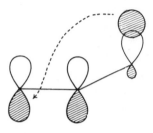

Fig. 4-87. Antarafacial 1-3 shift.

The resulting "antarafacial" 1,3-sigmatropic rearrangement is allowed in principle, but usually very difficult. In larger systems, however, this remains a definite possibility (Fig. 4-88). A special case is provided when the migrating

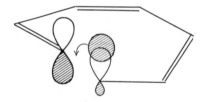

Fig. 4-88. Antarafacial 1-7 sigmatropic shift.

group is linked to the molecule by a p hybrid orbital. The back lobe of this orbital has then the right symmetry (Fig. 4-89) to attach to the proper lobe of

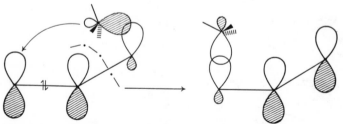

Fig. 4-89. 1-3 sigmatropic shift with inversion of configuration.

the π orbital and the reaction may then occur, although this is not anticipated to be an easy process.[49] The interesting feature about this reaction is that inversion of configuration of the migrating group should take place in the product.

Finally, one should realize that 1-3 shifts do occur in many cases involving a heteroatom. The best known of all is probably that of the keto-enol tautomerism:

$$\underset{\overset{|}{\underset{|}{\text{H}}}}{-\text{C}}-\text{C}\!\!\overset{\displaystyle\nearrow\text{O}}{\diagdown} \;\;\rightleftharpoons\;\; \diagup\text{C}\!\!=\!\!\text{C}\overset{\displaystyle\nearrow\text{OH}}{\diagdown}$$

However, although the reaction along the path described in this section is forbidden, and probably occurs most economically via a catalysed process, it is possible to provide a reasonable concerted mechanism for the reaction. The clue lies in the fact that the system involves a heteroatom (oxygen) possessing an occupied p orbital perpendicular to the $p\pi$ plane (Fig. 4-90). The reaction can be visualized as involving two fragments, a butadiene

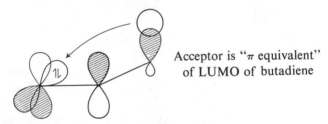

Acceptor is "π equivalent"
of LUMO of butadiene

Donor is p orbital
of oxygen

Fig. 4-90. Possible mechanism of intramolecular 1-3 sigmatropic shift on a heteroatom.

equivalent and a p orbital and consists of a $1 + 4$ intramolecular donor acceptor interaction.

$$\underset{\overset{|}{\underset{|}{\text{H}}}}{-\text{C}}-\text{C}\!\!\overset{\displaystyle\nearrow\text{O}}{\diagdown} \;\longrightarrow\; \diagup\text{C}\overset{\overset{\displaystyle\text{H}}{\diagdown}}{\!\!\cdots\!\!}\text{C}\!\!\overset{\displaystyle\nearrow\overset{\oplus}{\text{O}}}{\underset{\ominus}{\diagdown}} \;\longrightarrow\; \diagup\text{C}\!\!=\!\!\text{C}\overset{\displaystyle\nearrow\text{OH}}{\diagdown}$$

c. Case V. 1-2 Addition on a Double Bond. The concerted cis addition to an olefinic bond is a forbidden process and does not usually occur (Fig. 4-91).

Fig. 4-91. Forbidden intermolecular $(2 + 2)$ $(\pi + \sigma)$ addition.

Instead, the addition takes place in a stepwise fashion of which an electrophilic attack is the rate-determining step (see Section III.A.2) (Fig. 4-92). The

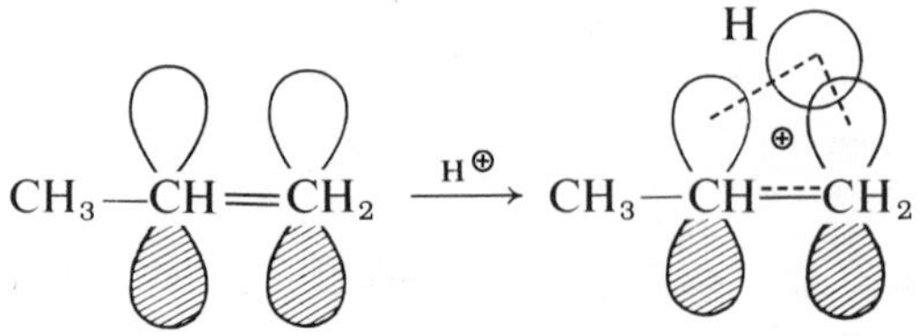

$$CH_3—CH{=}CH_2 \xrightarrow{\;H^{\oplus}\;} CH_3—CH{=\!=\!=}CH_2$$

Fig. 4-92. Front-side attack on a double bond by a proton.

more or less polarized carbocation then combines with the geigen ion to yield the trans product (Fig. 4-93). This mechanism is often prompted by the

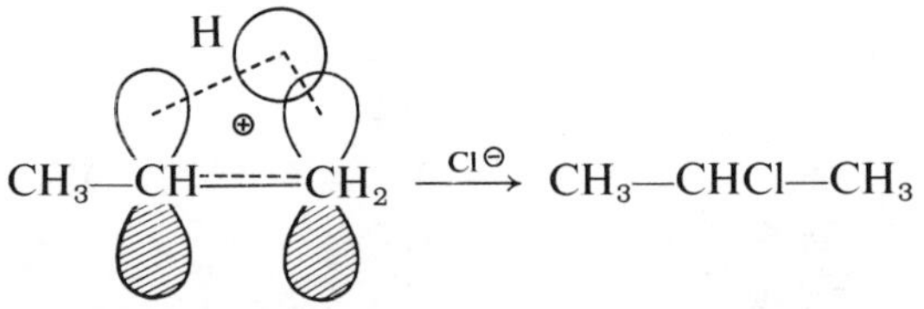

$$CH_3—CH{=\!=\!=\!=\!=}CH_2 \xrightarrow{\;Cl^{\ominus}\;} CH_3—CHCl—CH_3$$

Fig. 4-93. Product formation from cyclic intermediates.

fact that the reagent (e.g., HCl) is predissociated by the solvent into an electrophile (H$^+$) and a nucleophile (Cl$^-$). Alternatively, the reagent may interact with the π electrons via the back lobe of its antibonding orbital and its dissociation then accompanies the partial transfer of electrons into the antibonding orbital, for example, bromination (Fig. 4-94).

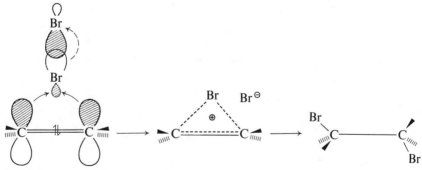

Fig. 4-94. Bromination of olefins.

In any event, the rate-determining step involves a 1 + 2 donor acceptor interaction identical to that described for general electrophilic attack. This mechanism is well substantiated by experimental findings such as in the example given in Fig. 4-95. In this example, stepwise trans addition is proven

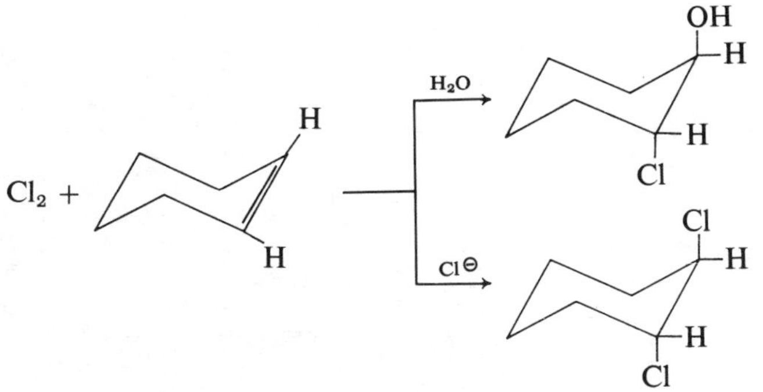

Fig. 4-95. Reaction of chlorine with cyclohexene in the presence of a trapping nucleophile.

by the fact that the intermediate carbocation can be trapped by either Cl^- or water to yield chlorocyclohexanol and 1,2-dichlorocyclohexane and that both products are exclusively trans substituted. Actually, this experiment proves only that the two chlorine atoms of the same chlorine molecule do not add in a concerted fashion but not necessarily that the nucleophilic and electrophilic step do not occur in a concerted fashion; that is, the following mechanism is still possible:

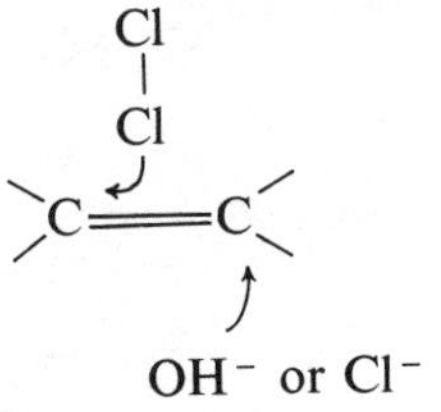

The direction of attack, that is, $\overset{\searrow}{C}=C$ or $C=\overset{\nearrow}{C}$ is determined by the relative contribution of each of the carbon p orbitals to the π molecular orbital. The Markownikoff rule can easily be deduced from a consideration of the inductive effects of the group attached to the double bond (see Section III.A.2 on electrophilic attack).

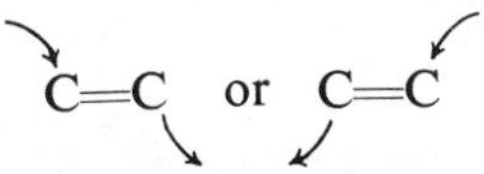

An important question that remains to be answered is why cis addition, which could occur via attack of the carbocation by the nucleophile, does not take place.

The main incentive for trans addition is probably provided by the fact that it yields the product in a staggered conformation, whereas cis addition conducts to an eclipsed conformation (Fig. 4-96).

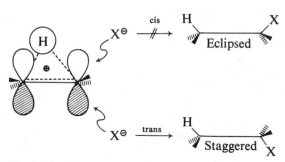

Fig. 4-96. Cis and trans addition to olefins.

This explanation lies on the same basis and is entirely consistent with that given to explain trans elimination (see Section III.B.1). Here again we may anticipate that cis addition will occur predominantly in molecules where the double bond is in such an environment that the product must remain in an eclipsed form. This was actually found experimentally for acenaphthylene[50a] and norbornene[37] where in contrast to other 1,2 addition, overwhelming cis addition* is observed (Fig. 4-97).

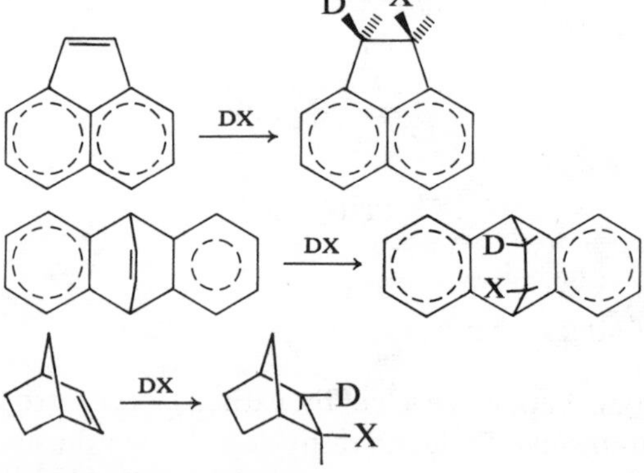

Fig. 4-97. Cis additions.

* For acenaphthylene, it was postulated that cis addition occurs because of the stabilization of a localized carbonium ion by resonance with the aromatic ring.[50b]

An interesting "apparent" exception to the above conclusions is the case of the borohydration of double bonds, which occurs exclusively cis. An examination of the relevant orbitals in this case is however quite explicit in showing the mechanism by which the reaction must proceed (Fig. 4-98).

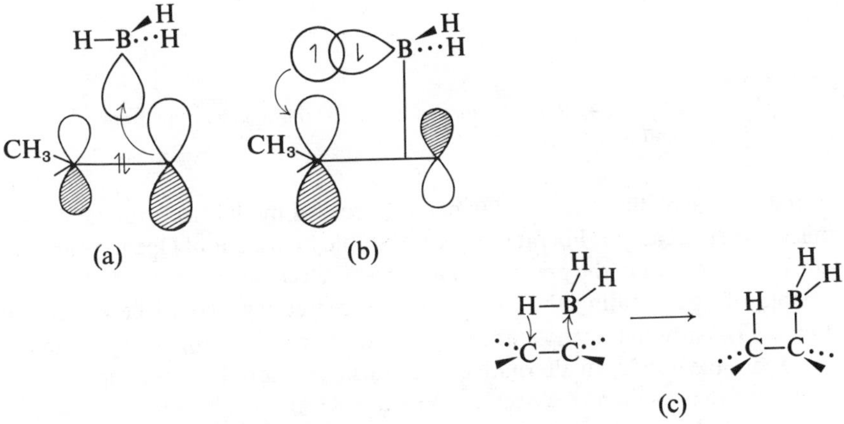

Fig. 4-98. Borohydration of alkenes.

The overwhelming feature of borohydrides is the unusual presence of an empty p orbital on the boron atom. This orbital undoubtably plays a leading role in the reaction and is believed to act as a powerful electrophile toward the double bond. (Fig. 4-98a). This is entirely consistent with the mechanism suggested for the rate determining step in all other electrophilic additions. However, due to the fact that the borohydride is an uncharged species and that the geometry of the intermediate is favorable, a return electrophilic attack is likely to occur in a more or less concerted fashion. Thus, one of the lobes of the π antibonding orbital may perform an electrophilic attack on one of the B—H bonds, and in the process, recuperate some of the electrons lost in the primary step (Fig. 4-98b).

This is not an exchange reaction since the two more or less concerted donor acceptor steps involve two orthogonal sets of orbitals on the borohydride.

The major driving force is the interaction of the empty orbital of BH_3 with the bonding orbital of the double bond, and it is not surprising therefore that the addition occurs in an anti-Markownikoff manner and leads exclusively to the cis derivative (Fig. 4-98c).

Interestingly enough, this picture also explains the structure of diboranes which can be visualized as stable intermediates of the kind discussed above but where the C=C bond is replaced by a B—H bond of another borane.

Before leaving this subject, let us notice that unlike other 1,2 additions, the addition of H_2 onto a double bond does not usually occur, whether in a concerted or stepwise fashion. As with the other cases, the exchange process is forbidden since the hydrogen molecular orbitals cannot match the adequate

Fig. 4-99. Possible intermediate in the catalytic hydrogenation of olefins.

orbitals of the olefin. The reaction cannot proceed by the electrophilic mechanism either because it liberates a very unstable hydride ion. One may speculate, however, that in the presence of a very soft electron acceptor (Pt, Hg, Au) capable of coordinating the hydride ion, the reactivity could be enhanced (Fig. 4-99). Although many soft acceptors act as catalysts for the hydrogenation of double bonds, an alternative mechanism seems to be more appropriate to explain the reaction (see Section IV, Case XIX).

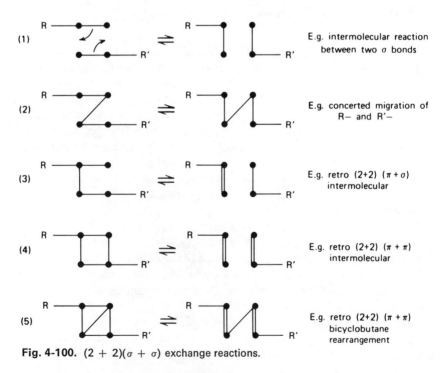

Fig. 4-100. $(2 + 2)(\sigma + \sigma)$ exchange reactions.

3. (2 + 2)(σ + σ) Exchange Reactions. Exchange reactions involving two σ bonds can, in principle, be of several kinds (Fig. 4.100). However, they are forbidden as concerted process and thus expected to occur, if at all, in a stepwise fashion. The only exception is reaction (5), where due to the transversal bond, the orbitals are held in such a position as to provide a configuration complementary to that shown in Fig. 4-75. In this case, the concerted reaction is possible as was proven by the fact that the dimethyl-

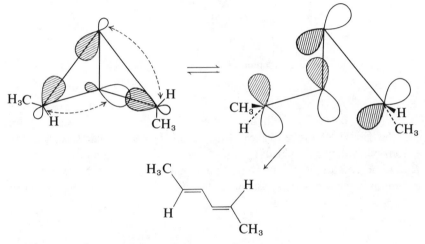

Fig. 4-101. Rearrangement of bicyclobutane derivatives.

bicyclobutane represented in Fig. 4.101 rearranges exclusively[51] into *trans-trans*-1-4 dimethyl butadiene.

4. (2 + 2) (π + n) Exchange Reactions

a. Case VI. Addition of Carbenes to Double Bonds. The addition of carbenes to olefins may occur in a stereospecific fashion if it results from a concerted interaction of the carbene orbital with both p_π orbitals of the double bond (Fig. 4-102). If, on the other hand, the addition occurs in two consecu-

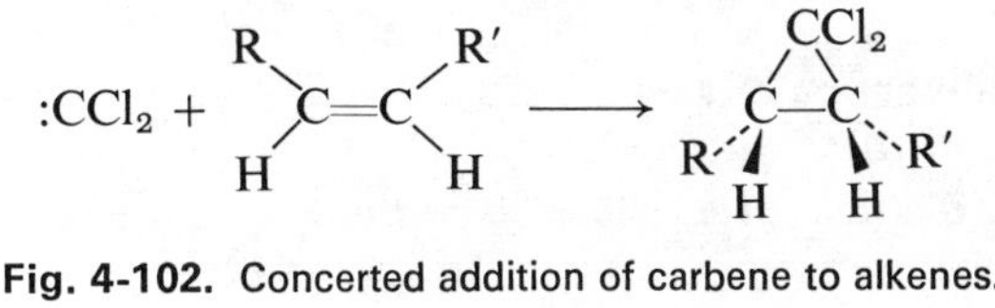

$$:CCl_2 + \underset{H}{\overset{R}{\diagdown}}C{=}C\underset{H}{\overset{R'}{\diagup}} \longrightarrow$$

Fig. 4-102. Concerted addition of carbene to alkenes.'

tive steps two stereoisomers could be obtained (Fig. 4-103). Neither path involves charge transfer; the concerted path consists of a double-site reaction and the stepwise path of a single-site radical reaction. In order to find whether

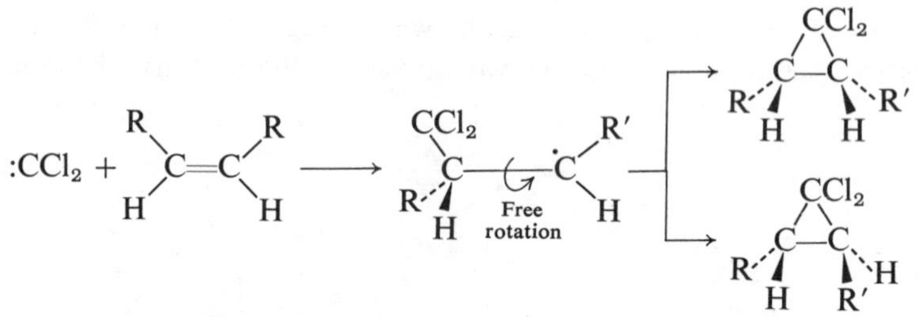

Fig. 4-103. Stepwise addition of carbene to alkenes.

the concerted path is allowed, let us consider the orbitals that are involved in the reaction and determine the perturbation between them. Let us assume that the orbitals of the central carbon atom of the carbene can be represented as four equivalent sp^3 hybrids.* Two of these are used to bind the two groups R_1 and R_2, chlorines in the present case, and the other two are left to accommodate the remaining two electrons (Fig. 4-104). We may transform the

Fig. 4-104. Orbitals of dichlorocarbene.

two available orbitals into two orthogonal orbitals of essentially the same energy by taking their positive and negative linear combination:

$$\psi_1 = \frac{1}{\sqrt{2}}(\phi_A + \phi_B) \quad \text{and} \quad \psi_2 = \frac{1}{\sqrt{2}}(\phi_A - \phi_B)$$

These two orbitals must accommodate two electrons and can do so in two ways, depending on whether the two electrons are paired or not. Thus carbenes may exist as singlets or triplets. In the singlet state the two electrons are paired and reside in either one of the available orbitals. The global picture of the interaction of singlet carbene with an olefin can thus be represented as in Fig. 4-105.

* This is not exactly true but simplifies the discussion and does not affect the results to any large extent.

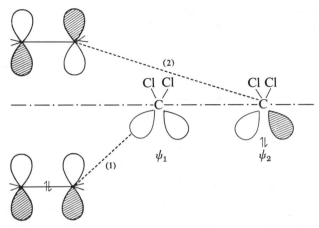

Fig. 4-105. Exchange interaction of singlet dichlorocarbene with double bonds.

The GP theory requires that the sum of the perturbations between the occupied orbital of one of the reagents with the unoccupied orbital of the other be evaluated. For a concerted process to be possible, it is necessary that the result be positive and this will be the case if all sites interacting orbitals can be made to match. In the present case such a match can be obtained; the reaction is thus allowed and proceeds via a concerted attack from the top of the double bond as illustrated in Fig. 4-106.

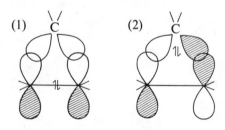

Fig. 4-106. Orbital match for the exchange interaction of singlet carbenes with double bonds.

This conclusion is borne out experimentally,[52] as singlet carbene generated by heating (or irradiating) diazomethane, has been shown to attack 2-butene in a concerted fashion to yield exclusively *cis*-1-2-dimethyl-cyclopropane.

$$H_2C_{(singlet)} + \underset{H}{\overset{H_3C}{\diagup}}C{=}C\underset{H}{\overset{CH_3}{\diagdown}} \longrightarrow \underset{H}{\overset{H_3C}{\diagdown}}\triangle\underset{H}{\overset{CH_3}{\diagup}}$$

The situation is quite different if the two electrons of the carbene molecule are unpaired. In this case the carbene is in a triplet state and the two electrons must lie in different orbitals. The diagram representing the interaction of a triplet carbene with an olefin now looks like that in Fig. 4-107.

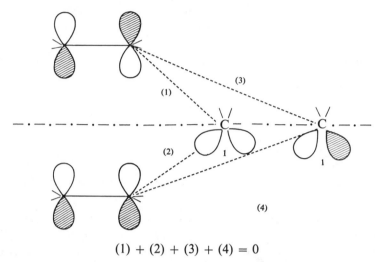

$$(1) + (2) + (3) + (4) = 0$$

Fig. 4-107. Forbidden interactions of triplet carbenes with double bonds.

Since each of the hybrid orbitals contains one electron, they may both interact with the occupied and unoccupied orbitals of the double bond, that is, total occupancy 3 and 1, respectively. Each of these interactions carries approximately the same weight, and it can be seen that they cancel each other. As a result the concerted reaction is forbidden. Here again, experiment confirms these conclusions as triplet carbenes generated by irradiating diluted diazomethane[53] in presence of a photosensitizer produce upon reaction with *cis*-2-butene a mixture of the *cis*- and *trans*-1-2-dimethylcyclopropane along a stepwise path:

$$H_2C_{(triplet)} + \overset{H_3C}{\underset{H}{\diagdown}}C{=}C\overset{CH_3}{\underset{H}{\diagup}} \longrightarrow \quad + \quad$$

b. Case VII. Adsorption of Olefins to Metals. Adsorption of olefins to metals may result as a consequence of favorable matching between their orbitals. Thus an empty *s* orbital may interact with the HOMO of a double bond to form a three-centered ring of the kind observed in the front-side electrophilic attack (Fig. 4-108). The ring interaction can be catalogued as a 1 + 2 donor-acceptor interaction since it results from a partial transfer of

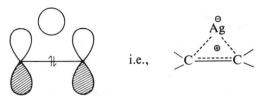

Fig. 4-108. Donor-acceptor interaction between a double bond and an *s* orbital of a metal.

electrons from the organic moiety to the metal. An alternate possibility is the interaction of an occupied *d* orbital of the metal with the LUMO of the double bond as in Fig. 4-109. Here again the process is a 1 + 2 donor-

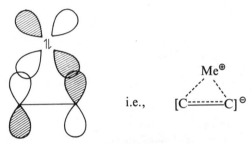

Fig. 4-109. Donor-acceptor interaction between a double bond and a *d* orbital of a metal.

acceptor interaction but the transfer of electrons now occurs from the metal to the organic moiety. Clearly the former process can occur only with a cation, whereas the latter may involve the metal itself. However, the transfer of electrons from the metal to a carbon moiety results in the formation of an unstabilized carbanion and is not very favorable.

One may consider combining both processes into a (2 + 2) exchange reaction in which two closely lying orbitals of the metal, a *d* and an *s* orbital interact in a concerted fashion with the π bond. The result of such interaction is simply adsorption on the metal surface with opening of the double bond. It does not involve significant transfer of electrons from one to the other fragment (Fig. 4-110), and can most properly be called a covalent metallic

Fig. 4-110. Exchange interaction between metals and double bonds.

bond. The metals that are suitable for such an interaction are those having closely lying a d and an s orbital, containing together two electrons. Depending on the environment the d orbital can be occupied and the s orbital empty, in which case adsorption of olefins could occur. Examples of such metals include essentially Pd, Pt, and Ni, and to a lesser degree Co, Fe, Mn, Rh, Ru, and Tc. These metals are precisely the ones known to catalyze some of the reactions of olefins (i.e., hydrogenation). This subject will be discussed in more detail in Section IV, as a 4 + 2 exchange reaction.

C. 2 + 3 Exchange Reactions (1,3-Dipolar Additions)

The π and "π equivalent" molecular orbitals for the two kinds of fragments of size 3 were shown previously (Fig. 4-43).

Fragments consisting of three orbitals are usually either anions or cations depending on whether the nonbonding orbital is occupied or not (Fig. 4-111).

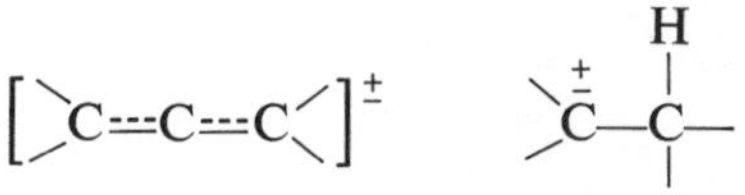

$$\left[\underset{/}{\overset{\backslash}{C}} \text{---} C \text{---} \underset{\backslash}{\overset{/}{C}} \right]^{\pm} \qquad \underset{/}{\overset{\backslash}{C}} \overset{+}{\underset{|}{\text{---}}} \overset{H}{\underset{|}{C}} \text{---}$$

Fig. 4-111. Size 3 fragments.

Their reactions toward external reagents are therefore mostly of the donor-acceptor type and proceed via single-site attack. This behavior is prompted by the fact that the charge on the fragment augments its ability to transfer electrons in a definite direction rather than to exchange electrons with another fragment. One notable exception to such behavior is that of the three orbital fragments involving one or more heteroatoms that contribute two electrons to the system (Fig. 4-112). In these cases, the molecule is not formally charged

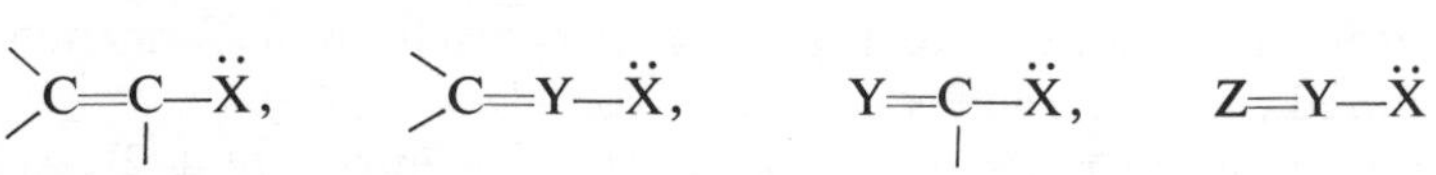

$$\underset{/}{\overset{\backslash}{C}}{=}\underset{|}{C}{-}\ddot{X}, \qquad \underset{/}{\overset{\backslash}{C}}{=}Y{-}\ddot{X}, \qquad Y{=}\underset{|}{C}{-}\ddot{X}, \qquad Z{=}Y{-}\ddot{X}$$

Where X, Y, and Z are O, N, or S

Fig. 4-112. Four electrons size 3 fragments.

and can engage in an exchange reaction (Fig. 4-113). Among the various possibilities, only the (2 + 3) concerted cis addition is allowed and of practical importance.

Numerous types of molecules belong to the family of formally uncharged species that could react in a concerted fashion with a double bond. However, only a few behave in such a fashion for the following two reasons. First of all

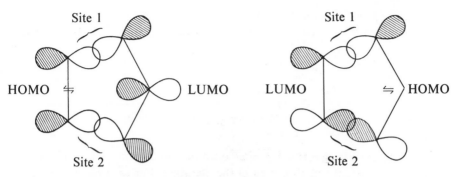

Fig. 4-113. Cis allowed (2 + 3) concerted exchange interaction.

most of the compounds of the general formulation given above exist as stable isomers under a rearranged form unsuited for the purpose, for example,

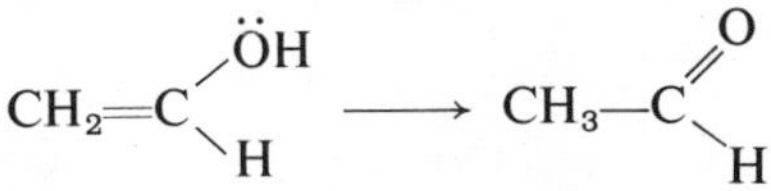

$$CH_2{=}C\overset{\ddot{O}H}{\underset{H}{\diagdown}} \longrightarrow CH_3{-}C\overset{O}{\underset{H}{\diagdown}}$$

Furthermore, the exchange process results in a separation of charge, always detrimental to the stability of the compound and therefore preventing the formation of product (Fig. 4-114). The only class of compounds[54] that could

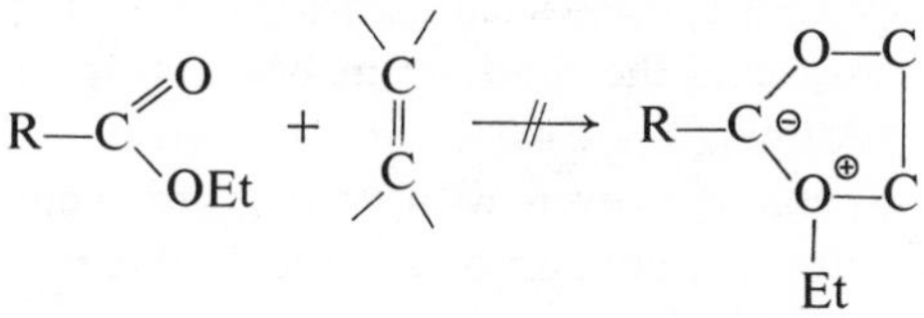

$$R{-}C\overset{O}{\underset{OEt}{\diagdown}} + \overset{\diagup C}{\underset{\diagdown C}{\parallel}} \xrightarrow{\;\;\not\longrightarrow\;\;} R{-}C^{\ominus}\overset{O{-}C}{\underset{\underset{Et}{\overset{\oplus}{O}}{-}C}{\diagdown}}$$

Fig. 4-114. Unfavorable (2 + 3) addition.

still react are those where charge separation already existed in the fragment and is lifted by the reaction, for example as in Fig. 4-115.

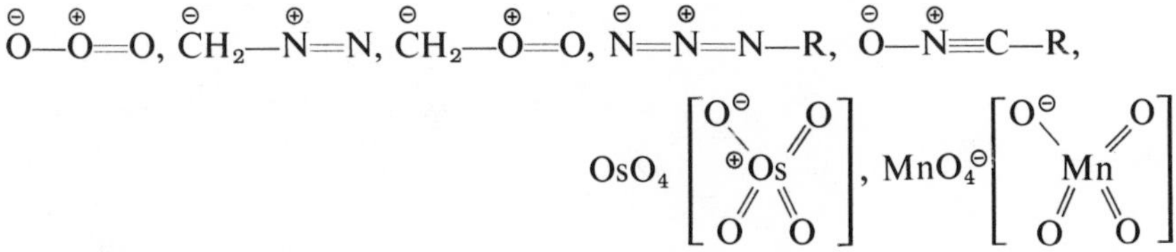

$$\overset{\ominus}{O}{-}\overset{\oplus}{O}{=}O,\ \overset{\ominus}{CH_2}{-}\overset{\oplus}{N}{=}N,\ \overset{\ominus}{CH_2}{-}\overset{\oplus}{O}{=}O,\ \overset{\ominus}{N}{=}N{=}\overset{\oplus}{N}{-}R,\ \overset{\ominus}{O}{-}\overset{\oplus}{N}{\equiv}C{-}R,$$

$$OsO_4 \begin{bmatrix} O^{\ominus} & O \\ {}^{\oplus}Os & \\ O & O \end{bmatrix},\ MnO_4^{\ominus} \begin{bmatrix} O^{\ominus} & O \\ Mn & \\ O & O \end{bmatrix}$$

Fig. 4-115. Candidates for favorable (2 + 3) addition.

a. Case VIII. Ozonolysis of olefins. The ozonolysis of olefins yields the ozonide[55] resulting from the breakage of a double bond (Fig. 4-116). This

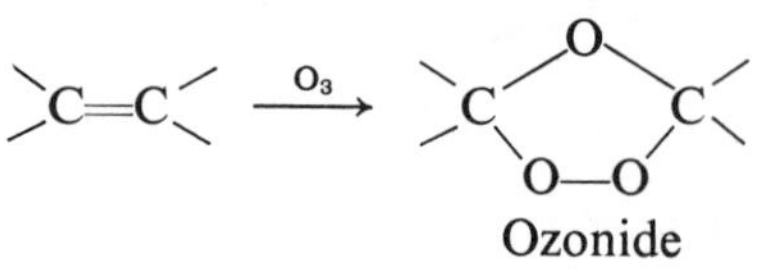

Ozonide   **Fig. 4-116.** Ozonolysis of a double bond.

apparently surprising formation of an ozonide, rather than a straightforward adduct, was extensively discussed in the literature. For example, it was suggested that the reaction proceeds via an initial cis addition of the double bond of ozone to the double bond of the olefin (Fig. 4-117). Although the (2 + 2)-

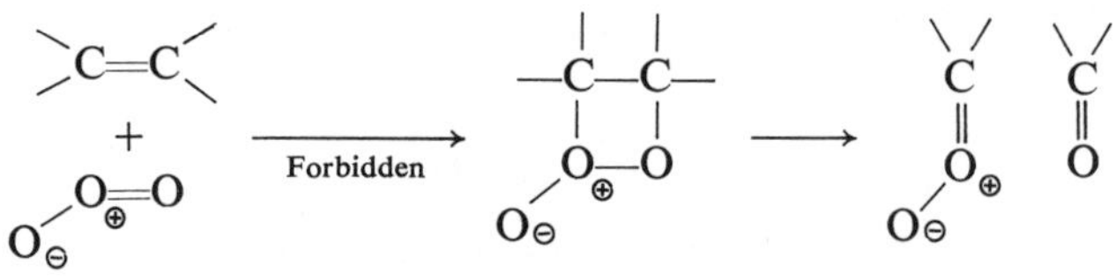

Fig. 4-117. (2 + 2) exchange mechanism for ozonolysis.

($\pi + \pi$) concerted addition is a forbidden process, such a mechanism may nevertheless occur in a stepwise fashion analogous to the electrophile addition via the intermediacy of a species such as described in Fig. 4-75. Far more reasonable, however, seems the mechanism below (Fig. 4-118) where each step proceeds via allowed (2 + 3)($\pi + \pi\sigma$) or retro (2 + 3)($\pi + \pi\sigma$) interactions. Interestingly enough, even with conjugated olefins, such as butadiene, the observed attack occurs on one of the double bonds and not, as in

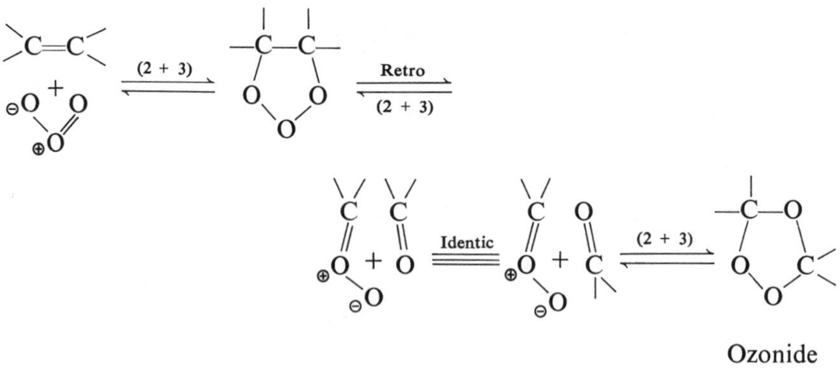

Ozonide

Fig. 4-118. (2 + 3) exchange mechanism for ozonolysis.

most other additions, at positions 1–4. This is easily understandable as no match is found for the 3 + 4 interaction (Fig. 4-119) between butadiene and

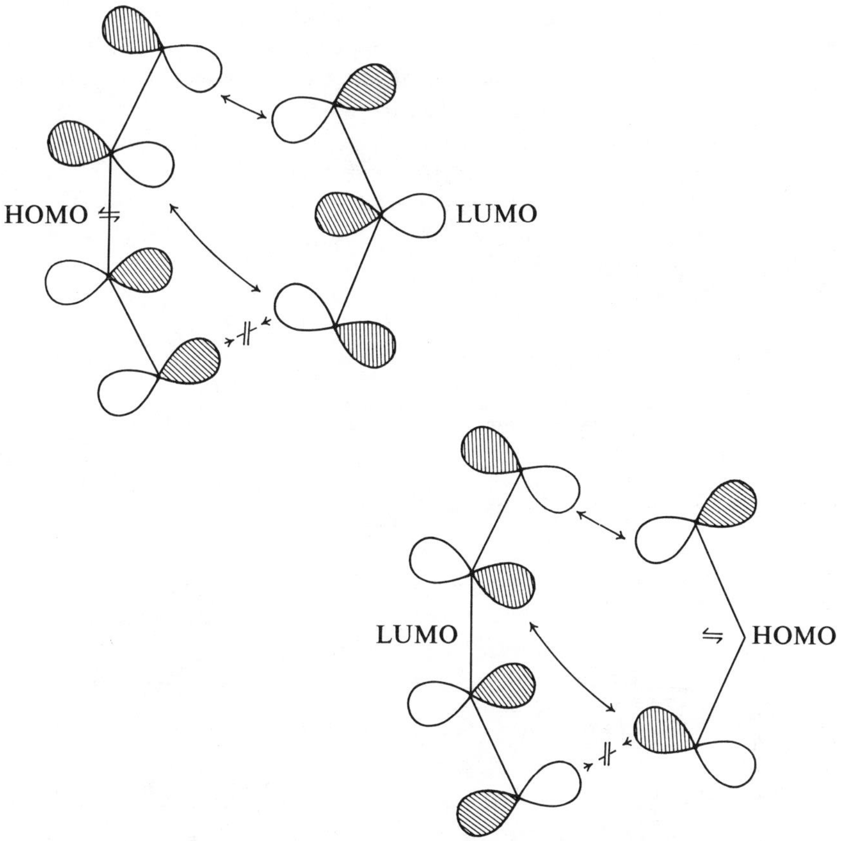

Fig. 4-119. Molecular orbitals pertinent to the ozonolysis of butadiene.

ozone. Instead, the 2 + 3 interaction is still allowed and leads to the product:

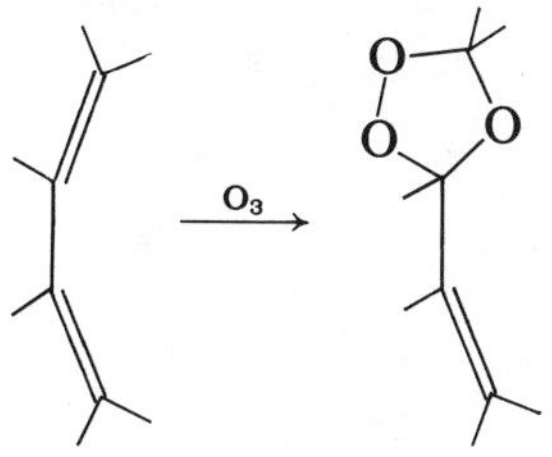

b. Case IX. Other 1-3 Dipolar Additions.[54]   Whereas the initial addition of ozone to a double bond produces an unstable species, this is not always the case for other 1-3 dipolar reagents. For example, diazomethane yields a stable adduct upon reaction with methyl-dimethyl-maleate and methyl-dimethyl-fumarate. This provides us with an experimental possibility to verify the correctness of the theoretical model. Thus, upon concerted cis addition of diazomethane to either methyl maleate or methyl fumarate, the product is expected to be stereochemically pure, whereas the same racemic mixture would be obtained from either acid if the addition occurred stepwise (Fig. 4.120). As predicted, the reaction gives in each case a pure product in perfect agreement with the theoretical conclusion that the $(2 + 3)$ exchange reaction is a concerted cis addition.

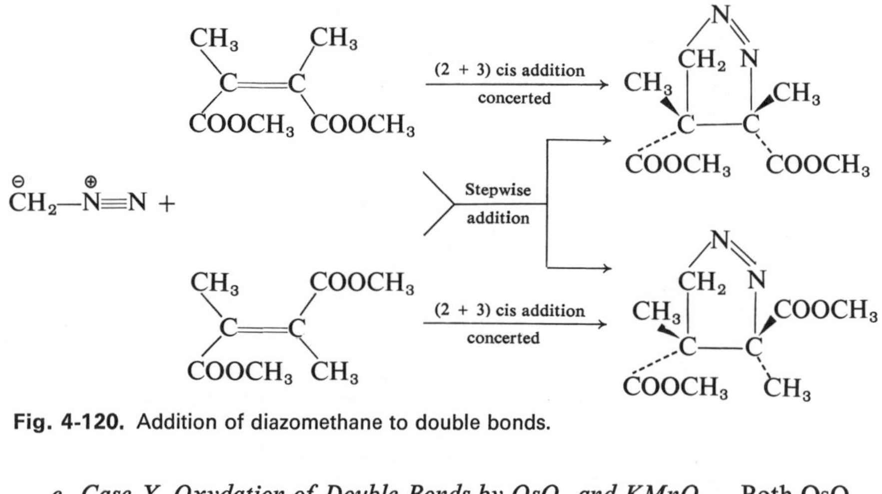

Fig. 4-120. Addition of diazomethane to double bonds.

c. Case X. Oxydation of Double Bonds by OsO_4 and $KMnO_4$.   Both OsO_4 and MnO_4^- may be represented as 4 electron, 3 orbital reaction fragments:

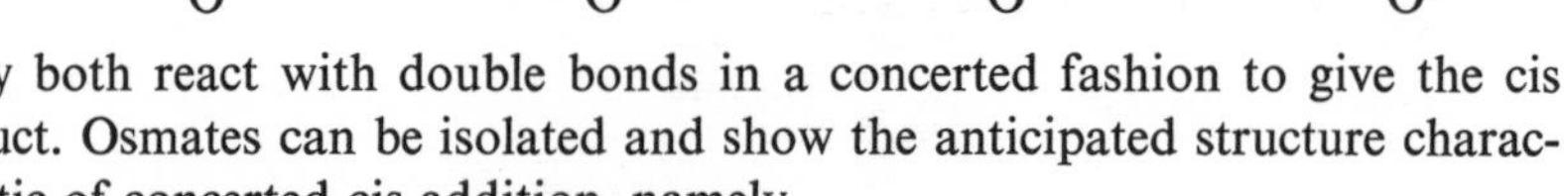

They both react with double bonds in a concerted fashion to give the cis adduct. Osmates can be isolated and show the anticipated structure characteristic of concerted cis addition, namely,

The permanganate adducts, on the other hand, are not usually stable but decompose to give the expected cis diols:

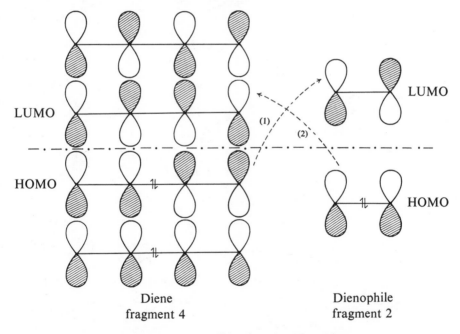

D. 2 + 4 Exchange Reactions

The 2 + 4 exchange reactions constitute the bulk of the known concerted reactions. The reason is that the process is favorable and usually allowed to proceed along a geometrically unstrained path.

The molecular orbitals relevant to this mechanism and the interactions that take place are represented in Fig. 4-121 for the most common of them, the $\pi + \pi\pi$ interaction illustrative of the Diels-Alder reaction. Analogous systems involve size 4 fragments of the type $\pi\sigma$ or even $\sigma\sigma$ (Fig. 4-122).

Diene
fragment 4

Dienophile
fragment 2

Fig. 4-121. Molecular orbitals in the (2 + 4) exchange reactions.

An examination of the interaction pattern of Fig. 4-123 reveals some interesting properties to be associated with the 2 + 4 reactions. Thus, we can see that process 1 involves a very favorable configuration since both sites are

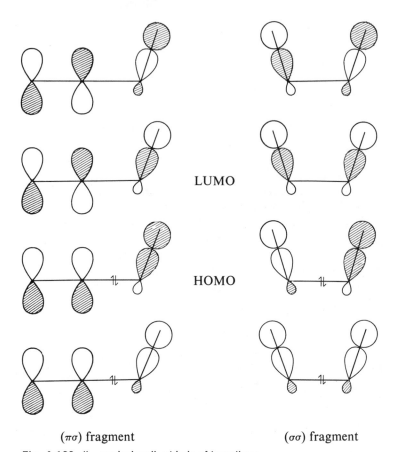

LUMO

HOMO

($\pi\sigma$) fragment   ($\sigma\sigma$) fragment

Fig. 4-122. "π equivalent" orbitals of butadiene.

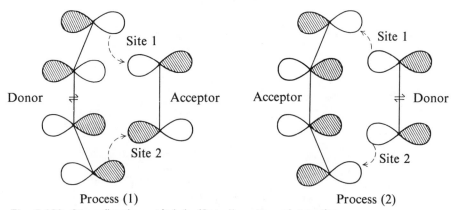

Site 1

Donor $\rightleftharpoons$ Acceptor   Acceptor $\rightleftharpoons$ Donor

Site 2

Process (1)   Process (2)

Fig. 4-123. Stereodirecting orbitals in (2 + 4) exchange interactions.

typically front-side type interaction intermediates, that is, the acceptor's orbitals (fragment 2) reach for electrons at the center of the bonds of the donor (fragment 4). On the other hand, process 2 is not very satisfactory as the acceptor's orbitals (fragments 4) are not reaching at the center of the donor's bond (fragment 2).

We may therefore deduce that the exchange reaction is most favorable when process 1 is predominant, that is, the main contributor. As a result, the reaction will be most likely to occur if the double bond (fragment 2) behaves basically as an electron acceptor.

1. (2 + 4) (π + ππ) Exchange Reactions

a. Case XI. The Diels-Alder Reaction. The Diels-Alder addition is the reaction that takes place between a diene and a dienophile. The orbital diagram (Fig. 4-123) suggests an allowed concerted pathway occurring via an intermediate whose geometry is dictated by the most favorable orbital overlap (Fig. 4-124). This mechanism is amply substantiated by the fact that

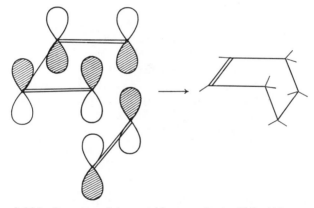

Fig. 4-124. Geometry of the transition state in the Diels-Alder reaction.

unsymmetrical dienes maintain their configuration in the product,[56] for example,

According to our observations of the interaction pattern (see p. 146), we expect the reaction to be the most favorable when the dienophile behaves as an electron acceptor. It is not surprising, therefore, to find that, usually, for the reaction to proceed, the dienophile has to be activated by electron withdrawing substituents. The main contributor to the rate determining step is the transfer of electrons from the diene to the dienophile, that is, the stereo-directing orbitals are the HOMO of the dienophile and the LUMO of the diene (Fig. 4-125). This, however, is not always so as cases are known where

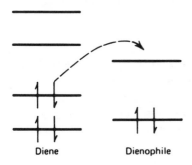

Fig. 4-125. Rate-determining donor-acceptor step in (2 + 4) exchange reactions.

the reaction is activated by electron donating substituents on the dienophile and electron withdrawing groups on the electrophile.[57]

In cases where several alternative products can be expected from the reaction, such as if the reagents are unsymmetrically substituted, the most stable single site interaction intermediate will usually provide the clue as to which product is formed[58] (see p. 115).

For example, an acceptor fragment of size 2 has its largest LUMO coefficient at position 2 with respect to electron-withdrawing substituents, and consequently will tend to coordinate the donor in this position during the primary step (Fig. 4-126).

In order to determine the preferred position of attack it is necessary to determine first the donor or acceptor character of the reagent and the position at which the highest orbital coefficient is found in the appropriate stereo-directing orbital.

Although these properties depend on the structure of the specific reagents, it is possible to establish a chart indicating the properties associated with the general configurations of various reagents.

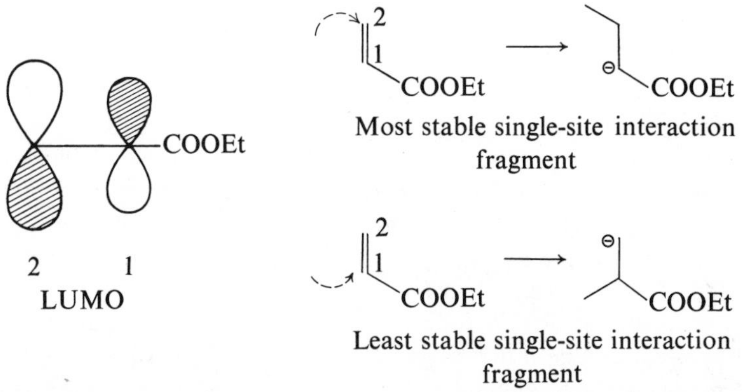

Most stable single-site interaction
fragment

LUMO

Least stable single-site interaction
fragment

Fig. 4-126. Alternative sites of attack on a fragment of size 2 attached to an electron-withdrawing substituent.

Table 4-12. Donor-Acceptor Properties of Reagent

Position 1—2—3—4		Highest Coefficient in HOMO LUMO At Position		HOMO Energy	LUMO Energy	Primary Single-Site Process
1	C=C	1	1	1.0	−1.0	Donor or acceptor at position 1
2	C=Ẋ	2	1	1.618	−0.618	Acceptor of position 1
3	C=C—Ẍ	1	2	0.69	−1.17	Donor at position 1
4	C=C—R	1	2	0.78	−1.28	Donor at position 1
5	C=C—C=C	1	1	0.61	−0.61	Donor or acceptor at position 1
6	C=C—C=Ẋ	1	1	1.0	−0.35	Acceptor at position 1
7	C=C—Ẋ=C	1	4	0.74	−0.47	Acceptor at position 4
8	C=C—C=C—R	1	4	0.45	−0.81	Donor at position 1
9	C=C—C=R with R	4	1	0.54	−0.68	Donor at position 4
10	C=C—C=C—Ẍ	1	4	0.42	−0.75	Donor at position 1
11	C=C—C=C with X:	4	1	0.52	−0.66	Donor at position 4
12	C=Ẋ—Ẋ=C	1	1	1.0	−0.41	Acceptor at position 1

$$\alpha_C = \alpha \qquad \beta_{CC} = \beta_{CX} = \beta$$
$$\alpha_{\dot{X}} = \alpha + \beta \qquad \beta_{CR} = 0$$
$$\alpha_{\ddot{X}} = \alpha + 2\beta$$
$$\alpha_{C(-R)} = \alpha - 0.5\beta$$
$$\alpha_R = 0$$

149

This is done in Table 4-12, where a simple Hückel calculation was performed for several general structures containing a doubly bonded heteroatom, $=\ddot{X}$, (e.g., $=\ddot{O}$, $=\ddot{N}H—$), a singly bonded heteroatom, $—\ddot{X}$ (e.g., $—\ddot{O}R$, $—\ddot{N}H_2$, $—\ddot{C}l$), or an electron donating substituent, R (e.g., alkyl group). The donor-acceptor ability of the reagents is judged on the basis of which of the stereodirecting orbitals (HOMO or LUMO) is closest to 0, and the preferred position of attack is then given by the highest coefficient in it. The results of Table 4-12 provide the necessary information to determine the structure of the product in most Diels-Alder reactions.

This is illustrated in the following examples (the arrows represent the donor acceptor ability of the reagent as determined from Table 4-12, and the numbers refer to the compound of Table 4-12, used as a model for the reagent).

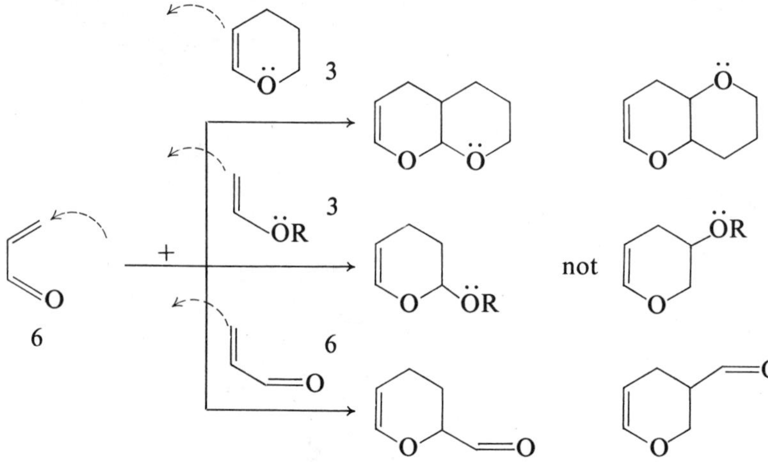

(Ref. 59)

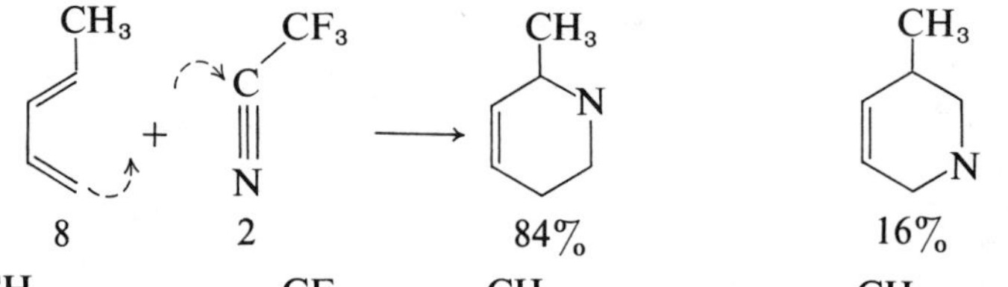

(Ref. 60)

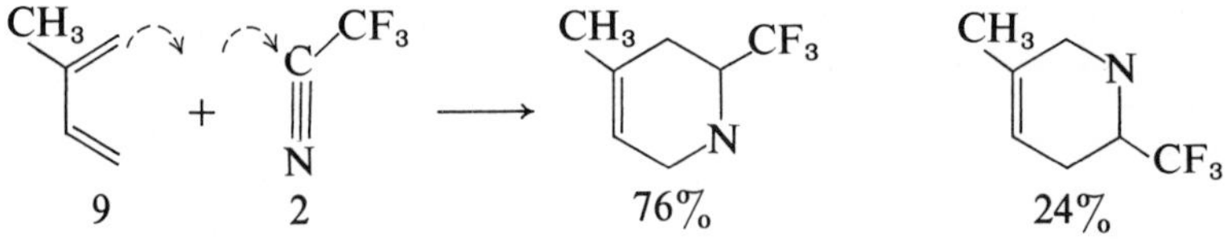

(Ref. 60)

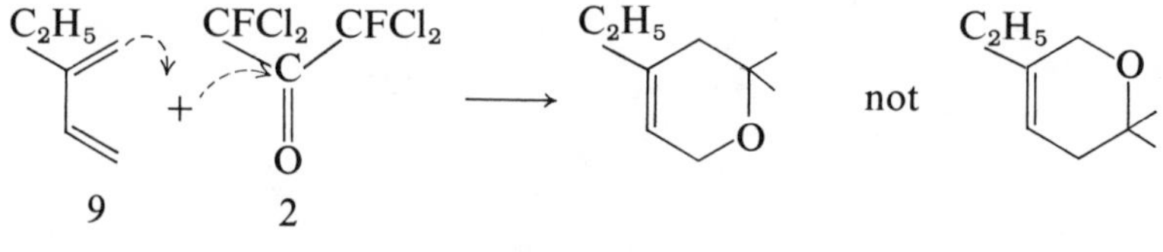

(Ref. 61)

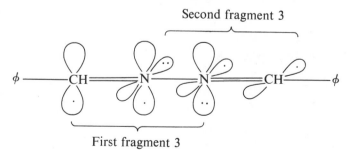

Here the reagents, being both good acceptors, cannot find a suitable path to interact. However, the reaction is not forbidden and would have occurred if another, much easier path were not available.

Fig. 4-127. The π system in benzylideneazine.

Indeed, it can be seen that due to repulsion between the adjacent nitrogen lone pairs, the benzylideneazine molecule can easily rotate to form two perpendicular π systems (Fig. 4-127). These consist of two fragments of size 3 each containing four electrons. One of these can easily react with maleic anhydride via a 1,3 dipolar addition to give

The resulting compound in turn can react with another molecule of maleic anhydride via a 1,3 dipolar addition to give the observed final product.

2. (2 + 4) (σ + ππ) Exchange Reactions

a. Case XII. 1,5-Sigmatropic Shift in Polyenes. 1,5-Sigmatropic rearrangements can be visualized as $(2 + 4)(\sigma + \pi\pi)$ exchange interactions; see Fig. 4-128. As mentioned before (see p. 115) a σ fragment of size 2 cannot engage in exchange reactions unless already attached to the other fragment. This is the case here and the process can be described in the same terms as the $(2 + 2)(\pi + \sigma)$ process illustrative of the 1,3-sigmatropic shift.

As was done in the latter case, the reaction can be studied either as a 2 + 4

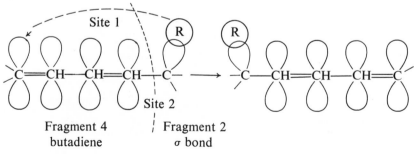

Fragment 4   Fragment 2
butadiene   σ bond

Fig. 4-128. 1,5-Sigmatropic shift in polyenes.

addition or as a self-perturbation of a hexatriene π equivalent. Both treatments lead to the same result, namely that the migration of the R group is allowed in a suprafacial mode (Fig. 4-129). This result has been amply

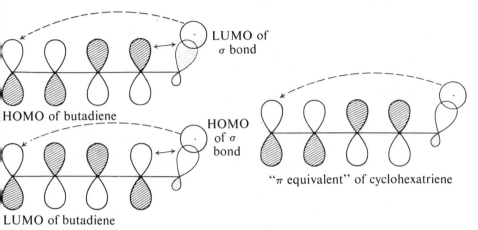

LUMO of σ bond

HOMO of butadiene

HOMO of σ bond

"π equivalent" of cyclohexatriene

LUMO of butadiene

Fig. 4-129. Alternative representations of the molecular orbitals for 1,5-sigmatropic shifts.

substantiated, as it was shown that, in contrast with the forbidden 1,3 shifts, 1,5 rearrangements readily occur. For example,[64] the active 2D, 6-methyl-2-4-octadiene rearranges exclusively to the indicated stereospecific products.

3. (2 + 4)(π + σπ) Exchange Reactions. Reactions involving $\pi\sigma$ fragments, as a class, include in principle a large body of known organic reactions. However, because the fragment involves a σ bond, it is not always clear whether the reaction is an exchange reaction proceeding via a cyclic intermediate or a donor-acceptor process occurring with a predissociated reagent.

This distinction is a subtle one and depends essentially on whether one of the bonds is formed at a substantially faster rate than (1) the other bond is formed, and (2) the σ bond is broken; that is,

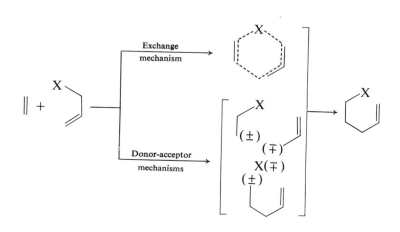

It is clear however, that if an exchange mechanism is available, it will always enhance the rate of the reaction since it provides assistance to the rate-determining step.

Thus, in the $\pi + \sigma\pi$ interaction, the stereodirecting π equivalent orbitals of the $\sigma\pi$ fragments (Fig. 4-130) can easily match the complementary orbitals of a π fragment and as a result the exchange mechanism is a favorable one.

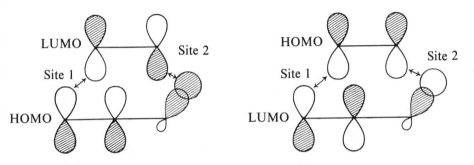

Fig. 4-130. Exchange mechanism of the $(\pi + \sigma\pi)$ interaction.

a. Case XIII. The Claisen Rearrangement. In the Claisen rearrangement, an alkyl ether rearranges as shown below:

The migration is allowed exclusively to the ortho position. Cases are known, however, where the product is a para phenol. This occurs when the ortho positions are blocked by alkyl groups. In these cases, however, it has been shown that the para derivative is obtained via two successive $2 + 4 (\pi + \sigma\pi)$ interaction steps.[65]

The product is found experimentally to be exclusively the one shown above and is different from the one that would have resulted from a direct $4 + 4$ migration, that is,

b. Case XIV. The Cope Rearrangement. The Cope rearrangement[66] is similar to the Claisen rearrangement except that the oxygen atom of the fragment 4 is replaced by a carbon atom, for example,

The conclusions reached for the Claisen rearrangement remain thus valid here.

An interesting problem arises when carbons 3 and 4 are each center of chirality. Depending on whether the starting material is the racemic or the meso derivative, one might obtain a different product.

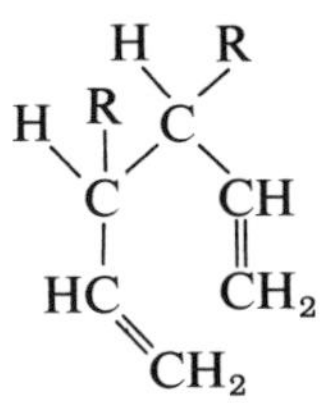

$$
\begin{array}{cc}
\text{H} \quad \text{R} & \text{H} \quad \text{R} \\
\text{H} \overset{\text{R}}{\underset{|}{\text{C}}} \overset{\diagup}{\underset{\diagdown}{\text{C}}} & \text{H} \overset{\text{R}}{\underset{|}{\text{C}}} \overset{\diagup}{\underset{\diagdown}{\text{C}}} \\
\overset{|}{\text{C}} \quad \overset{||}{\text{CH}} & \overset{||}{\text{C}} \quad \overset{}{\text{CH}} \\
\text{HC} \quad \text{CH}_2 & \text{HC} \quad \text{CH}_2 \\
\text{CH}_2 & \text{CH}_2 \\
\end{array}
$$

meso or racemic 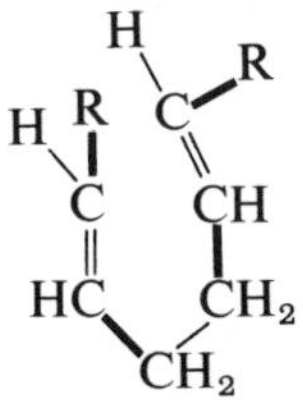   cis-trans
cis-cis
or trans-trans

This problem can easily be solved by determining the structure of the most stable transition state with respect to (a) the most favorable overlap situation between the reacting site, and (b) the less strained configuration of the intermediate.

Two structures fulfill the first requirement and are represented in Fig. 4-131 for the meso derivative. However, it can be seen that structure I is far

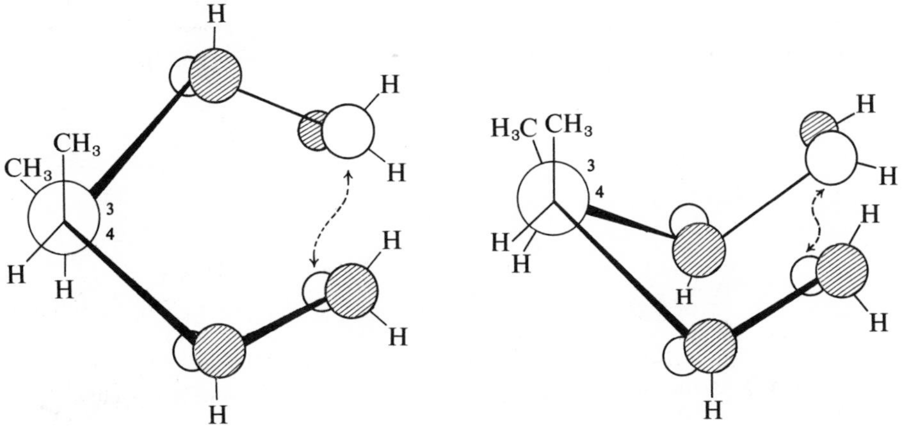

I
Gives cis-trans

II
Gives cis-cis

Fig. 4-131. Molecular orbitals in the Claisen rearrangement of the meso derivative of 3-4 dimethyl-1-5-hexadiene.

more stable than structure II (staggered versus eclipsed group on atoms 3 and 4, cyclic chair intermediate versus cyclic boat intermediate) and as a result, the reaction will proceed in that direction. Thus the meso derivative gives preferentially the cis-trans product, and it can be shown in an analogous way that the racemic mixture should yield a 50–50 mixture of trans-trans and cis-cis product. These are exactly the results obtained experimentally.[66]

c. Case XV. Addition of Allylic Grignard Reagents to Carbonyl Groups. This reaction, which may involve a cyclic intermediate, occurs mostly when the ketonic bond is hindered, for example,

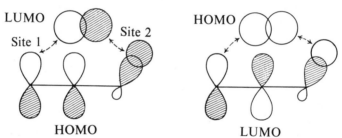

However, the mechanism of this reaction is not well established. It may be that, due to the weak C—Mg interaction, the reaction proceeds via a donor-acceptor path.

4. (2 + 4)(π + σσ) (or σ + σπ) Exchange Reactions. In this case again, good matching of the orbital pairs can easily be obtained between the two fragments (Fig. 4-132), leading to an allowed process. However, the reserva-

LUMO   Site 2   HOMO

Site 1

HOMO   LUMO

Fig. 4-132. Stereodirecting orbitals in (2 + 4) (σ + σπ) exchange interactions.

tions made in the previous section concerning the concertness of the process apply even more to the present case.

a. Case XVI. Intermolecular Cis Elimination: The Chugaev Reaction.[67] In the Chugaev reaction, a Xanthate is pyrolyzed and gives the following products:

Fragment 2 (π)

Fragment 4 (σσ)

$$S=C \begin{array}{c} SCH_3 \\ \\ O \end{array} \quad \xrightarrow{200} \quad H \begin{array}{c} S-C \begin{array}{c} SCH_3 \\ \\ O \end{array} \\ + \\ C=C \end{array} \quad (\longrightarrow HSCH_3 + COS)$$

The reaction can be viewed as resulting from an intramolecular $(2 + 4)$ $(\pi + \sigma\sigma)$ exchange interaction where the adequate fragments are as illustrated in Fig. 4-133. This description is a general one and applies to many

Fragment 2, Ethylene

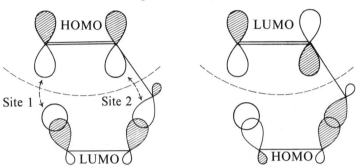

Fragment 4, "π equivalent" of butadiene

Fig. 4-133. Stereodirecting orbitals in the Chugaev reaction.

other intramolecular cis eliminations such as

$$\begin{array}{c} O=C \overset{R}{\diagup} \\ H \quad \diagdown NH \\ \overset{|}{C}-\overset{|}{C} \\ \diagup \diagdown \diagup \diagdown \end{array} \longrightarrow \begin{array}{c} O-C \overset{R}{\diagup} \\ H \quad + \quad \diagdown NH \\ \diagdown \diagup \\ C=C \\ \diagup \diagdown \end{array} \left(\longrightarrow O=C \overset{R}{\underset{NH_2}{\diagup}} \right)$$

In each of these cases the favorable $(2 + 4)(\pi + \sigma\pi)$ return reaction is prevented by the fact that one of the reagents reorganizes into a form unsuited for further reaction.

b. Case XVII. Enolization of Carbonyl Groups by Grignard Reagents. Stetically hindered ketones sometimes fail to add in the usual fashion with Grignard reagents. In these cases, and depending on the nature of the reagents, a less strained transition state may be provided by $2 + 4$ additions, such as

$$\begin{array}{ccc} O \longrightarrow MgX & & OMgX \\ \| \quad CH_2 & \xrightarrow[\pi+\sigma\sigma]{\text{Reduction}} & | \quad CH_2 \\ C \longleftarrow \; + \; | & & C-H \; + \; \| \\ R \quad CHR \quad CH_2 & & R \quad CH_2R \quad CH_2 \\ \diagdown H \quad \diagup H & & \end{array}$$

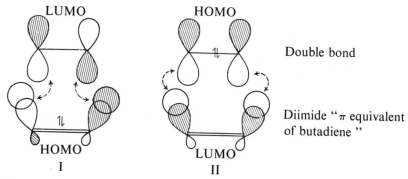

These reactions can be described as allowed ($\sigma\sigma + \pi$) exchange interactions for the reduction reactions and ($\sigma\pi + \sigma$) exchange* interactions for the enolization reaction.

c. *Case XVIII. The Reaction of Diimides with Alkznes.*[43]   The reaction of diimide with ethylene to give nitrogen and ethane provides a good example of intramolecular cis elimination via a $(2 + 4)(\pi + \sigma\sigma)$ exchange reaction.

The reaction is viewed as occurring between the π system of the ethylene fragment and the σ system of the diimide; see Fig. 4-134. As with the $\pi + \pi\pi$

LUMO   HOMO

Double bond

HOMO   LUMO

Diimide "π equivalent of butadiene"

I   II

Fig. 4-134. Molecular orbitals in the reaction of diimides with alkenes.

reaction, it can be seen that process I provides a more favorable pathway for the transfer of electrons than process II. It is therefore anticipated that, as with the Diels-Alder reaction, electron withdrawing groups attached to the ethylene moiety will enhance the rate of the reaction.

d. *Case XIX. Catalytic Reduction of Alkenes.*   It was suggested in Section IV, case VII, that the metals having closely lying an s and a d orbital, containing together two electrons, can interact with double bonds to form a "covalent metallic" bond.

---

* ($\sigma\pi + \sigma$) reactions are equivalent to $\sigma\sigma + \pi$ reaction providing that the alternative fragments are chosen. This is why they are grouped with the $\sigma\sigma + \pi$ reactions.

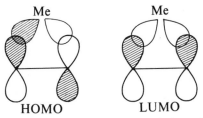

Fig. 4-135. Metal alkene adduct as a π equivalent of butadiene.

The orbital picture of the resulting adducts can be visualized as π equivalents of butadiene where the two ends are the metal sd hybrids (Fig. 4-135). The interesting property of these molecular orbitals is that the symmetry of the ethylenic π orbitals is reversed from that of the isolated double bond. As a consequence, the reaction pattern of the double bond is very strongly altered.

Indeed, all the conclusions concerning the exchange reactions of the double bond are now to be reversed. In particular, the reaction equivalent to a $(2 + 2)(\pi + \pi)$ reaction of the uncomplexed alkenes should now be allowed (Fig. 4-136). This is observed experimentally; for example,

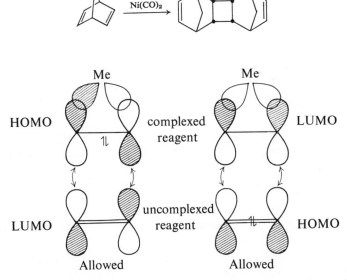

Fig. 4-136. Molecular orbitals in the reaction of alkenes with metal alkene adducts.

We may envision the reaction occurring with a σ bond in a similar fashion (Fig. 4-137). When the sigma bond is a hydrogen molecule, the reaction

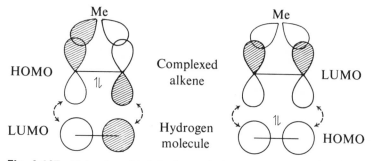

Fig. 4-137. Molecular orbitals in the catalytic hydrogenation of alkenes.

yields the reduced hydrocarbon, for example,

$$H_2 + R—CH=CH—R' \xrightarrow{Me} R—CH_2—CH_2—R'$$

An equally interesting alternative occurs if it is the σ bond (hydrogen) that is first adsorbed by the metal and then reaction occurs with the double bond (Fig. 4-138).

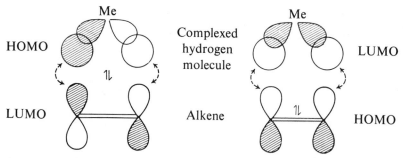

Fig. 4-138. Alternative mechanism of the catalytic hydrogenation of alkenes.

V. CONCLUDING REMARKS

In the preceding pages, we have described some of the areas of application of the generalized perturbation theory.

We have attempted a classification of the reagents in terms of (1) their ability to react in a particular fashion, and (2) the size of the reacting fragments. We are far from having exhaused all the possible applications of the method and need many more experiments to prove or disprove the predictions of the theory.

No attempt was made in this chapter to compile all the data relevant to the various examples as this would have covered the whole field of chemistry and require several volumes. Furthermore we limited ourselves to systems involving fragments no larger than 4. The interpretation of data involving larger reactive fragments could however have been carried out on the same basis.

The predictive ability of the GP theory is not restricted to reagents. It also covers the conditions under which experiments are performed. However, most of the experimental data necessary to test or corroborate the theoretical results are still lacking in the current literature and a major effort might be undertaken to close this gap.

One aspect of particular interest that is only marginally covered in the above treatment at this stage is the influence of the leaving group on the rate of reaction. The GP theory should be able to include this factor and this might represent a further area of development of the theory.

References

1. For a review see G. Klopman and B. O'Leary, *Fortschr. Chem. Forsch.* **15**, 4 (1970).
2. (a) W. J. Hehre, R. F. Stewart, and J. A. Pople, *J. Chem. Phys.* **51**, 2657 (1969); (b) L. Radom and J. A. Pople, *J. Am. Chem. Soc.* **92**, 4786 (1970).
3. (a) L. Salem, *Twenty-third International Congress of Pure and Applied Chemistry*, Butterworths, London, 1971, Vol. 1, p. 197; (b) N. Bodor and M. J. S. Dewar, *Tetrahedron* **25**, 5777 (1970); (c) M. J. S. Dewar and S. Kirschner, *J. Am. Chem. Soc.* **93**, 4290, 4291 (1971); (d) E. Clementi, J. Mehl, and W. Von Niessen, *J. Chem. Phys.* **54**, 508 (1971).
4. (a) G. A. Olah, G. Klopman, and R. H. Schlosberg, *J. Am. Chem. Soc.* **91**, 3261 (1969); (b) R. Sustman, J. E. Williams, M. J. S. Dewar, L. C. Allen, and P. von R. Schleyer, *J. Am. Chem. Soc.* **91**, 5350 (1969); (c) L. Radom, J. A. Pople, V. Buss, and P. von Schleyer, *J. Am. Chem. Soc.* **94**, 311 (1972).
5. (a) D. T. Clark and D. R. Armstrong, *Theor. Chim. Acta* **13**, 369 (1969); (b) L. Radom, W. J. Hehre, and J. A. Pople, *J. Am. Chem. Soc.* **93**, 289 (1971); (c) L. Radom, W. J. Hehre, and J. A. Pople, *J. Chem. Soc.* (A) 2299 (1971); (d) A. Dedieu and A. Veillard, *Chem. Phys. Letters* **5**, 328 (1970).
6. A. Streitwieser, Jr., *Molecular Orbital Theory for Organic Chemists*, Wiley, New York, 1967.
7. M. J. S. Dewar, *J. Am. Chem. Soc.* **74**, 3341, 3345, 3350, 3353, 3357 (1952).
8. R. B. Woodward and R. Hoffmann, *J. Am. Chem. Soc.* **87**, 395, 2046, 2511 (1965).
9. H. C. Longuett Higgins and E. W. Abrahamson, *J. Am. Chem. Soc.* **87**, 2045 (1965).
10. R. B. Woodward and R. Hoffmann, *The Conservation of Orbital Symmetry*, Verlag Chemie, Berlin, 1970.

11. (a) R. F. Hudson and G. Klopman, *Tetrahedron Lett.* 12, 1103 (1967); (b) G. Klopman and R. F. Hudson, *Theor. Chim. Acta* 8, 165 (1967); (c) G. Klopman, *Sigma Molecular Orbital Theory*, Sinanoğlu and Wiberg, Eds., Yale Univ. Press, New Haven, 1970.

12. L. Salem, *J. Am. Chem. Soc.* 90, 543, 553 (1968).

13. (a) K. Fukui, T. Yonezawa, and H. Shingu, *J. Chem. Phys.* 20, 722 (1952); (b) K. Fukui, T. Yonezawa, C. Nagata, and H. Shingu, *J. Chem. Phys.* 22, 1433 (1954).

14. G. Klopman, *J. Am. Chem. Soc.* 90, 223 (1968).

15. (a) L. Pauling, *J. Am. Chem. Soc.* 54, 3570 (1932); (b) R. S. Mulliken, *J. Chem. Phys.* 2, 782 (1934); 3, 573 (1935).

16. G. Klopman, *J. Am. Chem. Soc.* 86, 1463 (1964); *J. Chem. Phys.* 43, S124 (1965).

17. R. G. Pearson, *J. Chem. Ed.* 45, 581, 643 (1968).

18. J. O. Edwards, *J. Am. Chem. Soc.* 84, 22 (1962).

19. J. O. Edwards, *J. Am. Chem. Soc.* 76, 1540 (1954).

20. W. P. Jencks and J. Carriulo, *J. Am. Chem. Soc.* 83, 1743 (1961).

21. J. O. Edwards and R. G. Pearson, *J. Am. Chem. Soc.* 84, 16 (1962).

22. R. E. Davis and A. Cohen, *J. Am. Chem. Soc.* 86, 440 (1964).

23. G. Klopman, K. Tsuda, J. B. Louis, and R. E. Davis, *Tetrahedron* 26, 4549 (1970).

24. F. G. Bordwell, *Organic Chemistry*, Macmillan, New York, 1963, p. 218.

25. A. J. Parke, *Adv. Phys. Org. Chem.* 5, 173 (1967).

26. R. F. Hudson and G. Klopman, *J. Chem. Soc.* 5 (1964).

27. E. M. Kosower, *J. Am. Chem. Soc.* 78, 3497 (1956).

28. R. G. Pearson and J. Songstad, *J. Am. Chem. Soc.* 89, 1827 (1967).

29. (a) J. B. Hendrickson, D. J. Cram, and G. S. Hammond, *Organic Chemistry*, McGraw-Hill, New York, 1970, pp. 394–395; (b) N. Kornblum, R. A. Smiley, R. K. Blackwood, and D. C. Iffland, *J. Am. Chem. Soc.* 77, 6269 (1955).

30. P. A. Chopard, R. F. Hudson, and G. Klopman, *J. Chem. Soc.* 1379 (1965).

31. A. R. Katrizki and C. D. Johnson, Chemical Society Anniversary Meeting at Exeter (1967).

32. M. J. S. Dewar, *The Molecular Orbital Theory of Organic Chemistry*, McGraw-Hill, New York, 1969, p. 364.

33. (a) PNDO: M. J. S. Dewar and G. Klopman, *J. Am. Chem. Soc.* 89, 3089 (1967); (b) CNDO: J. A. Pople and G. A. Segal, *J. Chem. Phys.* 44, 3289 (1966).

34. G. A. Olah, *Acct. Chem. Res.* 4, 240 (1971), and references given therein.

35. See ref. 4a.

36. S. J. Cristol and N. L. Hause, *J. Am. Chem. Soc.* 74, 2193 (1952).

37. (a) S. J. Cristol and E. F. Hoegger, *J. Am. Chem. Soc.* 79, 3438 (1957); (b) H. C. Brown, *J. Am. Chem. Soc.* 89, 3400 (1967); (c) H. Kwart, *J. Am. Chem. Soc.* 86, 2601, 2606 (1964); (d) C. H. De Puy, *J. Am. Chem. Soc.* 84, 1314 (1962); 87 2421 (1965).

38. See ref. 29a, p. 722.
39. S. H. Pine, *J. Chem. Ed.* **48**, 99 (1971).
40. (a) T. S. Stevens, *J. Chem. Soc.* 2107 (1930); (b) T. Thomson and T. S. Stevens, *J. Chem. Soc.* 55 (1932).
41. R. S. H. Liu, *J. Am. Soc.* **89**, 112 (1967).
42. K. Fukui, *Acct. Chem. Res.* **4**, 57 (1971).
43. R. G. Pearson, *Acct. Chem. Res.* **4**, 152 (1971).
44. H. Yamazaki and R. J. Cvetanovic, *J. Am. Chem. Soc.* **91**, 520 (1969).
45. R. Breslow, *Organic Reaction Mechanisms*, Benjamin, New York, 1969, p. 135.
46. R. Huisgen and P. Otto, *J. Am. Chem. Soc.* **90**, 5342 (1968), and references therein.
47. R. Huisgen and P. Otto, *Tetrahedron Lett.* 4491 (1968).
48. (a) R. E. K. Winter, *Tetrahedron Lett.* 1207 (1965); (b) R. Criegee, D. Seebach, R. E. Winter, B. Börretzen and H. A. Brune, *Chem. Ber.* **98**, 2339 (1965).
49. M. R. Willcott and V. M. Cargle, *J. Am. Chem. Soc.* **89**, 723 (1967).
50. (a) S. J. Cristol, F. R. Sternitz, and P. S. Romney, *J. Am. Chem. Soc.* **78**, 4939 (1956); (b) M. J. S. Dewar and R. C. Fahey, *J. Am. Chem. Soc.* **85**, 2215 (1963), *Angew. Chem. Intl. Ed.* **3**, 245 (1964).
51. G. L. Closs and P. E. Pfeffer, *J. Am. Chem. Soc.* **90**, 2452 (1968).
52. R. Woodworth and P. Skell, *J. Am. Chem. Soc.* **81**, 3383 (1950).
53. K. Kopecky, G. Hammond, and P. Leermakers, *J. Am. Chem. Soc.* **84**, 1015 (1962).
54. R. Huisgen, *Angew. Chem. Intl. Ed.* **2**, 563, 633 (1963), and references therein.
55. P. Bailey, *Chem. Rev.* **58**, 925 (1958), and references therein.
56. J. G. Martin and R. K. Hill, *Chem. Rev.* **62**, 537 (1962), and references therein.
57. C. H. Depuy and O. L. Chapman, *Molecular Reactions and Photochemistry*, Prentice Hall, Englewood Cliffs, N.J., 1972, pp. 138.
58. R. Woodward and T. Katz, *Tetrahedron* **5**, 70 (1959).
59. J. Colonge and G. Descotes, in *1,4-Cycloaddition Reaction* J. Hamer, Ed. Academic, New York, 1967, p. 211, and references therein.
60. G. J. Janz and A. R. Monahan, *J. Org. Chem.* **29**, 569 (1964).
61. J. F. Harris, U.S. Patent 3, 136, 786, 1964.
62. L. S. Povarov and B. M. Mikhailov, *Izv. Akad. Nauk SSSR Otd. Khim. Nauk.* 955 (1963).
63. (a) T. Wagner-Jauregg, *Ber.* **63**, 3213 (1930); (b) J. V. Alphen, *Rec. Trav. Chem.* **61**, 892 (1942).
64. Ref. 10, p. 123.
65. Ref. 29a, p. 866.
66. S. J. Rhoads, in *Molecular Rearrangements*, P. de Mayo, Ed., Wiley, New York, 1963, Vol. I, p. 655.
67. Ref. 29a, p. 595.

CHAPTER 5

Nucleophilic Reactivity

R. F. Hudson
University of Kent at Canterbury
Canterbury, Kent, England

I. GENERAL NUCLEOPHILIC REACTIVITY

A. Introduction

Chemical reactivity is usually discussed in terms of transition-state theory, but this method is too complex to be applied to any but the simplest of reactions (e.g., $H + H_2$). There is consequently no satisfactory theory of chemical reactivity at this time,[1] and it is customary to relate the rate constant (or a relative rate constant) to various parameters, for example, bond energies, solvation energies, ionization potentials, redox potentials, polarizability, and equilibrium constants, in particular pK_a. These correlations can lead to useful descriptions of the reaction process. In particular, the Brönsted equation,[2]

$$\log \frac{k}{k^0} = \beta(pK_{a_1} - pK_{a^0}) \tag{1}$$

has been widely applied both in its original form to rate-determining proton transfers[3] (general base catalysis), and in the extended form to reactions at other centers[4] (*vide infra*).

This is one of a large number of linear free-energy relationships which relate reactivity to thermodynamic stability for *a series of closely related reactions*.[5] As first pointed out by Hammett,[6] the order of nucleophilic reactivity of halide ions toward alkyl halides is the reverse of the order of stability of the products. This rate order is given by the Swain equation,[7]

$$\log \frac{k}{k^0} = sn \tag{2}$$

where n is the nucleophilic parameter of the nucleophile and s is a constant for the reaction concerned.

This equation, like the Brönsted relationship, requires a common reactivity order for nucleophiles in all reactions, which is clearly not the case.

The Swain equation has been modified in various ways[8] to give essentially four-parameter equations, for example,*

$$\log \frac{k}{k^0} = yE_r + xpK_a \tag{3}$$

which in effect give a quantitative statement of the principle[9] of "hard and soft acids and bases." The recognition of different reactivity orders for different reactions (*vide infra*) is of considerable importance in the discussion of reaction mechanisms.[10]

The transition-state method[11] is based on the postulate of an equilibrium between ground state and transition state (an abnormal molecular configuration defined in a certain specific way), the concentration of which determines the reaction rate.

The quasi-thermodynamic system is then treated by classical statistical thermodynamics.[11] In recent years perturbation theory has been widely applied to chemical reactions,[12] and it has been particularly successful in interpreting electrocyclic, sigmatropic, and cycloaddition reactions.[13] The importance of orbital symmetry has greatly stimulated theoretical investigations of many chemical reactions.

B. The Perturbation Method[12]

This method attempts to calculate the energy changes in the initial stages of a reaction as the orbitals of the reacting molecules mutually interact (i.e., perturb each other), and to estimate the initial slope of the potential energy (PE) curve.[14] The perturbation method is particularly suited to the study of chemical reactions since the transition state can be regarded as a perturbed ground state, and hence the usual difficulties of calculating ground-state and transition-state energies are avoided. When two systems interact, the combined wave functions of the perturbed system are given approximately by combinations of the wave functions of the independent systems. Thus by considering the interaction of each pair of orbitals separately, the total perturbation energy can be found. As simple examples we consider the combination of two atoms, for example, $H\cdot + H\cdot$ and of two ions, for example, Na^+ and Cl^-.

In the first case the interacting orbitals are degenerate and combination gives a bonding orbital (B.O) an antibonding orbital (AB.O), and a bond energy of 2β (Fig. 5-1).

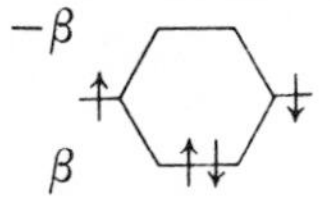

Fig. 5-1. Perturbation of degenerate levels.

* E_r is the oxidation-reduction potential of the nucleophile.

In the second case, the combination leads to essentially no covalent bond formation. Here the orbitals have widely differing energies (~ 5.1 and ~ 13 eV). It follows therefore that the perturbation energy decreases with the difference in energy of the interacting orbitals ($\alpha_j - \alpha_k$).

Thus in general we have for the interaction of two degenerate orbitals with coefficients C_1 and C_2, respectively,[12] $E_+ = \alpha + \beta$ and $E_- = \alpha - \beta$, where α and β are Coulomb and resonance integrals, respectively,

$$\beta = \langle \psi_1 | H | \psi_2 \rangle = \langle C_1\phi_1 | H | C_2\phi_2 \rangle$$
$$= C_1 C_2 (\phi_1 H \phi_2) = C_1 C_2 \beta_{1\cdot2}$$

For nondegenerate orbitals, the second-order term gives a bonding energy ΔE,

$$\Delta E_2 = \frac{2C_1{}^2 C_2{}^2 \beta_{1\cdot2}^2}{\alpha_j - \alpha_k}$$

For complex systems the perturbation energies are additive,[12] and hence in general we have for the interaction of atoms r and s

$$\Delta E_1 = 2 \sum C_{rj} C_{sk} \beta_{rs} \tag{4}$$

and

$$\Delta E_2 = \sum_j^{occ} \sum_k^{unocc} \frac{C_{rj}{}^2 C_{sk}{}^2 \beta_{rs}{}^2}{\alpha_j - \alpha_k}$$

This approach can be used in principle to analyze all chemical reactions since these involve orbital perturbation leading to energy changes ΔE_1 or ΔE_2. Figure 5-2 gives a general scheme for σ, π, and nonbonding orbitals.

These perturbations correspond to the molecular interactions, summarized in Table 5-1.

These interactions have been discussed in some detail in a recent review by

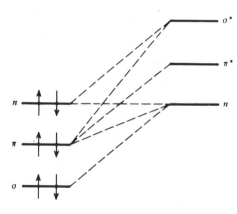

Fig. 5-2. Relative energy levels of σ, π, and nonbonding orbitals and their interaction with σ^* and π^* levels.

Table 5-1. Summary of Orbital Perturbations Leading to Molecular Interactions

Perturbation	Molecular System
1. $n \to a^a$	σ complexes, e.g., coordination compounds; neutral molecules
2. $n \to \pi^*$	Substitution at unsaturated systems; nucleophilic aromatic substitution
3. $n \to \sigma^*$	Nucleophilic aliphatic substitution; molecular complexes (e.g., I_3^-, XeF_2)
4. $\pi \to a$	π complexes; non classical carbonium ions
5. $\pi \to \pi^*$	Molecular complexes (e.g., picrates); cycloadditions (e.g., Diels-Alder)
6. $\pi \to n$	Electrophilic aromatic substitution; π complexes (e.g., Ag^+)
7. $\sigma \to \pi^*$	Intramolecular rearrangements, e.g., carbonium ions;
$\sigma \to \sigma^*$	hyperconjugation

[a] a refers to an acceptor orbital.

the author.[15] Here we shall restrict ourselves to a discussion of interactions normally encountered in nucleophilic substitutions.

π-Systems. The energies calculated by the perturbation method apply strictly to small interactions only. Chemical reactions proceeding through transition states, however, involve strong interactions with considerable energy changes. Since the perturbation method predicts the behavior of the system as the reaction begins, we have no way of ensuring that such predictions will be followed throughout the reaction, and that a sudden change in reaction path may not occur.

By the application of the perturbation method to electrocyclic reactions and cycloadditions,[13] the predictions based on a consideration of *either* end of the reaction path are the same as those given by correlating the orbital symmetries of reactants and products.[16] This must indicate that the initial slope of the energy curve along the reaction coordinate determines its general shape. Thus the Hoffmann-Woodward rules can be deduced from either the perturbation theory[13] or from the Longuet-Higgins and Abrahamson group-theoretical method.[16] The equivalence of the two methods depends on the following general rules which hold for reactions of a given type:

a. The initial symmetry is preserved through the transition state to the final products.

b. The initial perturbation is proportional to the energy changes along the reaction path.

The second rule, which enables reactivity to be considered,* is sometimes referred to as the noncrossing rule,[17] which is in fact a qualitative statement of a linear free-energy relationship. By way of illustration we shall consider electrophilic aromatic substitution since this reaction has been treated quantitatively more than any other.

The interaction of an aromatic compound with an electrophile $E^{\oplus}$ may be interpreted in two general ways. In the first treatment[18] the substitution is considered to proceed via a σ complex leading to a decrease in π-bond energy due to the removal of aromaticity, for example,

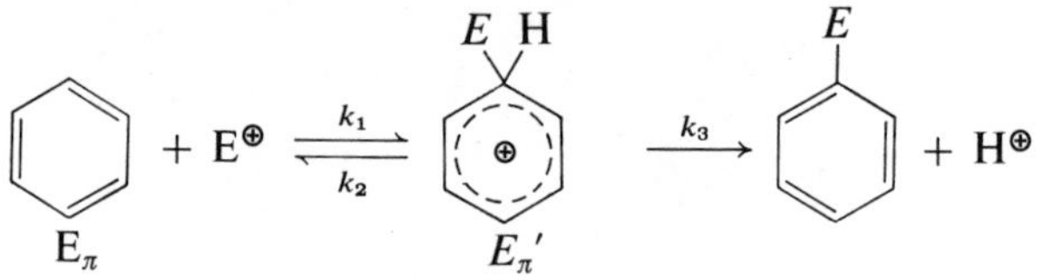

If $k_3 > k_2$ (as indicated by H isotope effects in most reactions),[19] k_1 is rate determining. The change in π-bond energy, termed the *localization energy, L,* is a measure of the activation energy if a given electrophile E reacts with a series of aromatic compounds. This can be obtained either by calculating[18] $E\pi$ and $E\pi'$, or by using a simple perturbation method[20] (PMO), developed by M. J. S. Dewar.

In a different approach,[21] the reaction may be assumed to involve the perturbation of the π system by E^+ at the point where the final substitution occurs, ΔE_p,

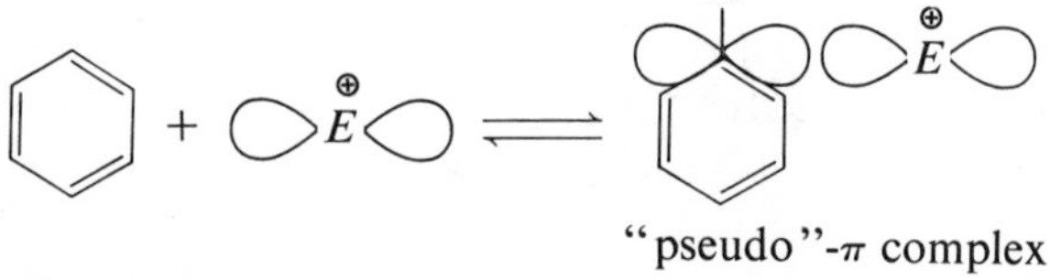

"pseudo"-π complex

The perturbation energy, ΔE_p, is given by

$$\Delta E_p = 2C_{sk}{}^2\beta_{rs}{}^2 \sum_{j}^{occ} \frac{C_{rj}{}^2}{\alpha_j - \alpha_k}$$

If $\alpha_j - \alpha_k$ is large (see Fig. 5-2), that is, if the antibonding orbital of the electrophile has a low energy (e.g., the electron affinity of bromine, chlorine

---

* The correlation diagram method does not give a measure of reactivity.

$\simeq 0$), $\alpha_j \gg \alpha_k$ and

$$\Delta E_p \simeq 2C_{sk}{}^2\beta_{rs}{}^2 \sum \frac{C_{rj}{}^2}{\alpha_j} \simeq 2C_{sk}{}^2\beta_{rs}{}^2(S) \qquad (5)$$

The term $\sum_j (C_{rj}{}^2/\alpha_j)$ was termed the *superdelocalizability* by Fukui,[21] and this is found to represent quantitatively the relative reactivity of a large number of polycyclic hydrocarbons with various electrophiles.

As shown by Fig. 5-3, S tends to increase with L, that is, the initial perturbation energy is approximately proportional to the reaction enthalpy.

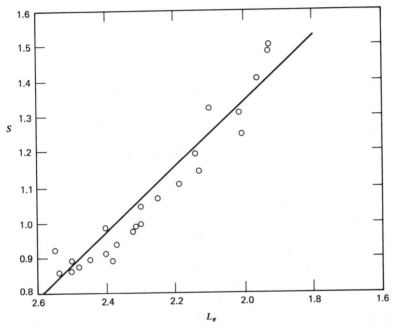

Fig. 5-3. Values of superdelocalizability and localization energy for reactions of alternant hydrocarbons.

The reason is that according to the noncrossing rule,[17] the PE profiles of similar reactions have the same form but are displaced by the enthalpy differences. If these curves are represented idealistically as in Fig. 5-4, the proportionality between ΔS, ΔL, and $\Delta\Delta E^*$ is evident. A general relationship between S and L has been derived analytically.[22]

It is now clear that the transition state of an aromatic substitution varies considerably with the nature of the electrophile and the reaction conditions.[23] This is shown by the relationship between rate (represented logarithmically) and localization energy for the reactions of various electrophiles with a series of alternant hydrocarbons.[24] If the transition state resembles the Wheland

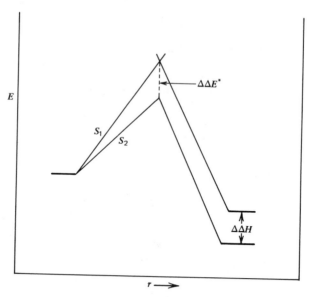

Fig. 5-4. Idealized P.E. Profiles for two closely related reactions.

intermediate,[18] then changes in $RT \ln k$ should follow changes in L. In all cases, however, the rate changes are much smaller, that is, $\Delta RT \ln k = x \Delta L$. x varies from about 0.1 to 0.5. Although part of the discrepancy may be due to solvation differences, the low values of x show that the structure of the transition state may lie close to that of the reactants (i.e., low x) and the rate still follows the localization energy.

Reactivity can be related to localization energy only when the noncrossing rule holds, as is usually the case when the reactions of similar nucleophiles (in this case, aromatics) are compared.

C. Linear Free-Energy Relationships

According to the above discussion, activation energy changes are proportional to localization energy changes for a series of similar reactions. This is a particular example of the linear free-energy relationship which states that for similar reactions

$$\Delta \Delta F^* = \beta \cdot \Delta \Delta F$$

This relationship, which has been derived parametrically[25] and from a consideration of energy profiles[26] (cf. Fig. 5-4), can also be derived by the perturbation method as follows.[15] The interaction of two *similar* nucleophiles of orbital energies α and α' with an acceptor orbital, Ψ^*, of energy α_s and coefficient C_s at the reaction site (Fig. 5-5) leads to a difference in perturbation energy of

$$\Delta\Delta E^* = \frac{2C_s{}^2\beta^2}{\alpha - \alpha_s} - \frac{2C_s{}^2\beta^2}{\alpha' - \alpha_s} \simeq 2C_s{}^2\frac{(\alpha' - \alpha)\beta^2}{(\alpha - \alpha_s)}$$

Similarly for the interaction of the same two nucleophiles with the corresponding atomic orbital, ϕ_s, of energy $\alpha_s{}^0$ leading to a covalent bond,

$$\Delta\Delta H \simeq 2\beta^2\frac{(\alpha' - \alpha)}{(\alpha - \alpha_s{}^0)^2}$$

so that

$$\Delta\Delta E^* \simeq C_s{}^2\left(\frac{\alpha - \alpha_s{}^0}{\alpha - \alpha_s}\right)^2 \cdot \Delta\Delta H \qquad (6)$$

that is,

$$\Delta\Delta E^* \simeq x\cdot\Delta\Delta H$$

Accordingly when $\alpha_s \ll \alpha_s{}^0$, $x \to 0$, and as $\alpha_s \to \alpha_s{}^0$ and $C_s \to 1$, $x \to 1$.

These are the boundary conditions for a generalized linear free-energy relationship, which is usually expressed[5] in the form

$$\Delta\log k \simeq x\cdot\Delta\log K$$

if changes in activation entropy and entropy of reaction are assumed to mutually cancel.

It is difficult to measure equilibrium constants for many organic reactions (*vide infra*), and for this reason the change in velocity constant k with the nature of the nucleophile is usually related to pK_a changes, that is, the changes in the affinity of the nucleophile for the proton.[2] This relationship is the Brönsted equation (1).

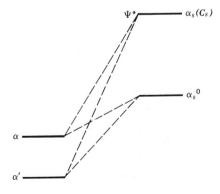

Fig. 5-5. Interaction of two nonbonding orbitals on similar nucleophiles with acceptor orbitals (antibonding and nonbonding levels).

1. The Brönsted Relationship. The Brönsted relationship[2] was originally restricted to general base (and acid) catalysis, that is, rate-determining proton transfers. A linear relationship is observed only for a restricted class of nucleophiles and in particular oxygen bases (alcohols, phenols, acids, oximes) and amines have been widely used. The results are affected by steric factors, and in general *m*- and *p*-substituted phenols and amines give the most reliable correlations.

Aromatic hydrocarbons are weak bases and as mentioned above their reactions with strong electrophiles have been widely followed. The proton affinities can be conveniently measured in HF solvent[27] and the rate of various electrophilic substitutions related[28] to the pK_a values measured in this way follow Eq. (1). The Brönsted coefficients lie within the prescribed limits of 0 and 1 (the value for proton-deuteron exchange is about 0.5 as expected for a symmetrical transition state), and show, as concluded in the previous section, that the transition state is not close to the Wheland intermediate. Values of β of 0.26 and 0.64 have been obtained for nitration in acetic anhydride[29] and for chlorination in acetic anhydride,[30] but the pK_a's in one solvent cannot be compared with rates given for a different solvent in general.

In studies of carbanion formation, Streitwieser[31] has followed the rates of ionization of the following hydrocarbons:

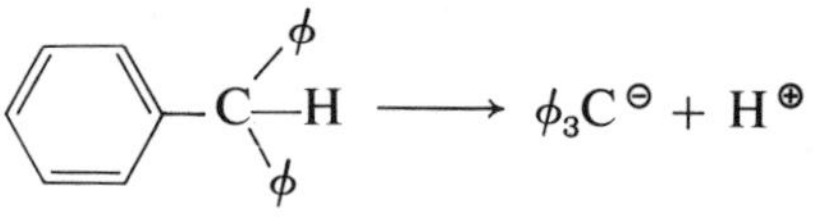

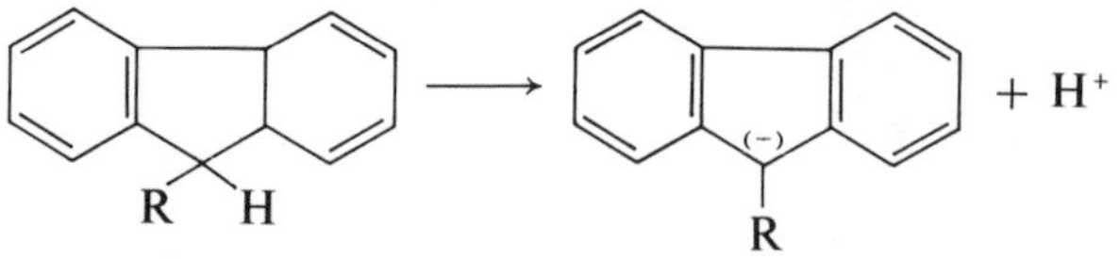

The Brönsted relationship holds over wide pH ranges although the values of β differ for these two similar systems (Fig. 5-6).

The correlation between hydrogen bonding (K_H) and acidity has been discussed by several authors,[32] and in general the Brönsted relation is followed closely (see Fig. 5-7) for variations in the acceptor ($x = b$) and in the donor ($x = a$). Since the coefficient x may be taken as a measure of the perturbation (see p. 175), it should vary in a predictable way with structural changes.

Thus, in Table 5-2, on comparing the values for donors (2) and (3) and for donors (1) and (4) it is seen that *a* increases with increase in the basicity of the donor. The value of *a* for combination of phenol and amine is close to 0.5, to

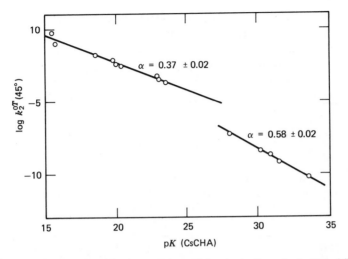

Fig. 5-6. Brönsted relationships for rate-determining ionization of substituted fluorenes (upper line) and polyarylmethanes (lower line).

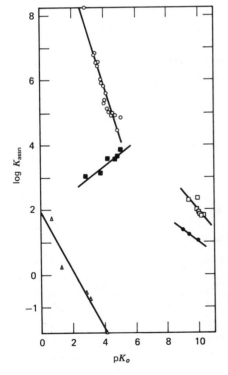

Fig. 5-7. Equilibria involving hydrogen bonding versus acidity: ○ benzoic acids with 1.3 diphenyl guanidine in benzene; △ $R \cdot COOH \cdots NC_5H_5$ = $RCOO^{(-)}$ $\overset{(+)}{HNC_5H_5}$ in carbon tetrachloride; □ phenols with triethylamine; (⊡ = naphthols); ● phenols with methyl acetate; ■ dimerization of carboxylic acids in benzene.

177

Table 5-2. Brönsted Coefficients *a* and *b* for Hydrogen Bonding[32]

	Donor	Acceptor	Solvent	
				a
(1)	Et₃N	ArOH	Cyclohexane	0.58
(2)	Et₃N	ROH	CCl₄	0.15
(3)	Et₂O	ROH	CCl₄	0.065
(4)	Et₂O	ArOH	CCl₄	0.25
				b
(5)	Pyridines	CH₃OD	—	0.17
(6)	Me-Benzenes	HCl	Heptane	0.044
(7)	Pyridines	CHCl₃	—	0.14
(8)	Cyclic ethers	CHCl₃	—	0.089

be expected for a symmetrical intermediate. Similarly, b increases with the basicity of the donor [cf. (7) and (8)]. In these cases an increase in basicity is due to a decrease in the magnitude of α_j [Eq. (4)] and an increase in perturbation as given by x.

A comparison of the data (1) and (2) and of (3) and (4) show that the coefficient a increases with acidity of the acceptor, that is, with an increase in the values of α_k.

These changes are the reverse of those observed for the Brönsted coefficients for rate-determining proton transfers as shown by the data given in Table 5-3 for the enolization of ketones,[33]

$$B^\ominus + H{-}CH_2{-}\overset{\overset{\displaystyle R}{|}}{C}{=}O \xrightarrow{\ k\ } BH + CH_2{\cdots}\underset{\underset{\displaystyle R}{|}}{C}{\cdots}O^\ominus$$

Here the value of β actually *decreases* from about 0.9 to 0.4, with increasing acidity of the ketone, a change which must be associated with a change in the nature of the transition state. The same conclusion is reached from the change in the rate constants of a series of pseudoacids with the basicity of the attacking nucleophile. Thus we can assume that a change in structure has little effect on the electronic distribution in the C—H bond in the ground state. Ionization, however, produces large changes in the energy and charge distribution in the displaced anion. Let us consider the reactions of two similar molecules with a given base leading to anions $X_1{}^-$ and $X_2{}^-$, the second having a much larger conjugation energy, E_c. The reactions may be represented by PE profiles relating energy changes to X—H separation distance

Table 5-3. Variations in the Brönsted Exponent β in the
Base Catalyzed Halogenation of Ketones[33]

Substrate	log R	β	pK_a
$CH_3 \cdot CO \cdot CH_3$	-8.56	0.88	20.0
$CH_3CO \cdot CH_2 \cdot CH_2 \cdot COCH_3$	-7.85	0.89	18.7
$CH_3 \cdot CO \cdot CH_2 \cdot Cl$	-5.29	0.82	16.5
$CH_3CO \cdot CH_2 \cdot Br$	-5.03	0.82	16.1
$CH_3CO \cdot CH \cdot Cl_2$	-3.78	0.82	14.9
$CH_2\!\!-\!\!CO \cdot CH \cdot COOEt$ $\diagdown(CH_2)_3$	-1.76	0.64	13.1
$CH_3CO \cdot CH_2 \cdot COOEt$	-1.06	0.59	10.5
$CH_2CO \cdot CH \cdot COOEt$ $\diagdown(CH_2)_4$	-0.60	0.58	10.0
$CH_3 \cdot CO \cdot CH_2 \cdot COC_6H_5$	-0.45	0.52	9.7
$CH_3 \cdot CO \cdot CH_2 \cdot CO \cdot CH_3$	-0.24	0.48	9.3
$CH_3 \cdot CO \cdot CHBr \cdot CO \cdot CH_3$	$+0.26$	0.42	8.3

(Fig. 5-8). An increase in conjugation energy in the leaving group lowers the right-hand side of the profile, producing a change in transition-state structure towards the reactants, that is, the less basic anion interacts more strongly with the proton in the transition state (leading to a larger value of β). This rule was deduced by Swain from kinetic isotope effect measurements.[34]

It is interesting to note that the value of β for a nitroparaffin[35] is greater than predicted from the values for ketones given in Table 5-3. This may be

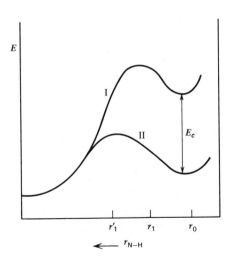

Fig. 5-8. Energy profile for the rate-determining ionization of an acid as a function of the distance between nucleophile and hydrogen atom in the transition state. Curve II is for an acid with a conjugated anion (conjugation energy E_c), curve I for a hypothetical compound with a nonconjugated anion.

due to the inductive effect of the NO_2 group in the *ground state* which tends to increase the value of β (*vide infra*).

Although the Brönsted equation is followed closely in most general base and acid catalyzed reactions, significant deviations are noted when the structure of the nucleophile is radically changed. Thus certain oxy anions, for example, oximes, are considerably more reactive than predicted by the equation, and this increased reactivity is well known in reactions at other centers (*vide infra*).[36]

Moreover, steric effects are observed, for example, as in the amine catalyzed decomposition of nitramine where the following results were obtained[37]:

Primary aromatic amines $\quad \log k = -5.39 + 0.64\, pK_a$
Secondary aromatic amines $\quad \log k = -5.04 + 0.64\, pK_a$
Tertiary aromatic amines $\quad \log k = -4.42 + 0.64\, pK_a$
Tertiary heterocyclic amines $\quad \log k = -3.49 + 0.64\, pK_a$

The parameter β is the same for the several series, the displacements of the $\log k$-pK_a lines being attributed to changes in solvation produced by steric factors.* A similar explanation has been advanced to explain the greater changes in pK_a of primary amines compared with secondary and tertiary amines on transfer from water to ethanol.[38]

More recently changes in β have been observed for the reactions of 2-methyl pyridines with ketones[39] (Table 5-4).

Table 5-4. Influence of Steric Factors on Brönsted β Values

	Acetone	diEt ketone	Pr^{i}Et ketone	Pinacolone	Cyclohexanone
Pyridines	0.73	0.77	0.77	0.75	0.58
2-Me pyridines	0.69	0.66	0.67	—	0.61
2,6-Me pyridines	0.59	0.46	0.65	—	0.50

The considerable steric hindrance in these reactions leads to decreases in bond formation, particularly prevalent in the case of cyclohexanone.

The influence of the basicity of the nucleophile on the rate of β elimination has been investigated along similar lines.[40] The general base catalyzed nature of E_2 elimination in ethanol solution and the change in transition-state with structure were established. The values of Table 5-5 were found for the Brönsted coefficients for β elimination (β_E) and concurrent substitution (β_s).[40,41]

The values of β_E of Table 5-5 show considerable variation from 0.17 for the tertiary chloride, where the transition state is probably close to the E_1

* Recent gas phase pK_a data support this interpretation.

$$\ominus B \cdots H - \overset{\overset{\displaystyle H}{|}}{\underset{\underset{\displaystyle H}{|}}{C}} - \overset{\overset{\displaystyle Me}{|}}{\underset{\underset{\displaystyle Me}{|}}{C}} \oplus Cl^{\ominus}$$

$$E_1$$

$$\overset{\ominus}{B} \cdots H \cdots \overset{\overset{\displaystyle H}{|}}{\underset{\underset{\displaystyle H}{|}}{C}} \cdots \overset{\overset{\displaystyle H}{|}}{\underset{\underset{\displaystyle H}{|}}{C}} \cdots Cl^{\ominus}$$

$$E_2$$

$$E_{1CB}$$

structure, to 0.88 for DDT. Here the two conjugating groups promote a transition state similar to that of a carbanion ($E_1CB\cdot$ mechanism); β_E is however significantly less than 1.0 and hence the carbanion is not fully formed. Stronger conjugating and carbanion stabilizing groups are required before the carbanion is liberated as a separate intermediate.

These data support proposals from other reactivity studies that the E_2 transition-state structure is highly variable, as in the case of S_N2 substitutions.[42]

It is interesting to note that in the absence of strongly conjugating substituents in the β position, β eliminations are only slightly more sensitive to increase in basicity of the nucleophile than substitutions. The selectivity is considerably greater in the corresponding reactions of n-propyl trimethyl

Table 5-5. Comparison of Experimental Values of β_E and β_S for Several Alkyl Compounds (in Ethanol Solution)

Compound	Temperature (deg)	β_E	β_S
Bu^{y}Cl[41]	45	0.17	—
Bu^{y}S$^+$Me$_3$[41]	25	0.46	0.27
Pr$_2$^{n}CHBr[40]	60	0.39	0.27
Bu^{n}Br[40]	60	—	0.37
Ph·CH$_2$·CH$_2$·Br[40]	60	0.56	0.35
p-NO$_2$·C$_6$H$_4$·CH$_2$CH$_2$Br[40]	60	0.67	—
Cl$_3$C·CH(C$_6$H$_4$·Clp)$_2$[41]	45	0.88	—

Table 5-6. Reactions with a Series of
Substituted Thiophenols[41]

Compound		Temperature (deg)	β_E
1,1-Cyclo-$C_6H_{10}XY$[41]			
X	Y		
H	OTs	35	0.27
H	Br	55	0.36
H	Cl	55	0.39
Br	Br	55	0.51
Cl	Cl	55	0.58

ammonium salts and of the sulphonium salt (Table 5-5) no doubt because of increased electrostatic attraction between nucleophile and incipient proton.

The relatively large value of β_E in a molecule where conjugation promotes further charge transfer from the nucleophile may also explain the considerable difference in behavior of piperidine which gives predominant elimination with α-bromoketones,[43] whereas morpholine gives predominant substitution as shown by the following reactions:

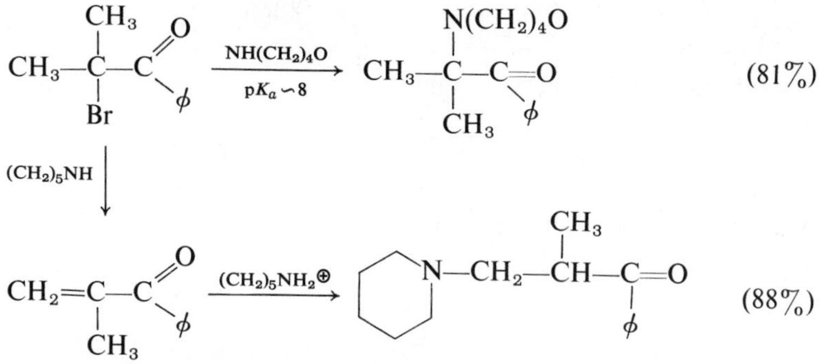

The above products would be obtained with a value of $\beta_E - \beta_S \sim 0.5$ using the pK_a values for the amines in water.

2. Deviant Brönsted Relationship. The Brönsted relationship (and other free-energy relationships) has been very widely used to provide a qualitative measure, β, of the extent of bond formation in the transition state. The success of this equation depends, however, on the assumption that a structural change in a reactant produces a continuous change in the interaction of this substituent with the reaction center as the reaction proceeds from a minimum at the initial state ($\beta = 0$) to a maximum ($\beta = 1$) at the final state.

According to this hypothesis β does not pass through extrema outside these limits but curved $\log k$-pK_a plots are to be expected (see Fig. 5-9) when a sufficiently wide range of bases is used.[44]

The change in a substituent in general produces changes in various energy factors, for example, charge distribution, solvation, steric interaction, ionization potential, and so on. Now for a series of closely related nucleophiles (e.g., p-substituted phenols, amines, or pyridines) steric influences are

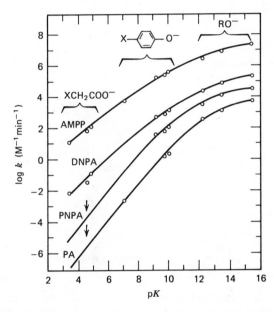

Fig. 5-9. Logarithmic plot of rate constants for the reactions of anionic oxygen nucleophiles with phenyl acetate (P.A.) p-nitrophenyl acetate (PNPA) dinitrophenyl acetate (DNPA) and 1 acetoxy-4-methoxy-pyridinium perchlorate (AMPP) against the basicity of the nucleophile at 25° in water.

constant, and it is assumed that the substituent produces proportional changes in all the energy factors.* We have considered this situation in detail elsewhere.[10] This may not be the case, and attention has recently been drawn[45] to reactions with carbanions which show large β values ($\beta > 1$) and negative β values. Thus nitromethane, nitroethane, and 2-nitropropane increase in acidity in this order but the rate of reaction with hydroxide ions decreases in the same order.[46] This can be explained by recognizing two main

* The usual "derivations" of the Brönsted law mask the relative contributions of various energy factors.

energy factors, the inductive and hyperconjugative effects of methyl groups, which act in opposition in this case (these effects augment in, for example, carbonium ion stability). The anion will be stabilized by hyperconjugation, in the planar form, whereas the transition state will be destabilized by a methyl group if the incipient carbanion is nonplanar and if the inductive influence dominates, namely,

The rate decreases are probably not due directly to steric hindrance, since large β values are observed for proton abstraction from the following nitroalkanes:

It is significant that β is greater for series (b) than for series (a) where conjugation is possible.

3. Extended Brönsted Relationship.[4] According to the derivation given on p. 175, the orbital of energy $\alpha_s{}^0$ may be any atomic orbital, and hence within the restrictions given above, rates or equilibria for reactions at one center can be correlated with rates or equilibria at another center. This conclusion has been widely confirmed in recent years.

We shall now consider the significance of the coefficient β for these correlations. Surprisingly few values of β greater than unity have been observed,[47] and it appears that $0 < \beta < 1$ for reactions at most centers. Few equilibrium comparisons have been made. The basicity of a series of carbanions[48] toward the $^{\oplus}CH_2$—OH group closely follows the change in proton basicity over a range of 10^{11} (Fig. 5-10). Also, the association constant[49] for a series of phenolate ions with the Fe^{3+} ion, a widely different process from the

$$HO—\overset{\oplus}{C}H_2 + \overset{\ominus}{C}(NO_2)_2X \; \xrightarrow{\;K\;} \; HO—CH_2—\overset{\overset{\displaystyle NO_2}{|}}{\underset{\underset{\displaystyle NO_2}{|}}{C}}—X$$

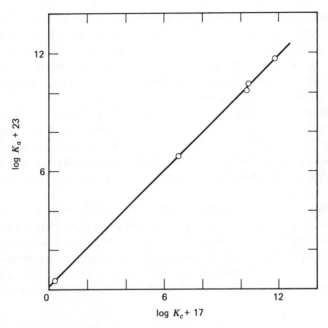

Fig. 5-10. Equilibrium constants, K_c, for the association of a series of nitrocarbanions with $HOCH_2^+$ related to their proton basicity, K_a.

addition to a carbon atom, is also related to the pK_a of the phenols with $\beta \simeq 1.0$ (Fig. 5-11).

$$Fe^{3+}(nH_2O) + \overset{\ominus}{O} - \underset{}{\overset{X}{\bigcirc}} \overset{K}{\rightleftharpoons} \overset{X}{\underset{}{\bigcirc}} - O - Fe^{2+}(n\text{-}1H_2O) + H_2O$$

On the other hand, the combination of a series of phenols and alcohols with the acyl group[50] leads to values of $\beta > 1$ (1.35 and 1.6, respectively) as shown by the data of Table 5-7. The greater slopes β in these reactions may be explained by conjugation in the esters formed.

The formation of the ester from a phenolate ion leads to conjugation between the benzene ring and carbonyl group,

$$\underset{X}{\overset{}{\bigcirc}} - \overset{\ominus}{O} + CH_3 - \overset{\oplus}{C} = O \longrightarrow$$

$$\underset{X}{\overset{}{\bigcirc}} - O - \underset{CH_3}{\overset{}{C}} = O \longleftrightarrow \overset{X}{\underset{}{\bigcirc}} - \overset{..}{O} - \underset{CH_3}{\overset{}{C}} = O$$

$\beta = 1.0$

The value of $\beta > 1$ thus implies the formation of a multiple bond between the nucleophile and reaction center. This cannot occur with a proton. Consequently β values in excess of unity are possible when a conjugating group is *released*, for example, in the hydrolysis of aromatic esters.[47] Care must therefore be exercised in the interpretation of β values for reactions proceeding through addition intermediates.

The Brönsted relationship proposed in 1922 was originally restricted to rate-determining proton transfers, and almost exclusively to general base catalysis.[3] In view of the outstanding success of this empirical rule, the converse relationship was widely used; that is, the demonstration of a Brönsted relationship for a particular reaction is taken as evidence for general base catalysis.

This view, however, is incorrect because the hydrogen atom *in a transition state* is not unique, and the field it exerts on an approaching nucleophile is small compared to that exerted by a proton. Thus cases have been found where the reactivities of two bases towards hydrogen are the reverse of the affinities towards a proton (*vide infra*). This means that *in general* the Brönsted law can be applied satisfactorily to a rate-determining proton transfer, pro-

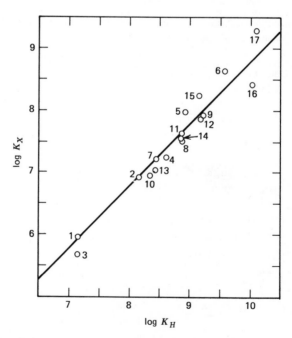

Fig. 5-11. Correlation of the equilibrium constants, K_x, for the association of phenoxide ions with Fe^{3+} in water and the pK_a of the corresponding phenol.

Table 5-7. Affinity of Oxygen Bases for the
Proton K_H and for Acyl-carbon K_C

	$K_H{}^a$	$K_C{}^a$
p-MeO·C_6H_4O—	1.5×10^{-6}	2.4×10^{-9}
p·MeC_6H_4O—	1.6×10^{-6}	1.6×10^{-9}
C_6H_5O—	9.0×10^{-7}	3.9×10^{-10}
p·Cl·C_6H_4O—	2.4×10^{-7}	7.2×10^{-11}
m-$NO_2C_6H_4O$—	2.2×10^{-8}	1.3×10^{-12}
p·$NO_2C_6H_4O$—	1.3×10^{-9}	1.7×10^{-13}
EtO—	9×10^{-1}	6.1
$MeOCH_2CH_2O$—	6×10^{-2}	1.7×10^{-1}
Cl CH_2CH_2O—	1.8×10^{-2}	1.7×10^{-2}
$Me_3\overset{\oplus}{N}CH_2CH_2O$—	9×10^{-4}	7×10^{-3}
$CF_3CH_2CH_2O$—	2.1×10^{-4}	5.5×10^{-6}

^{a} K_H refers to $H_2O + A^{\ominus} \overset{K_H}{\rightleftharpoons} HA + OH^{\ominus}$.

K_C refers to $ROH + A^{\ominus} \overset{K_C}{\rightleftharpoons} RA + OH^{\ominus}$.

viding that bases of similar structure (e.g., carboxylic acids, phenols, alcohols) are employed.

If we apply this same restriction to reactions at other centers, then the Brönsted equation should apply equally well. It is true, however, that in a proton transfer with large β, that is, when the transition state lies close to the reaction products, the reaction is effectively occurring with a proton, and hence a wider range of bases can be used in establishing a Brönsted relationship.*

In general, the range of bases is restricted. The success of the extended Brönsted relationship shows that the affinity of a series of similar bases for a given center follows the affinity for a proton. This is due to regular changes in bond dissociation energies for various bonds produced by a given substituent[51] (Fig. 5-12).

In spite of the generality of the Brönsted equation, it has been applied widely to centers other than hydrogen only in recent years. This is surprising since the reactions of several series of bases with alkylating and acylating agents were known to follow the Hammett equation.[52] The great success of

* In such cases, however, the lyate ion (OH^-) is highly reactive and it becomes experimentally difficult to follow the reactions of weaker bases, i.e., to differentiate between general and specific catalysis.

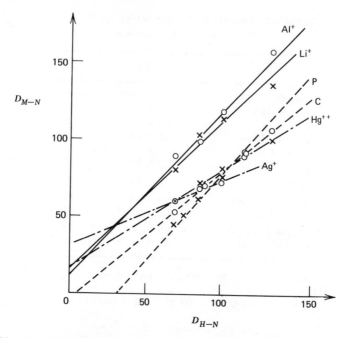

Fig. 5-12. The relationship between the dissociation energies of *M—N* bonds where *M* is Al^+, Li^+, P^{3+}, C, Hg^{2+} and Ag^+, and the corresponding bonds formed by the ligand *N* (halide ion) and hydrogen.

the Hammett law is its wide range of application. For nucleophilic displacements, however, it is limited since only *m*- and *p*-substituted aromatic compounds can be studied, and different substituent constants have to be used for different systems (e.g., for

$$\text{X}\!\!-\!\!\bigcirc\!\!-\text{CH}_2\!\!-\!\!\text{OH} \quad \text{and} \quad \text{X}\!\!-\!\!\bigcirc\!\!-\text{OH}$$

4. Applications

a. Carbonyl and Phosphyl Compounds. The differentiation between general base catalysis and nucleophilic catalysis is not easy. Kinetic isotope effects and linear free-energy relationships provide the main methods for attempting this differentiation.

Gold and Jefferson[53] noted that sterically hindered bases, for example, 2.6 lutidine and quinoline, react with acetic anhydride more slowly than predicted from their basicities, whereas 3- and 4-substituted pyridines and iso-quinoline obeyed the Brönsted relationship with $\beta = 0.92$. However, as discussed above, steric hindrance is also observed in some displacements on hydrogen, so these reactions could conceivably be general base catalyzed. However, formate ion reacts rapidly compared to the more basic acetate ion.[54] The slow reaction of the latter must be attributed to general base catalysis, namely,

$$CH_3-\overset{\overset{O}{\|}}{C}-O-\overset{\overset{O}{\|}}{C}-CH_3 + CH_3COO^{\ominus} \longrightarrow$$

$$CH_3\overset{\overset{O}{\|}}{C}-\overset{\ominus}{O}\cdots H-\overset{|}{\underset{H}{O}} \longrightarrow \overset{\overset{O}{|}}{\underset{CH_3}{C}}-O^{\overset{\ominus}{\diagdown}}\overset{CH_3}{\underset{}{C}}=O + CH_3COOH$$

Strong support for nucleophilic catalysis was provided by the work of Bender,[55] who showed that amines, including imidazoles* catalyzed the rate of hydrolysis of aromatic esters according to the Brönsted law, and in the case of imidazole itself direct evidence for intermediate formation was obtained.

$$R-\overset{\overset{O}{\|}}{C}-OAr + \underset{NH}{\overset{N}{\diagup\!\!\!\diagdown}} \rightleftharpoons \underset{}{\overset{HN-\diagdown}{\underset{\oplus}{N}}}-\overset{\overset{\ominus}{O}}{\underset{R}{C}}-OAr \longrightarrow$$

$$\underset{}{\overset{NH-\diagdown}{\underset{\oplus}{N}}}-C\overset{\diagup\!\!\!\diagup O}{\underset{R}{\diagdown}} + ArO^{(-)}$$

$$\downarrow H_2O$$

$$\underset{}{\overset{NH-\diagdown}{N}} + RCOO^{\ominus} + ArO^-$$

Further measurements by Bruice and Lapinskii[57] showed that all the catalytic bases cannot be accommodated by a single $\log k\text{-}pK_a$ line, the

---

* Williams has recently shown that imidazole is a general base in the hydrolysis of aryl diphenylphosphinates.[56]

different chemical types (e.g., anilines, pyridines) splitting to give a series of parallel lines (Fig. 5-13) with $\beta = 0.8$. This is similar to the value given above for the aminolysis of acetic anhydride and to that found for the reaction of a series of p-substituted phenols with ethyl chloroformate[58] ($\beta = 0.78$). A slightly higher value ($\beta = 0.9$) is found for the similar reaction of phenolate ions with acetylimidazole,[59] probably due to the positive charge on the reactant.

Several examples of general base catalysis in the reactions of esters and related compounds have been observed. The aminolysis of certain esters is found to be proportional to the square of the concentration of amine[60] indicating a mechanism of the kind

$$R'-\overset{\overset{\displaystyle O}{\|}}{C}-OR + NH_3 \;\rightleftharpoons\; R'-\overset{\overset{\displaystyle O^{\ominus}}{|}}{\underset{\underset{\displaystyle RO}{|}}{C}}-\overset{\oplus}{N}H_3 \;\xrightarrow{NH_3}\; R'-\overset{\overset{\displaystyle O}{\|}}{C}-NH_2 + NH_3 + ROH$$

$$R'-\overset{\overset{\displaystyle OH}{\diagup}}{\underset{\underset{\displaystyle RO}{|}}{C}}-NH_2$$

Tertiary amines and other bases, for example, hydroxide ions, may replace one of the ammonia molecules in the rate expression, and hence these reactions are general base catalyzed.[61]

The reaction of secondary amines with p-nitrophenyl phosphate[62] is similarly general base catalyzed, although here reaction occurs at the aromatic carbon atom.* Similarly, phosphorochloridates, cyanidates, and pyrophosphates react with n-butylamine to give the corresponding phosphoramidate, whereas the fluoridate gives the corresponding acid in a base catalyzed hydrolysis.[64] These examples show that good leaving groups promote nucleophilic attack whereas poor leaving groups require general base catalysis.

Some esters of strong carboxylic acids, for example, ethyl difluoroacetate, dichloroacetate, and oxamide hydrolyze with general base catalysis.[65] This is indicated by the decreased rate in deuterium oxide, an effect which is not observed in reactions proceeding by nucleophilic attack on carbonyl carbon.[66] Moreover, imidazole and the phosphate dianion have similar basicities and are similar catalysts, but the former is some 4×10^3 times more reactive

* See also the work of Bunnett.[63]

toward p-nitrophenyl acetate. Aniline is a strong catalyst, but does not form the anilide, but strong bases, for example, ammonia, do react preferentially at the carbonyl group.

The Brönsted coefficient for ethyl dichloroacetate[67] is 0.47, which is significantly lower than the values above referring to reaction at the carbonyl group. Ethyl trifluoroacetate[68] behaves similarly giving a value of $\beta = 0.33$. Again the reactivity in D_2O is less than in water (Long gives values of

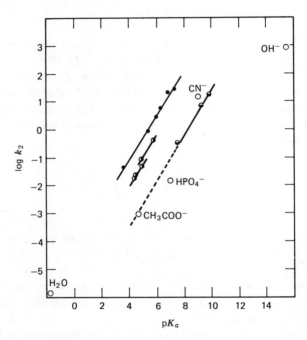

Fig. 5-13. A Brönsted-type plot of log k_2 versus pK_a for the reaction of p-nitrophenyl acetate with imidazoles, ●; pyridines, ◐; anilines, ◑; and phenolate ions, ◓.

k_D/k_H of 1.4–2.0 for esters in general). It is interesting to note that CH_3—ONH_2 produces a positive deviation in rate of approximately 10^2 although the product of reaction is trifluoroacetic acid only and the deuterium effect k_{H_2O}/k_{D_2O} is approximately 4.4, indicating a strong α effect for this nucleophile (*vide infra*).

There has been much discussion concerning the magnitude of β. According to the general treatment given on p. 175 and discussed in experimental terms on p. 176 the magnitude of β should lie between 0 and 1 unless additional conjugation affects the system. This is pertinent to reactions proceeding through addition intermediates as in acylations, namely,

For good leaving groups, that is, when $k_3 \gg k_2$ the rate measures k_1 directly and here values of β lie between 0 and 1.0. As $k_2 \to k_3$, that is, for poor leaving groups, the rate is measured by $(k_1/k_2)k_3$. Since changes in k_1/k_2 should lead to a value of $\beta \simeq 1.0$, the experimentally found value of β will be given by $1 + d\beta$ where $d\beta$ is the small (inductive) influence of Nu on k_3. Thus (limiting) values of 1.05 and 1.01 have been found[69] for the reactions of phenyl acetate and of

Large values of β may however be found for change in the nature of the leaving group, that is, variation in X of the above equation with a given $Nu^\ominus$.

Thus values of 0.42, 0.77 (0.83), and 1.3 are found for the alkaline hydrolysis, the nucleophilic catalysis by imidazole, and for the solvolysis of a series of p-substituted phenyl esters.[70]

As mentioned above, nucleophilic catalysis and displacement on carbon are usually subject to strong steric hindrance. This is shown by the extensive data by Hall[71] for the reactions of amines, and by the classical work of H. C. Brown.[72] Thus cyclic amines, for example, quinuclidine, are considerably more reactive than acyclic amines of comparable basicity,[72] and a similar reduction in steric hindrance may explain the high reactivity of azetidines[73] and aziridines toward phenyl acetate (Fig. 5-14).

In contrast, cyclic phosphites 2-alkoxy-1,3-dioxaphospholanes are considerably *less* reactive than trialkyl phosphites toward acylating agents[74]:

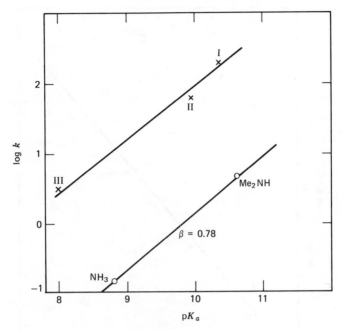

Fig. 5-14. Reactions of phenyl acetate in water at 30° with (I) and (II) and (III) aziridine.

This rate reduction has been ascribed to increase in ring strain on quaternization,[75] and since this should increase with the extent of bond formation in the transition state, it should change regularly with β (see Fig. 5-15).

This effect in principle provides a method, independent of linear free-energy relationships, of investigating transition-state structure.

The reactions of phosphylating agents with a wide range of nucleophiles have been followed kinetically, because of the importance of phosphorylation in biological systems. In particular, oximes and hydroxamic acids[76] have been studied with various phosphorus compounds (Table 5-8). The values of β are significantly higher than those obtained from a series of substituted phenols or amines. Thus β values for phosphinic and phosphate esters[77] lie in the range 0.35–0.40 (Table 5-8), and somewhat higher values (about 0.60) are

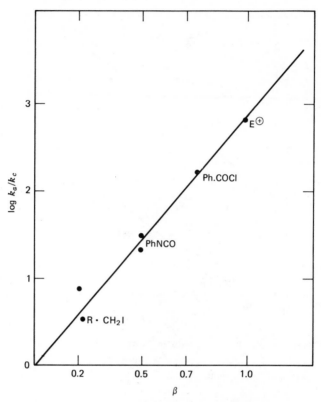

Fig. 5-15. Relationship between the relative reactivity of trimethyl phosphite and 2-methoxy 1.3 dioxaphospholane, k_a/kc toward various electrophiles and the Brönsted β for the reactions of these electrophiles with a series of p-substituted phenoxide ions (in water). The β value for PhNCO was obtained by interpolation.

Table 5-8. Brönsted β Values for Phosphoryl Compounds

Compound	Nucleophiles	β	Compound	Nucleophiles	β
$p \cdot NO_2C_6H_4 \cdot O \cdot P(O)O_2{}^{2-}$	Pyridines	0.13			
$NH_2P(O)O_2H^{(-)}$	Pyridines	0.22	$Et_2N(OEt)P(O)CN$	Hydroxamic acid	0.50
$Ph_2P(O)O \cdot C_6H_4 \cdot NO_2(p)$	Phenols	0.35	$[(EtO)_2P(O)]_2O$	Hydroxamic acid	0.70
$(EtO)_2P(O)O \cdot C_6H_4NO_2(p)$	Phenols	0.39			
$Pr^iO(Me)P(O)F$	Phenols	0.59	$Pr^iO(Me)P(O)F$	Hydroxamic acid	0.82
$Pr^iO(Me)P(O)F$	Cathechols	0.90	$Ph_2P(O)OC_6H_4NO_2p$	Hydroxamic acid	0.38

obtained for phosphorofluoridates.[78] The value is increased considerably to 0.90 if a series of substituted catecholates is studied,[79] an increase which is attributed to internal hydrogen bonding. This effectively increases the positive charge, and hence the electrophilic power of the phosphorus atom.

At the other end of the range, values of β are particularly small for anionic forms, decreasing from 0.39 for O-dialkyl p-nitrophenyl phosphates[77] to 0.15 for p-nitrophenyl phosphate.[80] For the latter, charged nucleophiles are generally unreactive, rendering uncharged nucleophiles such as amines and imidazoles relatively more reactive (*vide infra*). In these cases, evidence is

found for the intermediate quaternary ammonium compound, which rapidly hydrolyzes to the acid.

It may be concluded therefore from the examples given in Table 5-8 that the change in β is brought about primarily by a decrease in the electron density at the reaction center *in the ground state*, which increases the polarization of the nucleophile, and hence the charge transfer in the transition state.* The large values of β obtained for the reactions of hydroxamic acids[77] indicate a different reaction mechanism (by analogy with different β values for nucleophilic and general base catalyzed hydrolysis of similar esters). In agreement with this suggestion, the rate constants for the reactions of hydroxamate *ions* lie on the same $\log k$-pK_a line as those of *neutral* amidoximes[81] which react by intramolecular base catalysis similar to that proposed by Jencks[82] for the O acylation of NH_2OH by active esters, namely,

In view of the work of Exner[83] showing that hydroxamic acids ionize by dissociation of the N—H proton, a similar mechanism is possible for hydroxamate ions.

* Cf. hydrogen bonding and proton transfer (p. 177).

The same explanation has been advanced for the acylation[81] of hydroxamic acids, but here the β value is similar to the value obtained with phenolate ions, probably because these reactions involve the formation of a tetrahedral intermediate, and bond formation is almost complete in the transition state.

 b. Reactions at Saturated Carbon. In an important classical paper, G. F. Smith[84] has shown that the rate of reaction of anions with the α-chloroacetate ion follows the Brönsted law closely. By the inclusion of water and the hydroxyl ion, the range covered is extended to 3.5×10^4, and the deviations are no greater than those found for rate-determining proton transfers. Since the reaction is a substitution at carbon

$$ClCH_2CO_2^{(-)} + B^{(-)} \longrightarrow BCH_2CO_2^{(-)} + Cl^{(-)}$$

it was concluded that the transfer of a proton is not, as usually supposed at the time, the only reaction to which the Brönsted equation may be applied. Similar observations were made with the corresponding bromide, resulting in the following rate equations:

$$\log k = -6.09 + 0.203 \log K_b$$

$$\log k = -5.50 + 0.198 \log K_b$$

The coefficients for bromide and chloride are very similar to those in the case of β elimination (Table 5-9).

The available data for substitution at saturated carbon systems[58] are given in Table 5-9.

These data show clearly the decrease in the value of β with the tendency for

Table 5-9. Brönsted β Values for Alkyl Compounds

Compound	Nucleophiles	β^a	Compound	Nucleophiles	β
$MeOSO_2Me$	Pyridines	0.09	$n\text{-}Pr_2CHBr^b$	$R \cdot C_6H_4O^{\ominus}$	0.27
$EtOSO_2Me$	Pyridines	0.11	$n\text{-}C_8H_{17}Br^b$	$R \cdot C_6H_4O^-$	0.36
$MeOSO_3^{(-)}$	$R \cdot C_6H_4O^{(-)}$	0.16	$n\text{-}BuBr^b$	$RC_6H_4O^{\ominus}$	0.37
$Br \cdot (CH_2)_3 \cdot OH$	$R \cdot C_6H_4O^{(-)}$	0.22	$Ph(CH_2)_2Br^b$	$RC_6H_4O^{\ominus}$	0.35
$Cl \cdot CH_2COO^{(-)}$	$R \cdot COO^{(-)}$	0.20	$Ph \cdot CH_2Br^c$	$RC_6H_4S^{\ominus}$	0.20
$Br \cdot CH_2COO^{(-)}$	$RCO_2^{(-)}$	0.20	$CH_2{=}CH{-}CH_2Br^d$	Pyridines	0.37
$\overset{\displaystyle \frown}{\underset{\displaystyle }{O}}$ $CH_2{-}CH_2$	$R \cdot C_6H_4O^{(-)}$	0.32			

 a Water.
 b Ethanol.
 c Methanol.
 d Nitromethane.

the compounds to react by ionization S_N1 reactions, corresponding to a gradual change in the transition-state structure, as indicated by other investigations.[85] Thus sulphonate esters undergo S_N1 solvolysis more readily than bromides and chlorides,[86] and epoxides may be regarded as partially unsaturated, thus forming stronger bonds in the transition state.

This principle is widely used in the interpretation of the mechanism of substitution, first advanced by Hughes and Ingold[87] in a qualitative form in the famous concept of duality of mechanism.* It is of some interest to interpret the original postulate[87] in terms of the above approach. If the electronic distribution in the transition state for the reaction of a methyl halide is represented by structure I below, substitution of an electron releasing group, for example, CH_3, reduces $\delta+$ to $(\delta - \delta')+$, thus reducing the interaction with the nucleophile (II). The transition state is reached by a further extension of the C—X bond:

$$^{(-)}N \cdots\!\!\overset{H}{\underset{H}{\overset{\displaystyle\diagup}{\underset{\displaystyle\diagdown}{C}}}}\!\!\overset{\delta+}{} \cdots Cl^{(\delta-)}$$

(I)

$$^{(-)}N \cdots\!\!\overset{H\ \ H}{\underset{R^{\delta+}}{\overset{\displaystyle\diagdown\ \diagup}{C}}}\!\!\overset{(\delta-\delta)^+}{} Cl^{(\delta-)}$$

(II)

$$^{(-)}N \cdots\!\!\overset{H\ \ H}{\underset{R^{\delta+}}{\overset{\displaystyle\diagdown\ \diagup}{C}}}\!\!\overset{(\delta-\delta'+\delta'')^+}{} Cl^{(\delta+\delta'')-}$$

giving an increased charge δ'' (smaller than δ') on the carbon atom which now carries the charge $(\delta - \delta' + \delta'')$. The net effect is that although the charge on the central carbon atom $(\delta - \delta' + \delta'')$ of the substituted compound (and hence the selectivity) is less than that on the methyl carbon atom (δ), the charge on the group as a whole $(\delta + \delta'')$ is greater than that on the methyl group, that is, the carbonium ion character of the transition state is increased.† This gives an interpretation of the gradual change in mechanism $(S_N2 \rightarrow S_N1)$, based on considerations of nucleophilic reactivity.

The gradual change in the transition-state structure of displacement reactions of benzyl bromides[88] has been examined in a systematic way. That these transition states are highly dependent on reaction conditions is suggested by the changes in rate with substitution in the benzene ring for reactions with different nucleophiles. Generally, both electron-attracting and -donating

* The use of k_{OH^-}/k_{H_2O} as a criterion of mechanism is analogous to the use of β.

† It is not always realized that an increased S_N1 tendency does not usually increase the positive charge on the α carbon atom.

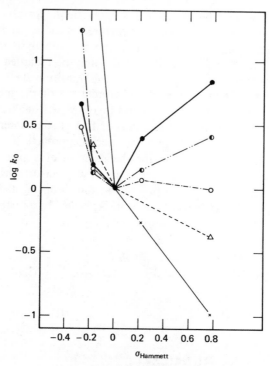

Fig. 5-16. Hammett relationships for the reactions of p-substituted benzyl halides with various nucleophiles: ◑ RBr + $C_6H_5S^-$ in methanol at 20°; ◐ RBr + $CH_3O^{(-)}$ in methanol at 20°; ○, RCl + $(CH_3)_3N$ in benzene at 100°; △, R-Cl + C_5H_5N in 90% aqueous alcohol at 20°, and X R-Cl in 50% aqueous acetone at 70°.

substituents increase the reactivity leading to a rate minimum for the unsubstituted compound[88] (Fig. 5-16). With weaker nucleophiles in highly polar solvents, however, the observed rate order corresponds to that of an S_N1 ionization process.

The changes in reactivity produced by changing R' for a given value of R in the following reactions of thiophenylate ions with p-substituted benzyl bromides have been measured in methanol solvent[89]:

$$R \bigcirc -CH_2Br + R'-\bigcirc -\overset{\ominus}{S} \longrightarrow$$

$$R-\bigcirc -CH_2-S-\bigcirc -R' + Br^{\ominus}$$

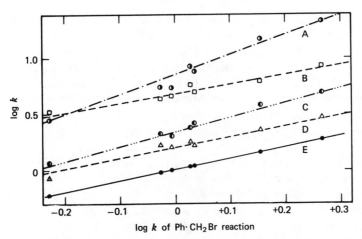

Fig. 5-17. Plot of log k for reaction of p-C_6H_4·CH_2Br with p-Y-$C_6H_4S^{(-)}$ against log k for the reaction of Ph·CH_2Br with pY-$C_6H_4S^{(-)}$ in methanol at 25°.

In these reactions the Brönsted relation is not accurately obeyed, and a tendency for the selectivity to increase with the pK_a of the thiophenol is observed. An approximate value of $\beta = 0.2$ for the unsubstituted compound can, however, be estimated. Since the deviations in the present case are of the same form for each bromide, linear relationships of the Hammett kind were obtained by plotting the log k values for the various substituents R' for a given substituted benzyl bromide against the corresponding values for the unsubstituted compounds (Fig. 5-17). The slope of each line ρ_N is found to be proportional to the Hammett substituent constant σ_R for the group R (Fig. 5-18). Since σ_R represents the change in the charge density at the electrophilic center produced by R, this provides a quantitative demonstration of the change in transition-state structure with substitution in the reactant.

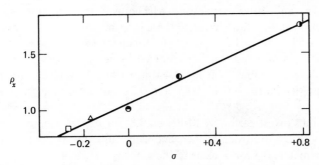

Fig. 5-18. Plot of ρ against Hammett's σ constant for the following substituents (points reading from right to left) : NO_2, Br, H, Me, MeO.

This mechanistic change is also brought about by a change in solvent as shown for example by the value of $\beta = 0.36$ for primary bromides in ethanol compared with $\beta = 0.22$ in water (Table 5-9). Further, benzyl bromide has the low value of 0.20 in methanol, whereas alkyl bromide has the value of 0.37 in nitromethane.[90] These values indicate that bond making is less advanced, and consequently bond breaking is more advanced in the solvating medium methanol than in nitromethane.

It should, however, be mentioned that the rate values for the reaction of allyl bromide in nitromethane are compared with pK_a values in water. Strictly the rate and pK_a values must be compared for identical reaction conditions, as in the case of the other data in Table 5-9. Solvation differences in the alkylation of tertiary amines in different media are probably unimportant (in view of the large size and wide dispersal of charge in the cation), and hence the allyl bromide value is included in Table 5-9.

c. Substitution in Aromatic Compounds. Nucleophilic substitutions of aromatic compounds are extremely slow in the absence of electron-withdrawing substituents. These render the aromatic carbon highly electron deficient in the transition state leading to extensive bond formation, for example,

In some cases tetrahedral intermediates may be isolated as in Meisenheimer complexes in reactions with alkoxide ions[91] or Janofsky complexes[92] with enolate ions and other carbanions. Mechanistic investigations suggest that the transition state is fairly close to these intermediates. For example, the rate order F > Cl > Br > I for the group X suggests strong bond formation in the transition state, and this is supported by recent observations of general base catalysis.[93] Values of β for 2.4 dinitro fluoro- and chloro- and iodobenzene are found to be 0.42, 0.45, and 0.52 with amines as the nucleophilic series.[94]

A previous investigation by Miller et al.[95] for the reaction of *p*-substituted phenols with 2.4 dinitrochlorobenzene using aqueous alcohol led to a value of approximately 0.5 for β, close to the value given above.

Systematic studies have been made of the reactions of a series of *p*-substituted thiophenols with several 2-halogenobenzthiazoles in ethanol which are

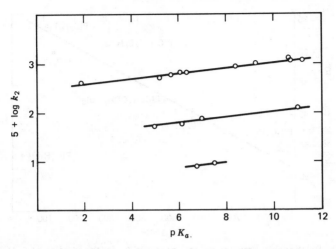

Fig. 5-19. Brönsted-type plots for the reaction of substituted pyridines with Pt(bipy)Cl$_2$. The lower two lines refer to 2-substituted and 2·6-substituted pyridines.

very reactive toward nucleophilic reagents.[96] As in the case of the reactions of thiophenols with benzyl bromide, the Brönsted relation is not obeyed particularly well, although the approximate value of $\beta = 0.7$ is found. The deviations are largely common for the fluoride, chloride, and bromide, and using the fluoride as a standard, values of the slope of 1.0, 0.84, and 0.70, respectively, are obtained indicating a gradual decrease in bond formation with decrease in the electronegativity of the halogen. (This sequence is very similar to the variation in the Brönsted coefficient observed for several phosphoryl compounds; see p. 194.) The iodide does not give a linear relationship, the more basic thiols being relatively more reactive than thiols with electron-attracting substituents. The slope for the latter group is approximately 0.73, that is, close to that of the bromide. The enhanced reactivity of p-methyl and p-methoxythiolates may be due to an increase in C—I bond breaking, that is, a significant change in the transition-state structure.

d. Metallic Complexes.   Recently the Brönsted law has been extended to the reactions of several inorganic complexes, formed by soft or B-group metals.

Chloride ion is readily displaced from complexes of Pt^{II} by amines,[97] and application of the Brönsted equation leads to the very low value of 0.06 for the complex Pt-(bipy)Cl$_2$ in methanol. Again 2-substituted pyridines give a similar value with rates about 10 times less (see Fig. 5-19). This low value suggests that bond formation is extremely weak in the transition state leading

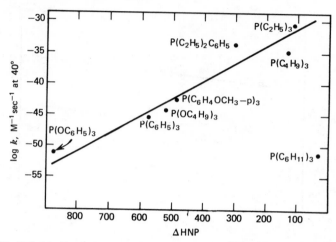

Fig. 5-20. Relationship between the rate constant, k_1, for the reaction of $C_5H_5Rh(CO)_2$ with phosphines and basicity of the phosphines, ΔHNP, determined by electrometric titration in nitromethane.

to the formation of a pentacovalent intermediate. Larger values of β are obtained for other Pt^{II} complexes.[98]

Similar observations[99] were made for complexes of Au^{III} although the selectivity is higher than for Pt^{II}. Thus a value of $\beta = 0.15$ was obtained for reaction of amines with $AuCl_4^{(-)}$. The considerably greater value of $\beta = 0.5$ was obtained for the cation $[Au(bipy)Cl_2]^+$ in acetone solution, in view of the

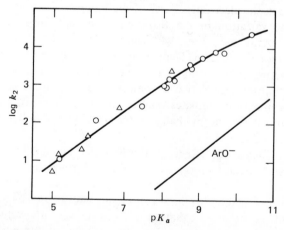

Fig. 5-21. Bimolecular rate constants, k_2, plotted against the pK_a of the conjugate acid for the reactions of hydroxamate anions (◯), amidoximes (△) and hydroxylamines (◯) with p-nitrophenyl acetate in water at 25°.

greater nuclear charge. The value for the corresponding complex of phenanthroline is, however, considerably less (0.22), showing the sensitivity of the gold atom to changes in electron density of the ligands.

The displacement of carbon monoxide from a cyclopentadienyl complex of rhodium by a series of phosphines and phosphites in toluene solution has been followed kinetically.[100]

With the exception of a few highly hindered phosphines, particularly the tricyclohexyl derivative, the rate constants follow the pK_a values measured in nitrobenzene fairly closely (Fig. 5-20). On the other hand, no reaction with pyridine could be detected after prolonged reaction showing the "soft" nature of the rhodium atom in this complex. The rhodium probably carries a formal positive charge in view of the high conjugation energy of the cyclopentadienyl ion, and the complex thus has a vacant orbital for bond formation,

The fact that phosphites lie on the same line as phosphines suggests that d_π bonding is negligible in the transition states. Tertiary phosphines are found to be more reactive toward $(NO)Co(CO)_3$ as in the case of the reaction of rhodium. On the other hand, Meriwether and Fine[101] have found that tertiary phosphines are displaced more rapidly than the less basic phosphites from nickel complexes of the kind $(CO)_2Ni(R_3P)_2$, thus providing evidence for strong d_π bonding in the ground state of these complexes.

It is evident that studies of this kind can provide important information in the nature of the transition state and reaction intermediates in reactions of organometallic compounds and metal complexes, and also on the type of bonding.

II. ENHANCED REACTIVITY

Some nucleophiles, for example, HO_2^-, ClO^-, RS_2^-, NH_2OH, N_2H_4, oximes, and hydroxamic acids appear to be more reactive than ions or molecules of comparable basicity, and thus show positive deviations from the Brönsted law.[102] This enhanced reactivity is particularly large in reactions at unsaturated centers as in acylation and phosphorylation (*vide infra*). Thus the hydroperoxide ion is less basic than the hydroxide ion owing to the inductive effect of the adjacent electronegative oxygen atom, $HO \leftarrow O^{\ominus}$. This leads to a reduced pK_a (e.g., $CH_3O_2^{\ominus}$ is $10^{4.5}$ times less basic than $CH_3O^{(-)}$). The hydroperoxide ion (in contrast to the hydroperoxide *radical*) is, however, more reactive than the hydroxide or alkoxide ion in all displacement reactions so far examined (Table 5-10). Similar rate enhancements are observed in reactions at the silicon atom, and with transition-metal complexes.[108]

Table 5-10. α-Effect of $HO_2^{\ominus}$ for Several
Electrophiles

Substrate	$k_{HO\bar{2}}/k_{OH}^-$	
Ph·CN	6600	(103)
$pNO_2C_6H_4O\cdot COMe$	200	(104)
$(EtO)_2P(O)OP(O)(OEt)_2$	100	(105)
$Pr^iO(Me)POF$	47	(106)
Ph·CH$_2$·Br	~50	(107)

Many explanations have been advanced,[109] most of which ascribe the rate increases to the influence of lone pairs of electrons on the atom adjacent to the nucleophilic center (the α effect). The position is, however, a complex one. Thus Jencks[81] showed that hydroxylamine in the neutral form is O-acylated by p-nitrophenyl acetate at a rate approximately 10^7 greater than expected from the basicity of the nucleophilic oxygen. On the other hand, hydrazines and hydroxylamines are not abnormally reactive towards alkyl halides,[110] and in simple proton abstraction reactions.[111]

Hydrazine itself shows enhanced reactivity towards esters,[112] isocyanates, and CO_2. Hydroxamate ions react at an enhanced rate with acylating[113] and phosphorylating[114] agents at a rate which increases logarithmically with pK_a, according to the extended Brönsted relation (Fig. 5-21) with a β comparable to that obtained for "normal" nucleophiles, for example, phenols and amines. On the other hand the reactivity of oximate ions changes in an irregular way (see Fig. 5-22) with the pK_a of the oximes,[115] the more acidic oximes generally showing the largest positive deviations from the Brönsted law.

It is highly likely that several factors contribute to enhanced reactivity,[109] and the differentiation of the contribution of these various factors is extremely difficult. Of the factors which have so far been suggested, lone-pair–lone-pair repulsion,[116] intramolecular catalysis,[81] stabilization of the transition state,[109] and differential solvation[109] would seem to be the most im-

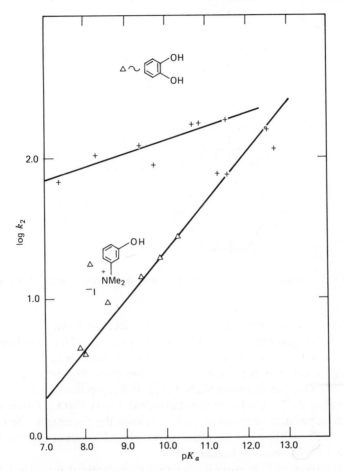

Fig. 5-22. Brönsted-type plot for the reactions of phenolate ions, △, and oximate ions, +, with *p*-nitrophenyl diphenylphosphinate in water at 25°.

portant. Moreover, differences in steric interaction (see p. 193) may also modify the reactivity, and the quantitative estimation of steric hindrance is particularly difficult.

There has, however, been much interest in recent years in the α effect, and in the following sections, we shall concentrate on the most recent work which seems to have led to a significant clarification of a confused situation.

A. Lone-Pair Repulsions[116]

There is considerable evidence to show that adjacent lone pairs on neighboring atoms may interact in such a way as to alter the physical properties of

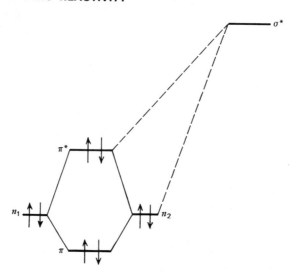

Fig. 5-23. Interaction of π and π^*, occupied "lone-pair" orbitals, with an acceptor, σ^* orbital.

the molecule or ion considered, for example, decreased ionization potential, reduced redox potential, and modified UV spectra.[117] Thus overlapping of lone-pair orbitals in a molecule leads to the absorption of light at longer wavelengths, for example, the color of the halogens (highest occupied orbitals are π^*). Overlap of two orbitals, for example, p-orbitals, leads to orbital splitting (Fig. 5-23), and hence the electrons in the HOMO are more readily removed on physical or chemical excitation than the electrons in the original AO (n). Such splitting can be observed by photoionization spectra, for example, 3.3 eV for *trans*-azomethane.[118]

This influence of lone pairs is also shown in chemical properties, for example, the acidity of the hydrogen atoms of nitrogen heterocycles.[119] The base catalyzed deuterium exchange of pyridine shows exchange at the 4-position, although Hückel charge densities suggest that the 2-position is more acidic.[120] Extended Hückel calculations, however, show that the 2-carbanion is destabilized as a result of lone-pair–lone-pair repulsion.[121] This is of interest in connection with the stability of the corresponding arynes.[119]

Relative Kinetic Acidities

Similarly the rate of base catalyzed deuterium exchange of diazines[120] can be explained readily by invoking lone-pair repulsions between electrons on nitrogen and on the incipient carbanion.

Conversely, a nucleophile of this kind, for example, pyridazine, should be more reactive than a comparable nucleophile, for example, pyrimidine or pyrazine since the lone-pair repulsion should decrease on the formation of a σ bond of the product.[122] It is generally assumed that the interaction of lone pairs with the electrons of σ bonds is less than the mutual interaction of two lone-pairs (cf. the Sidgwick-Powell and Gillespie-Nyholm rules[123]). In agreement with this prediction, the pK_a of pyridizine is significantly greater (2.44) than the pK_a of pyrimidine (1.23) and pyrazine (0.65) in spite of the greater electronegativity of the adjacent nitrogen atom. Moreover, the photoionization spectra of these molecules have been investigated in detail recently.[124] Although there is still some controversy over the assignment of the bands, it is now generally agreed that for pyridine, pyrazine, and pyrimidine the first ionization potential (IP) refers to a π electron, but the first ionization of pyridazine involves an essentially nonbonded electron; see Table 5-11. It is seen that the first IP of pyridazine is approximately 0.5 eV less than that of the other heterocycles in spite of the adjacent electronegative nitrogen atom, which should increase the IP.

The energy change produced by lone-pair interaction can be estimated from the orbital energies of the corresponding isolated atoms as follows,[125] (assuming essentially nonbonding electrons). The orbital energies E_+ and E_- on the combination of two atomic orbitals are

$$E_+ = \frac{\alpha + \beta}{1 + S} \quad \text{and} \quad E_- = \frac{\alpha - \beta}{1 - S} \tag{7}$$

where S is the overlap integral. For a system of two electrons in each orbital, the change in energy on this combination (to give an α nucleophile) is

$$\Delta E_r = 2E_+ + 2E_- - 4\alpha$$

$$= \frac{4\alpha S^2 - 4\beta S}{1 - S^2}$$

Table 5-11. Ionisation Potentials of Nitrogen Aromatics

IP, π	9.22	9.42	9.36	9.31
(eV) n.b.	8.90	9.64	9.51	9.51

If we apply the Mulliken approximation, $\beta = kS$, then

$$\Delta E_r \simeq -4(k - \alpha)S^2 \qquad (8)$$

This gives a measure of the repulsion energy of closed shells of electrons in terms of the overlap integral S. For example, assuming a value* of $E_n = 2$ eV (as estimated from the ionization potentials of water and hydrogen peroxide), and $S = 0.07$ for the O—O bond, $\Delta E_r = 0.28$ eV. This corresponds to a rate difference of $10^{4.6}$, which is similar to the maximum value observed for the α effect of $HO_2^{\ominus}$.

As a result of recent CNDO/2 calculation, S. Oae and collaborators [126] have suggested that an electrophile approaches the β atom of $HO_2^{\ominus}$ in a three-center interaction. Simple considerations show that the acyclic three-center four-electron system is destabilized relative to a two-center two-electron system. However, interaction of the electrophile with the π^* orbital produced by splitting may well lead to a three-center orbital since the coefficient on the α atom for the π^* orbital will be appreciable (see diagramatic representation below). According to this simple representation for large interatomic dis-

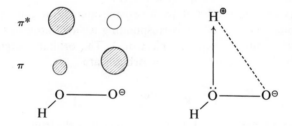

tances, covalent interaction proceeds mainly with the β oxygen atom, Coulombic interaction with the α oxygen atom.

This intriguing situation obviously requires further examination.

The magnitude of the overlap integral S depends on the conformation of the nucleophile, and in some cases it adopts a conformation which removes lone-pair repulsions, that is, when the corresponding doubly occupied orbitals are orthogonal. This cannot be the case for certain anions, for example, $RO_2^{\ominus}$, $RS_2^{\ominus}$, $ClO^{\ominus}$, as S remains large irrespective of orientation.

On the other hand the dihedral angles of NH_2—NH_2, NH_2OH, and R—S—S—R in their ground states are such as to lead to zero (or small) values of the overlap integrals, as shown by the Newman projections,[116]

---

* E_π is the orbital splitting energy.

R—S—S—R θ 90°   N$_2$H$_4$ θ 90°   NH$_2$OH θ = 180°

The ground-state conformations and rotational barriers of molecules of this kind including peroxides, F_2O_2, disulphides, hydrazine, N_2F_4, diphosphine, and hydroxylamine have been examined in great detail and can be explained by simple rules[127] similar to that proposed here for their reactivity. According to the argument developed above, such molecules do not exhibit an α effect due to electron repulsion (for example, the refractivities and polarizabilities of N_2H_4 and NH_2OH are normal), and consequently rate enhancements found in certain reactions of these nucleophiles (for example, phosphorylation and acylation) must be attributed to other causes (*vide infra*). These nucleophiles show normal reactivity in alkylation reactions and in general base catalysis by following the Brönsted rule, but in these cases small α effects are to be expected (see p. 214).

The effect of conformation on reactivity is shown by the reactions of cyclic disulphides. Calvin[128] has found that lipoic acid of importance in the reaction of coenzyme A, is oxidized by ammonium persulphate much more rapidly than 1,2-dithian or diethyl sulphide. If a mechanism of the following kind is assumed (by analogy with the reactions of phosphines),[129]

the high reactivity can be explained by increased lone-pair (p_π-p_π) repulsion as the dihedral angle decreases. This has been quantitatively demonstrated for the reaction of certain disulphides (Table 5-12) with methyl fluoro-sulphonate.[130]

$$R \underset{S}{\overset{S}{\underset{|}{\big|}}} \underset{R}{\big\backslash} + MeO—SO_2F \xrightarrow{k_2} R \underset{S}{\overset{\overset{\oplus}{S}\diagup Me}{\underset{|}{\big|}}} \underset{R}{\big\backslash} + SO_3F^{\ominus}$$

It is assumed (with H. Bock[134]) that each sulphur atom is largely un-hybridized and hence the splitting refers to p_π-p_π interaction of the highest occupied orbitals. The overlap integral S changes with cos θ and hence the repulsion energy increases with $\cos^2 \theta$ [see Eq. (9)] and this should give a measure of the increase in reactivity. Reference to Table 5-11 shows that (for the limited data available) log k_{Me} does in fact increase linearly with $\cos^2 \theta$.

The photoionization and kinetic data may be correlated in the following way.[130] If the changes in reactivity given by k_{Me} are produced mainly by orbital splitting, the difference in the free energy of activation of cyclic and acyclic disulphides is given approximately by

$$- RT \ln k_{Me} \simeq \beta \cdot \Delta E_r \simeq \beta \cdot [-4(k - \alpha)S_\pi^2 \cos^2 \theta] \qquad (9)$$

The proportionality constant β is expected to increase with the extent of bond formation in the transition state as measured by the Brönsted coefficient or some other linear free-energy parameter.

From the orbital energies [Eq. (7)] the photoionization splitting ΔE_π is

$$\Delta E_\pi = \frac{\alpha + \beta}{1 + S} - \frac{(\alpha - \beta)}{1 - S} \simeq 2(k - \alpha)S$$

Since the overlap integral is proportional to cos θ we have

$$\Delta E_\pi = 2(k - \alpha)S_\pi \cos \theta \qquad (10)$$

Table 5-12. Comparison of the Reactivity and Photoionization Data with the Dihedral Angle (θ) of Cyclic Disulphides

EtS—SEt	θ	ΔE_π(eV)[134]	log $k_{Me}{}^a$	$\cos^2 \theta$
1,2-Dithian	84.7[131]	0.24	0	0.01
5-[3-(1,2-dithiolanyl)]	60.0[132]	0.95	0.52	0.25
Valeric acid	35.0[133]	—	1.57	0.67

^{a} k_{Me} is the ratio of k_2 for the cyclic compound to k_2 for EtS—SEt.

Combination of these two equations leads to

$$-\frac{RT \ln k_{Me}}{\cos^2 \theta} \simeq 2\beta \cdot S_\pi \left(\frac{\Delta E_\pi}{\cos \theta}\right) \tag{11}$$

From the data of Table 5-12 taking S_π for the $S—S$ bond to be 0.16 as recommended by Mulliken[135] for Slater orbitals, a value of 0.21 is obtained for β. This value is close to those normally obtained for alkylation (0.2–0.3); see p. 196.

On the other hand, the relative reactivities of five- and six-membered cyclic hydrazines* and hydroxylamines (Table 5-13) toward p-nitrophenyl

Table 5-13. Reactions of Secondary Amines with p-NO$_2$C$_6$H$_4$O(Me)PO$_2^{(-)}$ in Water at 60° (Brönsted $\beta \simeq 0.3$)

	NH (cyclobutane)	NH (cyclohexane)	NH-NH (5-ring)	NH-NH (6-ring)	NH-O (5-ring)	NH-O (6-ring)
pK_a	11.20	11.12	7.6	7.7	5.05	5.20
$10^4 k$	285	47.7	39	13	2.9	1.1

Value for MeNHNHMe = 4.0 (pK_a = 7.52)

methylphosphonate[136] are similar to the relative reactivity of pyrrole and pyrrolidine. There is therefore no evidence for lone-pair–lone-pair inter-actions here, a conclusion supported by the similar pK_a values of the five- and six-membered heterocycles. An examination of Dreiding models shows the dihedral angles of

$$\text{HN—NH (5-ring)}$$

and the chair[137] form of

$$\text{HN—NH (6-ring)}$$

to be similar to that of hydrazine (approximately 90°). The small rate changes shown in Table 5-13 are probably due to changes in steric hindrance (see p. 180).

* Recent photoionisation data show small splittings for hydrazines.

B. The α Effect

It is clear from the above discussion that lone-pair–lone-pair repulsions can increase the reactivity of a particular nucleophile, and that the enhancement of reactivity is conformation dependent. This repulsive effect also modifies the affinity for protons and hence the pK_a is affected. It is not clear, therefore, at first sight why the α effect, which is usually defined by the positive deviations from the Brönsted equation, should be due to this lone-pair repulsion.[138] In a general explanation of this problem we have suggested that Coulombic interaction with the proton is relatively more important than with the reaction center of a transition state.[139] This suggestion has been developed further, by deriving an expression for the α effect based on a polyelectronic perturbation treatment which has been used to interpret the reactivity of polar molecules.[140] Here the one-electron Hamiltonian used previously (p. 170) is replaced by one including electron-electron interaction explicitly, that is, $H = H' + e^2/R_{rs}$. The electron-electron interaction is given by

$$\int \phi_i(1)\phi_k(2)H\phi_j(1)\phi_e(2)d_{\tau 1}d_{\tau 2}$$

and the core-core interaction by $Z_R Z_s{}^2 R_{rs}$. For a central-field approximation, electron-electron repulsion and core-core repulsion become equated in Γ_{rs}, and the perturbation energy is given approximately by

$$\Delta E \simeq q_r q_s \Gamma_{rs} + 2 \sum_j^{\text{occ}} \sum_k^{\text{unocc}} \frac{C_{rj}{}^2 C_{sk}{}^2 \beta_{rs}{}^2}{\alpha_j - \alpha_k} \qquad (12)$$

It is seen that Eq. (12) reduces to the equation using a one-electron operator for neutral molecules, and the Coulomb term $q_r q_s \Gamma_{rs}$ appears as a first-order perturbation.

This equation can be written in the following simplified form,[139] when reactions involving a given nucleophilic atom are compared (constant β),

$$\Delta E_p = k_1 q_r q_s + \frac{k_2}{\alpha_j - \alpha_k}$$

where q_r and q_s are the formal charges on nucleophile and electrophile, respectively, and α_j and α_k the corresponding orbital energies (Fig. 5-23).

The difference in the perturbation energies ($\Delta\Delta E$) of two similar nucleophiles, one of which has a lone pair of electrons adjacent to the nucleophilic atom (e.g., $RO_2{}^-$ and RO^-) reacting with a given electrophile is given by

$$(-\Delta\Delta E) \simeq k_1 q_r + k_2 \alpha_j - \{k_1 q_r' + k_2(\alpha_j - \alpha_j')\}, \qquad \alpha_j \gg \alpha_k$$

$$\simeq k_1 \Delta q_r - \frac{k_2 \alpha_j'}{\alpha_j{}^2}$$

Similarly for the interaction of these nucleophiles with a proton, Eq. (13) holds:

$$(-\Delta\Delta H) \simeq k_1{}^H \Delta q_r - \frac{k_2{}^H \alpha'_j}{\alpha_j{}^2} \tag{13}$$

Assuming that $(-\Delta\Delta E) \simeq RT\Delta \log k$ and $(-\Delta\Delta H) \simeq RT\Delta \log K$, it follows that

$$\frac{\Delta \log k}{\Delta \log K} = \frac{k_1 \Delta q_r - k_2 E_\pi}{k_1{}^H \Delta q_r - k_2{}^H E_\pi} \qquad \text{where } E_\pi = \frac{\alpha'_j}{(\alpha_j)^2} \tag{14}$$

In the case of two nucleophiles, neither of which is an α nucleophile, that is, $\alpha'_j = 0$, Eq. (15) is obtained:

$$\frac{\Delta \log k_0}{\Delta \log K} = \frac{k_1}{k_1{}^H} \equiv \beta \qquad \text{(Brönsted)} \tag{15}$$

The observed "α effect" is given by

$$\Delta\Delta \log k = \Delta \log k_0 - \Delta \log k$$

Combination of Eqs. (14) and (15) leads to the following expression:

$$\Delta\Delta \log k = \frac{\beta \cdot k_2{}^H (m - 1)\alpha'_j}{\alpha_j{}^2}, \qquad \text{where } m = \frac{k_1{}^H k_2}{k_2{}^H k_1} \tag{16}$$

According to this general treatment, the α effect for the reaction at a given center (other than hydrogen) should increase regularly with the extent of bond formation, β, to reach a maximum when bond formation is complete. Here the proton basicity and basicity towards the other center are compared. Hine[141] has shown that the carbon basicity of $Bu^\gamma OO^{(-)}$ is about 10^7 greater than that of $HO^{(-)}$, by a thermodynamic estimate of the appropriate equilibrium constants, and Jencks[142] has measured the equilibrium constants of the following reactions directly:

$$\underset{\underset{\text{Me}}{\overset{\text{O}}{\parallel}}{Me-C-N}}{} \overset{O^{(-)}}{} + Me-\overset{\overset{O}{\parallel}}{C}-\overset{\oplus}{Im}H \rightleftharpoons Me-\overset{\overset{O}{\parallel}}{C}\overset{\overset{O}{\underset{\underset{Me}{N}}{\overset{\text{Me}}{\nearrow}}}{O}}{} + ImH$$

and

$$RO^{(-)} + Me-\overset{\overset{O}{\parallel}}{C}-\overset{\oplus}{Im}H \rightleftharpoons Me-\overset{\overset{O}{\parallel}}{C}-OR + ImH$$

leading to a difference of approximately 10^3 for the relative affinity of a hydroxamate and alkoxide ion for the carbonyl group. (This compares favorably with the values given in Table 5-14).

It also follows from this treatment that α effects should not normally be observed in general base catalyzed reactions. There are several reports in the literature of such rate enhancements (p. 203), but the possibility of intramolecular catalysis (*vide infra*) must not be overlooked in these cases.

Table 5-14. Variation of α-Effect with Brönsted β-Coefficient

	α Effect^{a,b}	β^c	α Effect^{a,d}
$Ph \cdot CH_2 \cdot Cl$	0.60	0.30	1.10
$NO_2C_6H_4 \cdot CH_2 \cdot Cl$	—	0.33	1.20
$Pr^iOP(Me)OF$	—	0.55	1.35
$(MeO)_2P(O) \cdot OC_6H_5 \cdot NO_2(p)$	1.05	0.40	1.37
$EtOCOCl$	—	0.78	3.0
$p \cdot NO_2 \cdot C_6H_4 \cdot O \cdot COCH_3$	2.0	0.80	2.20

^{a} Logarithmic reactivity difference.
^{b} $CH_3CO \cdot C_{Me} = N - O^{(-)}$.
^{c} For p-substituted phenols.
^{d} $R - CO \cdot N(Me)O^{(-)}$.

It follows from this treatment that an "α effect" due to electron-electron repulsion can be observed only when $m > 1$, that is, when $k_1^H/k_2^H > k_1/k_2$. This means that the Coulombic term for the interaction with a proton is *relatively* more important than that for a typical transition state. The following conclusions can be drawn from this equation:

(i) the α effect (given by $\Delta\Delta \log k$) increases with the magnitude of the Brönsted coefficient β, (ii) α effects depend on orbital symmetry [α_j is a function of the lone-pair–lone-pair overlap; see Eq. (9)], (iii) the magnitude of the α effect decreases with the magnitude of α_j, that is, the ionization potential of the nucleophile, and (iv) large (maximum) α effects are observed in equilibria with centers other than the proton.

In agreement with these conclusions, it is found that the α effect for alkylation reactions is much less than the α effect for acylation. This is shown by the data given in Table 5-14 and by the recent data given by Zoltewicz[143] for the reactions of cyclic diazines. This latter work leads to a rate enhancement of threefold for methyl iodide and 20–30-fold for the reactions of "α" diazines with 2-4 dinitrophenyl acetate. The corresponding β values are 0.35 and 0.89, respectively.

Other nucleophiles are found to obey this rule semiquantitatively. Available data for the reactivity of N-methyl p-methoxybenzhydroxamate ions[144] relative to that of a phenoxide ion of the same basicity are collected in Table 5-14.

It is noted that the rate enhancements for alkylation are significantly less than the corresponding values for the reactions with active esters and chloroformates. Rate enhancements for phosphorylation lie between these two sets of values, in agreement with intermediate β values. The correlation is, however, only semiquantitative, which is to be expected in view of the various factors appearing in Eq. (16). Electrostatic energies are particularly high in phosphorus compounds (as shown for example by the large P—O and P—F bond energies) and it thus appears that the P^V phosphorus atom resembles a proton. This may lead to a value of m approaching unity, thus reducing the α effect.

The rate enhancement produced by oximate ions is also much greater for acylating and phosphorylating agents than for alkyl halides. The position here is complex, since the α effect varies considerably[145] with acidity of the oxime (Fig. 5-22). Thus the most basic oximes, for example, ketoximes and amidoximes, appear to show no α effect in the anionized form to all the substrates so far examined (e.g., PNPA, benzoyl fluoride, ethyl chloroformate, phosphate esters, and fluorides and benzoyl chloride). The more acidic oximes, for example, $CH_3CO \cdot C(R){=}N{-}OH$, show large α effects with acylating agents, for example, 10^2 for PNPA, and benzyl chloride[146] shows a logarithmic rate enhancement of 0.6. Here again these rate increases increase almost in proportion with the β values (0.8 and 0.3, respectively).

The large change in the α effect with pK_a of the oxime appears to be general for alkylation, acylation, and phosphorylation, but the precise reason is not known. Various factors could contribute to a decrease in the rate enhancement with pK_a of the nucleophile. As shown in Fig. 5-9 for the reaction of p-nitrophenyl acetate the slope of the log k-pK_a curve for phenols and alcohols decreases[147] from approximately 1.0 to a low value ($\beta \simeq 0.2$) with increasing pK_a. This would lead to a decreasing α effect according to Eq. (16). Moreover, the acidic oximes give highly conjugated anions, and it may be that for the most basic oximes, the longer N—O bond is free rotating, namely,

sp^3 hybridized oxygen sp^2 oxygen

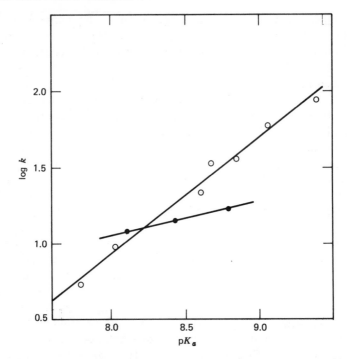

Fig. 5-24. Brönsted-type plots for the reaction of *p*-nitrophenyl acetate with hydroxamate ions, ○, and *N*-methyl hydroxamate ions.

The hybridization of oxygen could change with increasing conjugation energy, which would have the effect of decreasing the N—O distance, leading to greater values of the overlap integral S which determines the orbital splitting. The values of the α effect for N—Me hydroxamic acids also change with pK_a of the hydroxamic acid. Laloi et al.[148] give a value of approximately 0.2 for the Brönsted β value for the reaction of *p*-nitrophenyl acetate, compared with a value of 0.8 for the corresponding reactions of the unsubstituted acids. This is quoted as evidence for different reaction mechanisms in the two cases, the anions of the unsubstituted acids reacting by an intramolecular base catalyzed mechanism (Fig. 5-24).

C. Stabilization of the Transition State

It appears that some nucleophiles, for example, NH_2OH and N_2H_4, show large α effects in certain reactions (particularly acylation and phosphorylation) and small or negligible α effects in other reactions, particularly

Table 5-15. Second-Order Rate Constants for the Reaction of Various Substrates with Hydrazine, Glycylglycine, and Glycinamide

Substrate	$k_1(M^{-1}min^{-1})$			β	$\dfrac{k_{hydrazine}}{k_{glycoyl\ or\ glycinamide}}$ = α effect
	$k_{hydrazine}$	$k_{glycoyl}$	$k_{glycinamide}$		
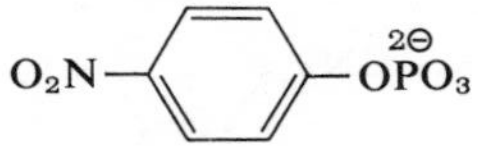 O_2N–⬡–$OPO_3^{2\ominus}$	6.3×10^{-5}		2.0×10^{-5}	0.15	3.0
 O_2N–⬡–$OSO_3^{\ominus}$	1.25×10^{-5}		2.0×10^{-6}	0.20	6.0
CH_3I	2.6×10^{-1}		5×10^{-2}	0.20	5.0
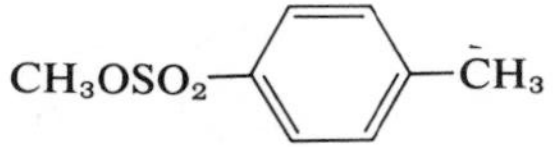 CH_3SO_2–⬡–$\bar{C}H_3$	0.39	0.42		0.27	0.9
ICH_2CONH_2	1.85×10^{-2}	5.6×10^{-4}		0.30	3.3
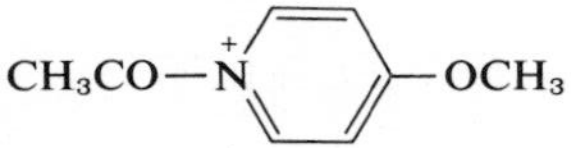 CH_3CO–$\overset{+}{N}$–⬡–OCH_3	4.9×10^5	2.9×10^4		0.32	17
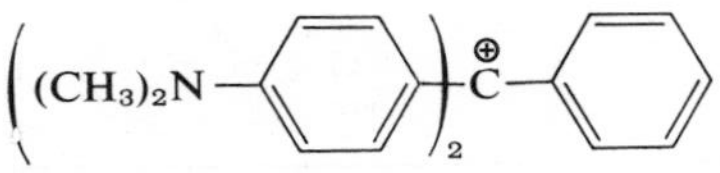 $\left((CH_3)_2N\text{–⬡–}\right)_2\overset{\oplus}{C}\text{–}C_6H_5$	3.75×10^3	1.33×10^2		0.41	28
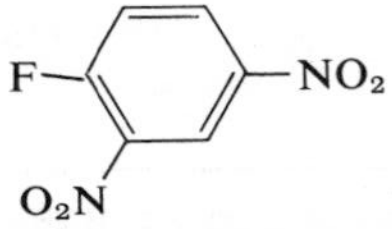 F–⬡(NO_2)(O_2N)	80.5	4.4	4.16	0.42	19

(continued)

217

Table 5-15. (*continued*)

Substrate	$k_1(\text{M}^{-1}\text{min}^{-1})$			β	$k_{\text{hydrazine}}/k_{\text{glycyl or glycinamide}}$ = α effect
	$k_{\text{hydrazine}}$	k_{glycyl}	$k_{\text{glycinamide}}$		
	8.2×10^{-2}	3.7×10^{-2}	3.85×10^{-3}	0.45	22
	0.216	7.0×10^{-3}	6.1×10^{-3}	0.52	35
	1.8×10^{4}	5.5×10^{2}		0.57	33
	7.2×10^{2}	4.3×10^{1}	3.1×10^{1}	0.64	23
	73	2.9		0.70	25.2

Structure				
CH_3COO–C$_6$H$_4$–NO_2	44	0.83	10.3	4.5×10^2
(N-acetyl pyrrolidine-dione benzylidene structure)	57	0.92	8.6	4.9×10^2
quinolin-6-yl $OOCH_3$	38	1.01	0.24	9.2
CH_3COO–C$_6$H$_5$	52	1.05	9.0×10^{-3}	0.47

alkylation. Part of the explanation may lie in changes in the extent of bond formation in the transition state as discussed in the previous section, but in some reactions, particularly the reaction of hydroxylamine with active esters,[82] the magnitude of the α effect is so large ($> 10^6$) as to indicate a different origin of the rate enhancement.

The position has been to some extent clarified recently by the detailed rate measurements of hydrazine by Bruice and Dixon.[149] By using glycinamide or glycyl-glycine as a reference the α effect of hydrazine is found to vary between 3 and 57 for 17 substrates (Table 5-15). Again the effect increases regularly with the magnitude of β.

The origin of this α effect is in some doubt as we have already noted that hydrazine shows no evidence of lone-pair splitting. Although intramolecular catalysis is possible (*vide infra*), hydroxylamine shows a similar effect when the reaction proceeds on nitrogen.[150] The explanation may lie in the original suggestion of Pearson[109] that the lone pair of electrons on the adjacent atom may stabilize the transition state, presumably by p_π bonding. Pearson[109] quoted the S_N1 ionization of α chloroethers as an analogy, namely

$$\text{R}\!-\!\overset{..}{\text{O}}\!\!\overset{\oplus}{-}\!\!\text{C}\!\!\begin{smallmatrix}\text{H}\\ \\ \text{H}\end{smallmatrix} \quad \text{cf.} \quad \text{R}\!-\!\overset{..}{\text{O}}\!-\!\overset{\oplus}{\text{N}}\!\!\overset{\text{H}}{\underset{\text{H}}{-}}\!\!\text{X}\!\cdots\!\overset{\ominus}{\text{Y}}$$

The two systems are not strictly comparable since a carbonium ion is electron deficient, a quaternary nitrogen atom is electronically saturated. The difference may well be one of magnitude, however, since high-energy orbitals ($3d$, $3p$, $4p$) can stabilize adjacent orbitals. For example, phosphorus and nitrogen ylids* are much more stable than the corresponding carbanion.

In support of this suggestion, it is found that in certain reactions (particularly when strong bonding in the transition state is evident) nucleophiles, for example, aniline, pyridines not normally considered to be α nucleophiles show some enhanced reactivity.[152] Thus aniline shows a similar "α effect" to hydrazine in acylation and to hydroxylamine in the reaction of ethyl chloroformate.

The reactions of hydroxylamine are, however, complicated by its ambient nature. Thus oxygen is the nucleophile in reactions with active esters and DFP and nitrogen in reactions with aryl, alkyl, and acyl halides. The large α effect observed for the reactions on oxygen is attributed to intramolecular catalysis discussed on p. 195. The smaller α effects found when the nitrogen atom is the nucleophile may be due to stabilization, as discussed above.

---

* The difference between phosphorus and nitrogen has probably been overemphasized, since their excited levels lie at similar energies.[151]

III. NUCLEOPHILIC ORDERS

A. General Introduction

We have so far been concerned with a comparison of the reactivity of nucleophiles of similar geometrical and electronic structures. Provided that the nucleophilic atom remains the same and that steric factors remain constant, quantitative relationships between reactivity and structure of the reactants can be developed.

In the general case, no simple relationship can be expected to hold, but a general pattern of behavior has emerged, which is described qualitatively[9] in the concept of hard and soft acids and bases (HSAB). This may be approached in a quantitative manner by classical thermodynamic methods.[10] Consider an ionic reaction of the following kind:

$$M^+ + N^- \underset{}{\overset{K_M}{\rightleftharpoons}} MN$$

The change in free energy ΔF° is given by

$$\Delta F^\circ = -RT \ln K_M = D_{M-N} + I_M - E_N - \Delta F_M^S - \Delta F_N^S + \sum \Delta S$$

where ΔF_M^S and ΔF_N^S are solvation free energies of the ions, I_M and E_N are the ionization potential and electron affinity of M and N, respectively, and D_{MN} is the dissociation energy of the product formed. $\sum \Delta S$ is the change in entropy. For a given series of nucleophiles reacting with a common center M^+, it follows that

$$\Delta F' = -RT \ln K_M \simeq D_{M-N} - (E_N + \Delta F_N^S) + \text{const} \qquad (17)$$

provided that differences in $\sum \Delta S$ are neglected. It is immediately apparent that the equilibrium is determined by two factors: (a) the interaction energy between M and N represented by the dissociation energy D_{MN},* and (b) the solution electron affinity given by $(E_N + \Delta F_N^S)$, which is a function of the nucleophile and solvent only. These terms are equivalent to† the driving force and reaction inertia discussed by Polanyi.[153]

It is instructive to calculate[154] values of $\Delta F'$ for different centers, for example, for a proton and Hg^{2+} ion (Table 5-16).

The equilibrium constant decreases in the first case and increases in the second with increasing size of the nucleophile. These orders are those associated with an A metal and B metal in the classification of Chatt and Ahrland,[155] and a "hard" acid and "soft" acid, respectively, in Pearson's nomenclature.[9]

---

* This includes a Coulombic term in the case of polar molecules.

† Repulsion energy is incorporated in D_{N-X}. For reaction rates this term may become a dominant one.

Table 5-16. Affinity of Halide Ions for H^+ and for Hg^{2+}

Ligand	$\Delta F_N{}^S$	E_N	$E_N{}^{Sa}$	D_{HN}	D_{HgN^+}	ΔF_{HN}	ΔF_{HgN^+}
F	121	83.5	204.5	134	100	70.5	104.5
Cl	93.5	88.2	181.7	102	81	80	100.7
Br	88.7	81.8	170.3	87	72	83.3	98
I	81.3	74.6	155.9	71	60	85	96

^{a} $E_N{}^S$ is $(E_N + \Delta F_N{}^S)$. All energies in kcal/mole.

These orders which have been widely used are general (although exceptions are found) for the following reasons. For a series of similar ligands, for example, halide ions or chalcogens, the energy of the polar bond M—N tends to increase[154] with the electronegativity of N, that is, increased Coulombic interaction. The term $E_N{}^S$ also increases with the electronegativity[154] of N. Thus the first order is observed when changes in D_{MN} are greater than changes in $E_N{}^S$, and the second order when the reverse situation obtains.* This behavior can be expressed quantitatively if $\Delta D_{MN} \propto \Delta E_N{}^S$ (as experimentally verified in some cases).

Although this is a rational approach it is of limited use since the energy terms are not readily available. For this reason empirical relationships have been proposed, of which the Swain-Edwards equation[8] is the most widely used,

$$\log \frac{k}{k^0} = xpK_a + yE_r \tag{3}$$

where E_r is the redox potential (or some similar physical property representing $E_N{}^S$ of the nucleophile).

A theoretical equation of this kind has been derived on the basis of a polyelectronic perturbation treatment.[140] The perturbation method was described briefly on p. 169, and this was limited to a one-electron treatment of the Hückel type.

If allowance is made for electron-electron and nuclear-nuclear interaction by incorporating the appropriate terms in the Hamiltonian, a general equation for the perturbation energy can be derived[140] which approximates to the following equation for the interaction of r and s:

$$\Delta E_p = -q_r q_s \Gamma_{rs} + \Delta \text{solv}(1) + \sum_j \sum_k \left[\frac{2C_{rj}{}^2 C_{sk}{}^2 \beta_{rs}{}^2}{\alpha_j - \alpha_k} \right] \tag{18}$$

* It should be noted that $E_N{}^S$ includes solvation energy and hence different orders may be found in different solvents. In the gas phase the hard-soft classification loses its significance.[156]

where Γ_{rs} is the Coulomb interaction between r and s with formal charges q_r and q_s in a central-field approximation.

Δsolv(1) is a partial desolvation energy on the approach of r and s. It is noted that the orbital interaction term is identical with that given in the equation (p. 170) for a one-electron treatment.

The limits of this equation,[157] determined mainly by the relative magnitudes of $\alpha_j - \alpha_k$ and the Coulomb term, lead to a theoretical derivation of the concept of HSAB. When $\alpha_j \gg \alpha_k$, that is, when the acceptor orbital has a low ionization potential and hence lies near the continuum, the last term of equation is small and the interaction energy ΔE_p is given mainly by the first term, that is, the Coulombic (Madelung) energy, and little charge transfer occurs. Such is the case when strongly electronegative and electropositive elements combine in a *charge controlled* reaction. This corresponds to a "hard-hard" interaction of HSAB theory.

On the other hand as $\alpha_k \rightarrow \alpha_j$ the interacting orbitals become degenerate, and the approximation on which the equation depends, no longer holds. Under these conditions the last term of Eq. (18) becomes large and equal to $2C_{rj}C_{sk}\beta_{rs}$. This term is then in general more important than the electrostatic term of Eq. (18) and the reaction is said to be *orbital controlled.*

This situation is found only for nucleophiles of low electronegativity and electrophiles of high electronegativity with good overlap of the interacting orbitals, corresponding to a soft-soft interaction. Other properties of the reactants modify the interaction energy and hence contribute to the chemical reactivity, in particular polarizability, solvation, energy, and orbital overlap.

The situation has been examined quantitatively by Klopman,[158] who showed that by making reasonable assumptions, the Chatt-Ahrland[155] classification could be derived from Eq. (18). Moreover, by making a regular (arbitary) change in α_k, several series of perturbation energies, which may be related to reactivity, were computed (Table 5-17).

Table 5-17. Reactivity Scales Computed from Equation for Centers of Changing Electrophilicity (α_k)[158]

	I$^-$	Br$^-$	Cl$^-$	F$^-$	SH$^-$	CN$^-$	OH$^-$
R_{rH} Å	1.60	1.41	1.27	0.92	1.34	1.06	0.96
$-2\beta_{rs}$ (eV)	3.16	3.60	4.10	4.48	4.0	3.8	4.50
$-\alpha_j$	8.31	9.22	9.94	12.18	8.59	8.75	10.45
$-\Delta E_p$ ($\alpha_k = -7$ eV)	2.52	1.75	1.54	1.06	2.64	2.30	1.49
($\alpha_k = -5$ eV)	1.07	0.98	0.97	0.82	1.25	1.17	1.01
($\alpha_k = +1$ eV)	0.45	0.48	0.52	0.53	0.55	0.56	0.58

These calculations were made with $q_r = -1$, $q_s = +1$, $C_r{}^2 = C_j{}^2 \equiv 1$, $\varepsilon = 80$, $\Gamma_{rs} = e^2/R_{rH}$, where R_{rH} is the bond distance in the molecule r—H. The β values are given empirically[159] by $\beta_{rs} = (\beta_{rr}\beta_{ss})^{1/2}$. Although this treatment involves several crude approximations, it shows that the nucleophilic order changes in a regular way with the change in orbital interaction. The order for $\alpha_k = -7$ eV corresponds to a "soft" nucleophilic order, for $\alpha_k = +1$ eV to a "hard" order. Examples of these different reactivity orders will be considered in the following sections. It is found that the order for $\alpha_k = -7$ eV is similar to that observed for peroxides and disulphides, and the order $\alpha_k = -5$ eV similar to that for attack on a saturated carbon atom discussed in the following section. The order $\alpha_k = +1$ which follows the pK_a closely is found for reactions at hard centers (e.g.

$$\overset{\diagup}{\underset{\diagdown}{P}}\overset{\textstyle O}{\diagdown X} , \quad -SO_2X, \ -COX, \ R_3SiX, \text{ and } R_2BX.$$

B. Reaction at the Saturated Carbon Atom†

Quantitative investigations of reactions at saturated carbon atoms (S_N2 displacements) and at acyl centers are more numerous than for other centers.* Hammett[6] pointed out originally that the relative reactivity of halide ions (in water) in S_N2 reactions (see Table 5-18) was the inverse of their thermodynamic stability. Thus iodide ion is the most nucleophilic and is also the most easily displaced of the halogens. The experimental data of Bathgate and Moelwyn-Hughes[160] for the equilibria

$$CH_3I + N^\ominus \underset{\xrightarrow{\hspace{1cm}}}{\overset{\Delta F}{\rightleftharpoons}} CH_3N + I^\ominus$$

are compared with the calculated thermodynamic data (Table 5-18).

Table 5-18. Affinity of Nucleophiles for the Saturated Carbon Atoms

	F⁻	Cl⁻	Br⁻	I⁻	HO⁻	HS⁻
$-\Delta F$ (kcal/mole)	0.7	−1.0	−1.54	0	a	a
$-\Delta F$ (calc)	5.2	1.3	−0.4	0	22.2	17.5
n^b	2.0	3.04	3.89	5.04	4.20	5.1

a Very large values.

b n is $\log k_{N^-}/k_{H_2O}$ for the reaction of MeI.

* In the gas phase the various nucleophiles NH_2^-, OH^-, F^-, H^- have similar reactivities.[161] The reactions may be encounter controlled ($\Delta H^* \simeq 0$).

† Stable intermediates, $CH_3X_2{}^{(-)}$ have been detected (X = Hal) in the gas phase.

Table 5-19. Affinity of Alkoxide and Thioalkoxide Ions
for Carbon Centers

Acceptor	Base	K_A	K_R
Me	$^-$OMe	0.11	12
Me	$^-$SMe	1.9×10^{-6}	4×10^4
Me	$^-$SPh	3.0×10^{-10}	2
Ph	$^-$OMe	0.11	1.5
Ph	$^-$SMe	1.9×10^{-6}	1.1
9-Ph-10 Me ^{a}	$^-$OH	0.11	$\sim 10^3$
Acridinium	$^-$SPh	1.9×10^{-6}	1
$CH_3C{=}O$	$^-$OEt	0.9	6.1
$CH_3C{=}O$	$^-$S—$(CH_2)_2$NHAc	2.7×10^{-7}	1.6×10^{-8}

$$H_2O + N^{\ominus} \xrightleftharpoons{K_A} HN + OH^{\ominus} \qquad ROH + N^{\ominus} \xrightleftharpoons{K_R} RN + OH^{\ominus}$$

^{a} Actual association constants.

The thermodynamic stability, like the rate constant, changes with structure, as shown for example by the relative affinity of alkoxide and thioalkoxide ions towards various carbon centers[10] (Table 5-19).

Bunnett, Hauser, and Nahabedian[162] measured the following equilibrium directly,

$$Nu^{\ominus} + \quad \xrightleftharpoons{K_R} \quad (Nu^{\ominus} = OH^{\ominus}, SPh^{\ominus})$$

and found a ratio of affinities of approximately 10^3 in favor of the hydroxide ion. Thermodynamic calculations however give a considerably lower value of 6 for combination with a methyl cation.[141] It is noted that the difference in K_R is greater than the difference in K_A for combination with an acyl group. This is probably due to differences in the conjugation energies of alkoxide and thiolate groups attached to a carbonyl center. Consequently these equilibrium constants are not relevant to the formation of a transition state formed from saturated or unsaturated compounds.* The comparable values of K_R for combination with a phenyl group indicate a specific stabilization of the C—S bond (this could be d_π-p_π bonding).

* See p. 185 for the influence of conjugation on Brönsted values.

Nucleophilic displacements of methyl bromide are usually taken as reference reactions for alkylation nucleophilic orders. The Swain relation,[7]

$$\log \frac{k}{k^0} = sn \tag{2}$$

represents the reactivity relative to the rate of hydrolysis in terms of a parameter n for the nucleophile, which is closely related to polarizability and redox potential, and a constant s for the particular substrate. Values of n are given in Table 5-20.

Table 5-20. n **Values for Reactions of Nucleophiles with Methyl[7] Bromide in Water at 25°**

Nucleophile	n	Nucleophile	n
H_2O	0	$(NH_2)_2C{=}S$	4.1
$p\text{-}Me\cdot C_6H_4SO_3{}^-$	<1.0	OH^-	4.20
$NO_3{}^-$	1.03	$PhNH_2$	4.19
Picrate	1.9	SCN^-	4.77
F^-	2.0	I^-	5.04
$SO_4{}^{2-}$	2.5	CN^-	5.1
$MeCO_2{}^-$	2.72	HS^-	5.1
Cl^-	3.04	$SO_3{}^{2-}$	5.1
$HPO_4{}^{2-}$	3.8	$S_2O_3{}^{2-}$	6.36
Br^-	3.89	$HPSO_3{}^{2-}$	6.6
$N_3{}^-$	4.00		

These values show that both terms of Eq. (13) contribute to the reactivity, the orbital term being generally the more dominant. This reactivity order holds approximately for many other alkylations, for example, the reactions of epoxides[7] including glycidol and epichlorhydrin, β-lactones,[7] sulphonate esters,[7] and phosphate esters. Similarly carbenes, which are known to be electrophilic, show the rate order[163] $I^- > Br^- > OH^- > Cl^-$, and the $\log k$ values for CCl_2 are proportional to n with $s = 0.54$. However, the reactivity order $F^- > Cl^-$ is found[164] for CF_2 showing the increased importance of the Coulombic term. The influence of this term is shown in the relatively large value of n for the OH^- and CN^- ions. As the formal positive charge on the carbon atom in the transition state increases, the relative importance of the Coulomb term increases and hence basic nucleophiles, for example, OH^- ions, become relatively more reactive, producing significant deviations from the Swain equation.

C. Organic Cations

Although the Swain equation predicts the nucleophilic reactivity of several types of alkylating agent reasonably well,[7] serious deviations are observed when the charge term makes an abnormally large contribution, for example, when the carbon atom is positively charged *in the transition state*. We must differentiate between two types of reaction, those in which an originally neutral molecule ionizes to give a highly electron-deficient reaction center (e.g., sulphonate ester), and reactions of charged species. There is much confusion in the literature regarding this situation.*

Thus, as shown by Davis,[165] basicity makes a particularly large contribution in the reactions of sulphonate esters, in which the transition state approaches the limiting S_N1 structure (cf. the Sneen[167] mechanism):

$$CH_3-O-SO_2R \longrightarrow \overset{\oplus}{C}H_3\overset{\ominus}{O}SO_2R \overset{N\ominus}{\longrightarrow} \overset{\ominus}{N}\text{---}\overset{\oplus}{C}H_3\text{---}\overset{\ominus}{O}SO_2R$$

Transition states of this kind are characterized by low β values† and large solvation entropy changes.[168] The nucleophilic order (Table 5-21) differs considerably from the alkylation order given by n, but is accurately represented by the Edwards equation or so-called oxibase scale[165] (Fig. 5-25).

Table 5-21. Nucleophilic Order for Ethyl *p*-Toluenesulphonate

Nucleophile	k_2	n	Nucleophile	k_2	n
H_2O	1.15×10^{-7}	0	$HS^{(-)}$	1.7×10^{-3}	5.1
$OH^{(-)}$	1.23×10^{-4}	4.20	$I^{(-)}$	6.5×10^{-5}	5.04
$S_2O_3^{2-}$	8.6×10^{-4}	6.36	$SCN^{(-)}$	6.8×10^{-5}	4.77
SO_3^{2-}	1.13×10^{-3}	5.1	$NO_2^{(-)}$	1.5×10^{-4}	
N_3^-	1.4×10^{-4}	4.0	ClO^-	1.0×10^{-3}	

The large rate constants for $ClO^\ominus$ show an appreciable α effect. Similarly some carbonium ions, for example, mustard cation[169] and chloroethers, where the positive charge is localized, show high reactivity towards $OH^\ominus$ and other basic nucleophiles.[7] For example, $OH^\ominus$ is about 40 times more reactive than given by the Swain equation, which is followed by most nucleophiles, in the reaction of mustard cation.[169]

However, when the charge in the cation is dispersed by conjugation as in

---

* Cf. the classical discussion on limiting S_N1 ionization.[166]
† Brönsted β values.

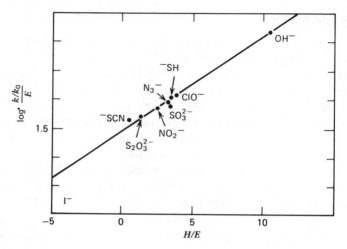

Fig. 5-25. Rate constants for reactions of ethyl tosylate with nucleophiles in water plotted according to the Edwards equation $\log k/k_0 = xH + yE$ where $H = pK_a +$ constant and E is the standard reduction potential of the nucleophile.

"stable" carbonium ions, for example, trityl cation and malachite green and in diazonium ions, the reactivity order

$$PhS^{(-)} > N_3^{\ominus} > CH_3O^{\ominus} > OH^{\ominus} > CN^{\ominus}$$

is anomalous in that $CN^{\ominus}$, a highly polarizable ion has a low reactivity, whereas the polarizable thiolate and azide ions have high reactivities.[170] This nucleophilic order does not follow the Swain-Edwards equation, and the selectivity appears to be independent of the nature of the cation, although the reaction rate for the cations studied with a particular ion changes by approximately 10^6.

This remarkable result shows that specific interactions between nucleophile and anion are absent, and hence the transition state is probably formed at a large internuclear separation. The approach of anion and cation in the gas phase should lead to a continuous decrease in the energy of the system. We assume therefore that the transition state is reached before charge transfer and orbital interaction occurs. Under these conditions Eq. (18) reduces to

$$\Delta E^* \simeq \{\alpha\Delta H_N{}^S + \delta\Delta H_E{}^S\} - \frac{q_N q_E}{r_N + r_E}$$

a. For a change in nucleophile from N_1 to N_2 ($q_N \equiv q_E \equiv 1$),

$$-RT\ln\frac{k_{N_1}}{k_{N_2}} \simeq \alpha(\Delta H_{N_1}{}^s - \Delta H_{N_2}{}^s) - \frac{(r_{N_1} - r_{N_2})e^2}{(r_{N_1} + r_E)^2} \simeq N_{\oplus}$$

If $(r_N + r_E)$ is taken to be constant for large internuclear separations, the relative nucleophilic order is independent of the nature of the cation (and is constant $N_\oplus$).

b. For a change in the electrophile from E_1 to E_2 in reactions with a given nucleophile

$$RT \ln \frac{k_{E_1}}{k_{E_2}} \simeq \delta \cdot \Delta E_E^s$$

that is, independent of the nature of the nucleophile. We suggest therefore that a constant nucleophilic selectivity as given by this equation is indicative of an ion-pair transition state.

D. Reaction at the Unsaturated Carbon Atom

Nucleophilic substitution at aromatic centers has been widely investigated and in certain stabilized systems, Meisenheimer complexes have been isolated.[91] There is little doubt therefore that the substitution proceeds through an addition intermediate, for example,

Brönsted coefficients of the order of 0.5 are usually obtained, and hence it appears that the transition-state structure is intermediate between ground state and addition intermediate. The nucleophilic order is thus determined by both terms of Eq. (3), the relative contribution depending on the initial charge distribution in the aromatic halide. Thus, as shown by the data of Table 5-22, PhS^- is considerably more reactive than $CH_3O^\ominus$ toward 2.4-dinitrochlorobenzene,[171] but the reactivities toward p-nitrofluorobenzene are similar.[172] This comparison shows the influence of the Coulomb term in the highly polar fluoride. Moreover, polarizable nucleophiles, for example, I^-, Br^-, are considerably less reactive than alkoxide ions toward the aromatic chloride (Table 5-22), showing the importance of this term in the transition state.

Table 5-22. Nucleophilic Reactivities Toward Aromatic Halides

	p-Nitrofluorobenzene (methanol 25°)				
	MeO$^-$	PhS$^-$	PhO$^-$	PhNH$_2$	Cl$^-$
k_2	1.8×10^{-4}	1.7×10^{-4}	$\sim 10^{-6}$	1.6×10^{-8}	1.7×10^{-14}

	2,4-Dinitrochlorobenzene (dioxan-water)				
	PhS$^-$	CH$_3$O$^-$	PhO$^-$	HO$^-$	C$_5$H$_{10}$N
k_2	1300	2.2	0.72	0.07	4.48

	2,4-Dinitrochlorobenzene (alcoholic solvents)				
	OEt$^-$	OMe$^-$	OPh$^-$	I$^-$	Br$^{(-)}$
k_2	4.95(a)^a	1.50(b)	0.90(a)	0.01(c)	0.003(c)

^{a} (a) In ethanol at 25°; (b) in methanol at 25°; (c) in ethylene glycol at 175°.

For substitution in the benzene ring of the more reactive benzene iodonium salts,[173] for example,

the rate order I$^-$ > Br$^-$ > Cl$^\ominus$ is found.

As in the corresponding reactions of aromatic halides, thiolate ions are very reactive toward vinyl halides[174] undergoing substitution* whereas alkoxide ions normally give elimination, namely,

$$\text{ROH} + \text{CH}\equiv\text{C—X} \xleftarrow{\text{RO}^\ominus} \text{X—CH}=\text{CH—X} \xrightarrow{2\text{RS}^\ominus} \text{RS—CH}_2=\text{CH}_2\text{—SR}$$

Modena and Todesco[176] give the rate order PhS$^\ominus$ > MeO$^\ominus$ > N$_3^\ominus$ for the *direct* substitution of the chlorovinyl sulphone, for example,

Fluoroolefins, however, react preferentially with the more basic ions, and Miller[177] has established qualitatively the order F$^-$ ≫ Cl$^-$ > Br$^-$, I$^-$ for

* These reactions are sometimes complex and proceed by elimination-addition mechanisms.[175]

$S_N'2$ displacements and nucleophilic addition, which normally proceed more rapidly than the alternative S_N2 displacement, namely

These reactions were, however, carried out in aprotic solvents in which the nucleophilic order toward a saturated carbon atom is $F^{(-)} \sim Cl^- \sim Br^- \sim I^-$. The selectivity is, however, much greater in the reactions of fluoroolefins, owing to the decreased electron density at the reaction center.

Changes in solvent change the nucleophilic order in a regular way,[178] such that the less polarizable (hard) strongly solvated ions become more reactive as the solvent polarity decreases. Hydroxylic solvents normally solvate ions strongly because of hydrogen bonding, compared with aprotic solvents and hence the rate order is similar in water alcohols and acids, but different in aprotic solvents, for example, acetonitrile and dimethyl formamide. The change is illustrated by data taken from the review by Parker,[178] for a typical alkylation and a typical aromatic substitution (Table 5-23).

Here the reactivity of the thiocyanate ion is taken as a reference and it is clearly seen that a change from alcohol to DMF produces a greater increase in reactivity for the more strongly solvated ions, for example, Cl^-, N_3^- than for the more polarizable ions, for example, Br^- and I^-.

Table 5-23. Effect of Solvent on Nucleophilic Order

	CH₃I		$NO_2\text{-}C_6H_3(NO_2)\text{-}I$	
	MeOH	DMF	MeOH	DMF
Cl^-	−2.25	+1.51	−5.56	−0.32
Br^-	−0.83	+1.20	−3.49	−1.89
N_3^-	−0.84	+1.62	+1.72	+4.36
I^-	+0.81	—		

In most of the work discussed in this chapter, reactivities in water or alcohols only are discussed.

Of other unsaturated centers, benzyne is particularly interesting, and the characteristic order $I^- > Br^- > Cl^-$ has been found for the reaction in ethanol.[179] Also, competition experiments with 9-10-dehydrophenanthrene[180] lead to the following (soft) reactivity order (in ether):

$$LiSPh > LiPh > LiNC_5H_{10} > LiNMePh > LiOPh$$

E. Reactivity of Hard Centers

As pointed out above, the Swain equation is reasonably successful in repre–senting the reactivity of a range of alkylating agents (and similar molecules), but changes in structure produce entirely different nucleophilic orders. When reactivity is determined mainly by the intensity of the electrostatic field* produced by the electrophilic atom *in the transition state*, the Coulomb term then dominates the rate equation.

Carbonyl and phosphonyl centers normally carry a large positive charge and the reactivity orders, for example, $HO^\ominus > NO_2^\ominus > F^- \gg I^-$, Br^-, $SCN, S_2O_3{}^{2-}$ for ethyl chloroformate[181] and $F^- > HO^\ominus > CH_3CO_2^\ominus \gg I^-$, Br^- for a phosphorochloridate[182] show clearly the influence of the Coulomb term. Attempts to explain these rate data by the Swains-Edwards and similar equations have not been successful. However, the relative reactivity of nucleo-philes toward chloroformate follows the thermodynamic value of ΔF (Table 5-18) semiquantitatively, suggesting that the transition state resembles a tetrahedral intermediate. This is also suggested by the large β values nor-mally observed for acylation.

Moreover, the rate constants for the reaction of a wide range of nucleo-philes with p-nitrophenyl acetate[183] show a rough logarithmic correlation with pK_a although various serious anomalies are observed (Fig. 5-26). If α nucleophiles are neglected (see p. 212) the most serious anomalies are the low rate of $OH^\ominus$ and the abnormally high rate of azide ions. The first is due to the abnormally high solvation energy of $OH^\ominus$ ion, as observed in other reactions. The nucleophilic order for δ-thiovalerolactone[184] is the same as for PNPA. Although specific effects are observed, the log k for many acylat-ing agents follows the log k for PNPA fairly closely.

Similarly in the reactions of phosphorus compounds, for example, O, O-diphenylphosphorochloridothioate, the rate follows the pK_a of oxygen and sulfur nucleophiles closely[185] (Fig. 5-27), although the reaction is carried out in t-butanol. The significance of the curvature of Fig. 5-27 is not known, although ion pairing may be significant in this low dielectric medium. The

---

* This of course is related to orbital occupancy.

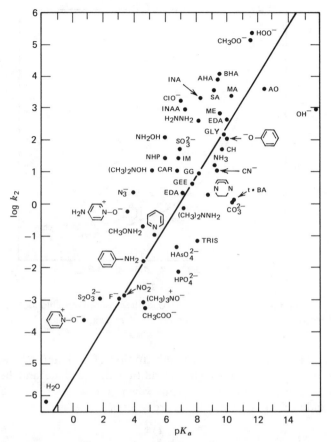

Fig. 5-26. Rates of nucleophilic reactions with p-nitrophenyl acetate in aqueous solution at 25° plotted against the basicity of the attacking reagent. GEE, glycine ethyl ester; GLY, glycine; GC, glycylglycine; IM, imidazole; AHA, acetohydroxamic acid; BHA, n-butyrylhydroxamic acid; CH, chloral hydrate anion (at 30°); INA, isonitrosoacetone; SA, salicyl aldoxime; MA, sodium mercaptoacetate; AO, adetoxime; ME, mercaptoethanol; INAA, isonitrosoacetylacetone; EDA, ethylenediamine; NHP, N-hydroxy phthalimide; BA, t-butylamine; CAR, carnosine; TRIS, tris-(hydroxymethyl)aminomethane. (Reproduced by permission of the American Chemical Society.)

high rate of reaction of fluoride ion with di-isopropyl phosphorochloridate[182] (in ethanol) must be due to the large Coulombic term in the P—F bond energy. The nucleophilic order toward p-nitrophenyl phosphate[186] is determined mainly by electrostatic repulsion owing to the large negative charge on the ester; for example, pyridine is 30 times more reactive than hydroxide ion.

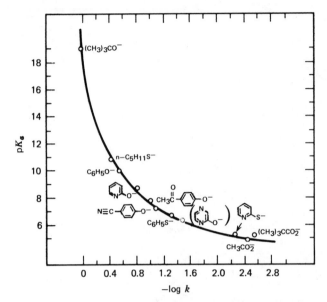

Fig. 5-27. Reactions of anions with O-O-diphenylphosphorochloridothioate in *t*-butanol at 58°. (Reproduced by permission of the American Chemical Society.)

It should be noted, however, that substitution of acyl, and perhaps phosphoryl, centers proceeds through an addition intermediate and hence the observed reactivity is composite. Thus when X is a poor leaving group, $k_3 \ll k_2$, and hence only strong bases are reactive,

$$R-C{\overset{O}{\underset{X}{\diagup}}} + Nu^{\ominus} \underset{k_2}{\overset{k_1}{\rightleftharpoons}} Nu-\underset{\underset{R}{|}}{\overset{\overset{O^{\ominus}}{|}}{C}}-X \xrightarrow{k_3} R-\overset{\overset{O}{\|}}{C}-Nu + X^{\ominus}$$

Thus aliphatic esters hydrolyze with specific base catalysis.

Similar reactivity orders are found for compounds of boron and silicon, typical hard electrophiles. Trigonal boron, R_2BX shows the rate order[187]

$$HO^- > RO^- > NH_3 > R_2NH \simeq RS^-$$

that is, the basicity order, modified by steric hindrance (cf. H. C. Brown). Tetrahedral boron,[188] for example, $NH_3 \cdot BF_3$ reacts in the order $OH^- > F^- > H_2O$ (Cl$^-$ no effect) as do acyl compounds, presumably by a dissociation-addition mechanism,

$$HN_3 \longrightarrow BF_3 \rightleftharpoons HN_3 + BF_3 \xrightarrow{N^{\ominus}} N-\overset{\ominus}{B}F_3 + NH_3$$

Pearson and Basolo[189] have recorded the following rate order for the reaction of the tris-acetylacetonate of silicon,

$$HO^- > HPO_4^{2-} > S_2O_3^{2-} > NO_2^- > H_2O$$

that is, the order of basicity. The peroxide ion and peroxide molecule show strong α effects. The change in nucleophilic order with oxidation state is shown clearly by recent measurements on sulfenyl, sulfinyl, and sulfonyl sulfur.[190] (Table 5-24) It has been known for some time that sulfonyl halides were more reactive toward basic reagents than toward polarizable nucleophiles,[191] and in the classical work of Foss[192] on chalcogens, sulfur nucleophiles were found to be strongly nucleophilic. Later work by Bartlett, Cox, and Davis[193] showed that Ph_3P, HS^-, and $HSO_3^{\ominus}$ are powerful nucleophiles for elementary sulfur.

Direct rate comparisons have been made recently of the reactions of a

Table 5-24. Nucleophilic Orders for Sulphur Atoms in different oxidation states

Nucleophile	Sulfenyl	Sulfinyl	Sulfonyl
F⁻	—	4.4	59
AcO⁻	—	9.0	(1.0)
Cl⁻	(1.0)	12	0.0016
Br⁻	35	65	0.0009
SCN⁻	5.4×10^3	1.7×10^2	—
I⁻	1.4×10^4	1.0×10^3	—
Thiourea	Very fast	3.5×10^3	—

series of nucleophiles in aqueous solution with R—S—SO$_2$R, R—S—SO$_2$R,

$$\underset{O}{\overset{\parallel}{}}$$

and RSO$_2$·SO$_2$R, thus maintaining the same leaving group in all cases.[190] A complete change in nucleophilic order from "soft" to "hard" is observed with the state of oxidation. Moreover, the nucleophilic order toward sulfonyl sulfur is very similar to that established for acyl carbon, as shown by the comparison in Table 5-25. The relative reactivities of F$^-$ and NO$_2^-$ are different for the two centers, which may reflect the high bond energies for bonds between

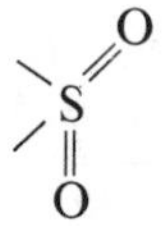

and F

(cf.

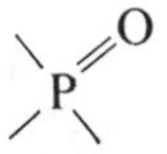

and F.)

Table 5-25. k_{Nu}/k_{OAc} **Values for a Di-sulfone and Acylating Agents**

	Ph·SO$_2$·SO$_2$Ph	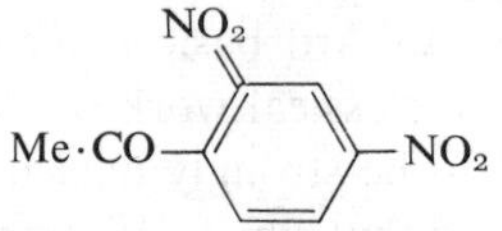 Me·CO—C$_6$H$_3$(NO$_2$)$_2$	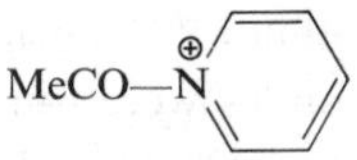 MeCO—N$^{\oplus}$ (pyridinium)
n-BuNH$_2$	5.9 × 10^3	2.1 × 10^5	3.1 × 10^3
N$_3^-$	3.3 × 10^2	1.7 × 10^3	3.5 × 10^2
F$^-$	59	5.6	3.3
NO$_2^-$	10	15	7.7
AcO$^-$	(1)	(1)	(1)
H$_2$O	3.0 × 10^{-5}	3.5 × 10^{-4}	2.0 × 10^{-4}

F. Reactivity of Soft Centers

In a previous section the relative reactivity of a series of nucleophiles toward the saturated carbon atom was shown to follow the polarizability of the nucleophile according to the Swain equation, modified in certain cases for Coulombic forces. Many other saturated centers, and some metallic complexes of B metals, show essentially the same rate or equilibrium order although specific differences are found. There is therefore no unique reactivity order corresponding to a "soft" acid. Examples are given of substitution of hydrogen peroxide[194] on electrophilic oxygen* and displacement at a Pt^{II} complex[195] (Table 5-26). The peroxide order is almost the same as the order toward sulfenyl sulfur although the selectivity is greater. Moreover, neutral molecules, for example, thiourea, tertiary phosphines, and olefins are very reactive toward peroxides. (Note: Peroxide rates are very sensitive to traces of Fe^{3+} catalyst.) This presumably is due to the influence of electron-electron repulsion between the lone pairs on oxygen and the attacking nucleophile. Consequently the Edwards equation does not fit the data for peroxides very well.

Table 5-26. Nucleophilic Orders for Various "Soft" Centers

	Peroxide	n	Pt^{II}	$HgMe^+$	$E_n{}^a$
Thiourea	Very fast	4.1	6.000	—	2.18
I^-	6×10^{-1}	5.04	107	8.60	2.06
$SO_3{}^{2-}$	2×10^{-1}	5.1	250	8.11	2.57
$S_2O_3{}^{2-}$	2.5×10^{-2}	6.36	9,000	10.90	2.52
$S \cdot (CH_2 \cdot CH_2OH)_2$	2.2×10^{-3}	—	—	—	—
CN^-	1.0×10^{-3}	5.1	—	14.1	2.02
SCN^-	5.2×10^{-4}	4.77	180	6.05	1.83
NEt_3	3.3×10^{-4}	—	(0.47)	—	2.0
Br^-	2.3×10^{-5}	3.89	3.7	6.62	1.51
$NO_2{}^-$	3×10^{-7}	—	0.68	—	1.73
Cl^-	1×10^{-7}	3.04	0.45	5.25	1.24
HO^-	$< 10^{-7}$	4.20	$\ll 0.1$	9.37	1.65

^{a} E_n is the Edwards polarisability parameter.

In the reactions of Pt^{II}, the low bond energies with NR_3 and $OH^{(-)}$ lead to particularly low rate values and this probably accounts for the low values for $SO_3{}^{2-}$ and $NO_2{}^-$. On the other hand $SCN^{(-)}$ has a relatively large value comparable to that of the $CN^\ominus$ ion. A wide range of Pt^{II} complexes has been investigated and in spite of large differences in selectivity, the same nucleophilic order is followed accurately.[196] In general this nucleophilic order follows the affinity of the nucleophiles for $MeHg^\oplus$ as given by Schwarzenbach's scale for soft bases.[197]

* Relative reactivity of HOCl and CH_3Cl toward I^- is about 10^{13} (probably due to bond energy difference).

The equilibrium constants for the association[198] of nucleophiles with I_2 show similar selectivity to the Swain n values, that is, the relative reactivities toward MeI.

	Cl^-	Br^-	SCN	I^-
$\log K\,(I_2)$	0.32	1.08	2.04	2.85
n	3.04	3.89	4.77	5.04

Moreover the $\log k$ values for the reaction

$$\begin{array}{c} R \\ \diagdown \\ S \\ \diagup \\ R \end{array} + \text{MeOSO}_2\text{F} \xrightarrow{\text{Me}_2\text{SO}_4} \begin{array}{c} R \\ \diagdown \oplus \\ S\text{—Me} \\ \diagup \\ R \end{array} + \text{SO}_3\text{F}^-$$

plotted against the $\log K$ values for the association with I_2

$$\begin{array}{c} R \\ \diagdown \\ S \\ \diagup \\ R \end{array} + I_2 \rightleftharpoons \begin{array}{c} R \\ \diagdown {}^{\Delta+} \\ S\text{---}I\text{---}\overset{\Delta-}{I} \\ \diagup \\ R \end{array}$$

are approximately linear with slope $\simeq 1.0$. These comparisons show that the iodine complex is a good model for S_N2 displacement on alkyl halides.*

G. Regular Changes in Transition-State Structure

It is apparent from the above discussion that no general reactivity order for nucleophiles can be devised and that nucleophilicity changes with the nature of the electrophile[10] in a manner given approximately by Eq. (3). Equations of this type provide a quantitative basis for the concept of HSAB, which, however, remains a qualitative principle.

Moreover, a given nucleophilic order is not found for a particular reaction center, unless the structure of the transition state is also considered. Thus in substitutions at carbon, as the bond extension, C—X, in the transition state increases in a heterolytic reaction, the charge on the electrophilic center, C, also increases and the energy of the acceptor σ^* orbital decreases, leading to increased perturbation. Thus the magnitude of both terms of equation 13 increases, and it is impossible to deduce which change makes the greater contribution to the activation energy, since the Coulombic term and orbital energy α_k refer to a hypothetical system (e.g., $\overset{\delta+}{C}\text{---}\overset{\delta-}{X}$) in which the reactive bond is partially dissociated.

---

* Recently Challis (private communication) has shown that the nucleophilic order toward 2,4-dinitrophenyl hydroxylamine (i.e., saturated nitrogen) is also given by E_n.

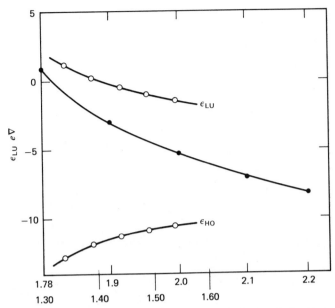

Fig. 5-28. Change in π and π^* orbital energies for ethylene ($\bigcirc$) and σ^* orbital energy for a C—Cl bond ($\bullet$), with bond separation (after Fukui).

The situation may be examined in a general way as follows. For substitution at a saturated carbon atom, the initial bond polarity is low, hence the reaction is orbital controlled. Moreover as the bond is extended the energy of σ^* decreases rapidly (see Fig. 5-28). Under such circumstances the increase in orbital term is greater than the increase in charge term. This means that the nucleophilic order, as represented by the relative reactivity of a typical "soft" nucleophile (e.g., $RS^\ominus$, I^-) to a "hard" nucleophile (e.g., $RO^\ominus$, $F^\ominus$) increases regularly with the extent of bond formation (as given for example by the Brönsted coefficient β). Such a case is found for the reaction of a series of p-substituted benzyl bromides with methoxide and thiophenoxide ions[188] (Table 5-27) and in the nucleophilic reactions of Pt^{II} complexes[196] (where the Coulomb factor remains small.

As seen from the data of Table 5-27, the value of $\log k_{Phs}/k_{Meo}$ increases regularly with the Hammett σ value for the p-substituent in the benzyl halide which gives a measure of the electrophilic power (positive charge density) on the reactive carbon atom.

On the other hand, the value of k_{Phs}/k_{Meo} decreases with the formal charge on the reactive carbon atom of substituted benzenes[171,172] (reaction B) and benzthiazoles (reaction C). Here the transition states probably resemble the corresponding addition intermediates (the values of β are certainly greater

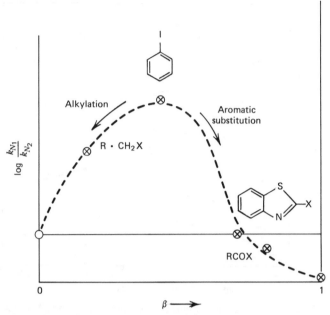

Fig. 5-29. General variation in the relative reactivity of "soft" and "hard" nucleophiles, k_{N_1}/k_{N_2} towards a carbon center, with the extent of bond formation as given by the Brönsted β parameter.

than the alkylation values). The inductive effect of the halogen atoms is exerted mainly in the ground state and hence the value of $k_{\mathrm{PhS}}/k_{\mathrm{MeO}}$ decreases regularly with χ_X, the Pauling electronegativity value of the halogen which determines the effective charge at the reaction center. Values of β suggest that the structure of the transition state is not sensitive to the nature of X; hence the value of $k_{\mathrm{RS^-}}/k_{\mathrm{RO^-}}$ changes with the Coulombic term. It should be noted that according to the calculations of Fukui,[199] the energy of the acceptor orbital (π^*) changes less rapidly with bond extension than the σ^* energy so that the transition state may be regarded as a perturbed initial state.

Although electron withdrawal from the reaction center appears to increase the $k_{\mathrm{MeO}}/k_{\mathrm{PhS}}$ ratio, substitution of electron-attracting groups in the ring, for example, the 6-nitro group, leads to considerable decreases in these ratios for the various halides.[201] This apparent anomaly is explained by a change in transition state produced by conjugation,

Table 5-27. Relative Reactivity of Alkoxide and Thioalkoxide Ions Towards several Unsaturated Centers

Substituent	Reaction A $\log k_{\text{PhS/MeO}}$	Substituent	Reaction B $\log k_{\text{PhS}}/k_{\text{MeO}}$	Reaction C	Reaction D
p-MeO (−0.27)[a]	2.70	I (2.5)[b]	5.22	0.34	—
H (0)	3.30	Br (2.8)	3.68	0	3.14
p-Br (0.23)	3.52	Cl (3.0)	3.29	−0.44	2.67
p-NO$_2$ (0.78)	3.72	F (4.0)	1.77	−1.40	1.19

[a] Hammett σ-values.
[b] Pauling electronegativity values. Reaction A. p-substituted benzyl bromides; 1-halogeno-2,4-dinitrobenzenes. Reaction C. 2-halogeno-benzthiazoles. Reaction D. Substitution of vinylic carbon atom of β-halo-p-nitrostyrenes.

As in the rate-determining ionization of ketones (see p. 179) conjugation stabilization of the final state (in this case the tetrahedral intermediate) produces a change in transition state structure toward that of the initial system.

As already discussed, acyl carbon, being a hard center, is normally more reactive to hard nucleophiles, that is, $k_{\text{RO}^-} > k_{\text{RS}^\ominus}$, and the affinity for a carbon atom is also in this order (Table 5-19). By considering all these data the variation in nucleophilic order (as represented by $k_{\text{RS}^-}/k_{\text{RO}^-}$) changes with bond formation in the transition state in a general way as shown in Fig. 5-29.

The inversion of nucleophilic order with change in transition-state structure is shown clearly by rate data for β elimination. Several workers have observed that contrary to anticipation, thioalkoxide ions are more reactive than alkoxide ions towards various alkyl halides, for example, t-butyl chloride[200] and 1,1-dimethyl 2-phenyl ethyl chloride.[202] However, as pointed out by Bunnett,[201] the rate of isomerization of menthone to iso-menthone is catalyzed by alkoxide ions several thousand times more strongly than by thioalkoxide ions. Further investigations showed that the ratio $k_{\text{PhS}}/k_{\text{MeO}}$ decreases with the decreasing lability of the leaving group:

$$\text{Ph·CH}_2\text{·C}\overset{\text{Me}}{\underset{\text{Me}}{-}}\text{X}$$

$\text{Ph·CH}_2\text{·C}\overset{\text{Me}}{\underset{\text{Me}}{-}}\text{X}$	Temperature	$k_{\text{EtS}}/k_{\text{MeO}}$
X = Cl	76°	6.5
SMe$_2^\oplus$	30°	0.86
SO$_2$Me	113°	∼0.05

The inference is that more charge is necessary on the β carbon atom to expel the group X when it is a poor leaving group and this requires more powerful N—H interaction, that is, stronger bonding in the transition state. The alkoxide ion thus becomes more reactive in view of the high HO bond energy (large Coulomb term).

The regular variation in the E_2 transition state has been investigated by England and McLennan,[41] enabling a direct comparison of k_{RS}/k_{MeO} with β to be made. Thus β for DDT, determined from rate studies with a series of substituted thiophenoxides, is 0.88 and k_{OEt}/k_{PhS} is 14,000. Here again strong bonding in the transition state leads to high alkoxide reactivity. At the other extreme, β for Bu$^\gamma$Cl is 0.17 and $k_{PhS}/k_{OEt} = 8.3$. An increase in β to 0.46 is produced by electrostatic interaction in the reaction of t-butyl dimethylsulfonium ions, with a corresponding decrease in k_{PhS}/k_{OEt} to 0.31. Similar regular variations are shown by the data for 1,1 substituted cyclohexanes (Table 5-28). Since the selectivity must be 1.0 when $\beta = 0$, it follows that the ratio must decrease again as the transition state becomes E_1, that is, the maximum value of k_{PhS}/k_{OEt} is found for very low β values in this case.

Table 5-28. Change in Nucleophilic Order with Transition
State Structure for β-Elimination

	X	Y	β	k_{PhS}/k_{OEt}
	H	OTs	0.27	7.0
	H	Br	0.36	1.6
	H	Cl	0.39	0.73
	Br	Br	0.51	0.63
	Cl	Cl	0.58	0.21

Alternative explanations have, however, been given of the change in reagent selectivity (e.g., as given by k_{RS}/k_{RO}). In particular, Bunnett[203] has drawn attention to the influence of α substituents and ortho groups on this ratio. For example, the ratio k_{RS^-}/k_{MeO^-} increases markedly with the size of the α-substituted halogen and α-substituted alkyl groups, as shown by the data of Table 5-29.

These results have been explained by postulating an interaction between the nucleophile and halogen atoms,[206] this polarization increasing in the order F < Cl < Br < I. This phenomenon is akin to symbiosis of ligands in inorganic complexes where ligands attract similar ligands.[204]

The large rate differences observed in the reactions of methylene halides, and the similar but smaller rate differences observed for o-substituted benzyl

Table 5-29. Change in Nucleophilic Reactivity with Structure

	k_{Phs}/k_{MeO}	k_I/k_{OMe}		k_{Br}/k_{Cl}	k_I/k_{Br}
EtBr	87	36	MeBr	22	—
$F \cdot CH_2Br$	17	6.0	MeI	58	—
$Cl \cdot CH_2Br$	288	92	EtBr	17	1.15
$Br \cdot CH_2Br$	749	186	EtI	36	4.03
$Cl \cdot CH_2 \cdot I$	855	—	Pr^iBr	14	0.77
$Br \cdot CH_2I$	2060	—	Pr^iI	28	1.43
$I \cdot CH_2 \cdot I$	5230	—			

halides (Table 5-29) may, however, be largely steric in origin. Alkoxide ions are heavily solvated and hence steric hindrance may be greater than in the case of the less solvated softer ions, for example, RS^- and I^- The latter would therefore tend to react relatively more rapidly with the more sterically hindered systems. The data of Table 5-30 show that the rate ratios are similar for o-Br and o-Me chlorides.[203] The values for the reactions of Me_3N show clearly the influence of steric hindrance. Since CH_3 and Br groups are of comparable size but quite different polarizabilities, it is questionable whether the ortho effect is caused by an attractive force between nucleophile and substituent. Steric hindrance to the approach of solvated reactants appears to be a more likely possibility.

Table 5-30. Reactions of o-Substituted
Benzyl Chlorides[203] k_Y/k_{MeO}
ratios

	Substituent		
Nucleophile	H	CH_3	Br
Li^+SPh^-	944	1740	2560
Me_3N	30.5	17.1	16.3
K^+I^-	24.0	78.9	103.7

IV. GENERAL CONCLUSIONS

The subject of nucleophilic reactivity is a very complex one, in the final analysis requiring the *ab initio* calculation of activation energies including solvent participation. In spite of considerable advances in the calculation of potential energy surfaces of organic reactions in the past few years, stimulated by detailed investigations of electrocyclic reactions, the theoretical treatment of heterolytic reactions in solution is a very formidable one.

For the next decade at least, mechanistic chemists will have to depend on empirical treatments and qualitative concepts for the interpretation of nucleophilicity.

The original contention of Ingold[87] that basicity is a satisfactory measure of nucleophilic reactivity is wide of the mark. His original differentiation of $S_N 1$ and $S_N 2$ reactions rested heavily on the significance of k_{OH^-}/k_{H_2O} rate ratios, and although this mechanistic scheme was completely substantiated by more sophisticated methods and analytical procedures, considerable controversy raged for many years over the significance of k_{OH^-}/k_{H_2O}. In retrospect, it is clear that such controversy was inevitable since the basic principles of nucleophilic reactivity were not understood. In fact, the realization that different substrates require different nucleophilic orders came very slowly, and it was only in the 1960s that the wide range of nucleophilic orders was appreciated.

The present account is based on a general approach, adapted from perturbation theory, which considers interaction between nucleophile and electrophile to involve three main factors: (1) Coulomb attraction, (2) electronic repulsion due to closed shells, and (3) orbital penetration. Two general situations can be recognized. First, a change in the nature of the nucleophile may produce a regular change in these three factors which affects the activation energy in the same direction. This leads to a direct relationship between the reactivity and affinity, usually represented as some kind of linear free-energy relationship. The Brönsted relationship is the most suitable of these for the treatment of the nucleophilic reactivities of a class of closely related nucleophiles, which normally follow the pK_a of their conjugate acids. Exceptions to the Brönsted equation must be examined on this basis, and interpreted in terms of steric factors, anomalous solvation or some other specific effect. It is realized of course that this equation cannot be theoretically derived without making gross assumptions, namely, the regular change in the three factors given above. There are many examples where such a situation obtains, but deviations from the law are to be expected.

Indeed, there is no reason to suppose that these factors will change in the same direction with changes in the nature of the nucleophile. In general, factors (1) and (2) will tend to change in the same direction (although solvation energies usually change in the opposite direction) whereas the third factor usually changes in the opposite direction. This situation leads to two extreme cases corresponding to a "soft" and "hard" nucleophilic order, respectively, with the possibility of many nucleophilic orders between these limits. The HSAB concept does therefore have a theoretical basis as provided by the classical work of Polanyi and Evans, who introduced the idea of a driving force (e.g., Coulomb and exchange forces) and reaction inertia (e.g., repulsion, charge transfer, and desolvation).

This general principle embodied qualitatively in the HSAB treatment, and quantitatively in various linear free-energy relationships of which the Swain-Edwards equation is the most well known, should be regarded as a guide line rather than a means of calculating reactivity in a parametric way. At the present time there is no possibility of deriving a satisfactory equation for nucleophilicity since this of necessity requires calculation of differences in reaction rates. Various limiting relationships, of which the Swain equation and Brönsted relationship are examples, may be used for a restricted range of reactions. Considerations of the deviations from these reference equations can lead to important conclusions with regard to transition-state structure. The HSAB concept has wide practical applications, and gives a qualitative guide to reagent selectivity, particularly in competitive reactions.

A consideration of the relative magnitude of Coulomb and orbital interaction leads to a general interpretation of the influence of neighboring atoms with lone pairs of electrons on the nucleophilic reactivity (sometimes termed the α effect). Quantitative correlations can be derived and the influence of the conformation of the nucleophile can be interpreted on the same basis. Steric hindrance, however, must be treated separately, as in most electronic treatments of reactivity.

The scope of this review has been severely limited, by necessity. No mention is made, for example, of ambient reagents and positional reactivity, and solvation is treated only very briefly. What we have tried to do is to develop a broad framework, based on the perturbation method, within which nucleophilic substitution can be treated, and to discuss some experimental results which illustrate various facets of this approach.

References

1. G. Porter, *Chem. Soc. Spec. Pub.* **16**, 1 (1962).
2. J. N. Brönsted, *Z. Phys. Chem.* **102**, 169 (1922).
3. R. P. Bell, *The Proton in Chemistry*, Methuen, London, 1959, p. 124.
4. J. F. Bunnett, *Ann. Rev. Phys. Chem.* **14**, 271 (1963).
5. E. Grunwald and J. E. Leffler, *Rates and Equilibria of Organic Reactions*, Wiley, New York, 1963.
6. H. R. McCleary and L. P. Hammett, *J. Am. Chem. Soc.* **63**, 2254 (1941).
7. G. C. Swain and C. B. Scott, *J. Am. Chem. Soc.* **75**, 141 (1953).
8. J. O. Edwards, *J. Am. Chem. Soc.* **76**, 1540 (1954).
9. R. G. Pearson, *J. Am. Chem. Soc.* **85**, 3533 (1963).

10. R. F. Hudson, *Chimia* **16**, 173 (1962).
11. S. Glasstone, K. J. Laidler, and H. Eyring, *Theory of Rate Processes*, McGraw-Hill, New York, 1941.
12. M. J. S. Dewar, *Adv. Chem. Phys.* **8**, 65 (1965); K. Fukui, in *Molecular Orbitals in Chemistry, Physics and Biology*, P. O. Löwdin and B. Pullman, Eds., Academic, New York, 1964, pp. 513–537.
13. R. B. Woodward and R. Hoffmann, *Angew. Chem.* **81**, 797 (1969); *Angew. Chem. Intl. Ed.* **8**, 781 (1969).
14. W. C. Herndon, *Chem. Rev.* **72**, 157 (1972); L. Salem, *J. Am. Chem. Soc.* **90**, 543 (1968).
15. R. F. Hudson, *Angew. Chem.* **85**, 63 (1973).
16. H. C. Longuet-Higgins and E. W. Abrahamson, *J. Am. Chem. Soc.* **87**, 2045 (1965).
17. R. D. Brown, *Quart. Rev.* **6**, 63 (1952).
18. G. Wheland, *J. Am. Chem. Soc.* **64**, 900 (1942).
19. L. Melander, *Chem. Soc. Spec. Pub.* **16**, 77 (1962).
20. M. J. S. Dewar, *J. Am. Chem. Soc.* **74**, 3341 (1952).
21. K. Fukui, T. Yonezawa, C. Nagata, and H. Shingu, *J. Chem. Phys.* **22**, 1433 (1954).
22. K. Fukui, T. Yonezawa, and C. Nagata, *J. Chem. Phys.* **26**, 831 (1957).
23. G. Olah, *Acct. Chem. Res.* **4**, 247 (1971).
24. M. J. S. Dewar, *Adv. Chem. Phys.* **8**, 65 (1965).
25. M. G. Evans and M. Polanyi, *Trans. Faraday Soc.* **32**, 1333 (1936).
26. R. P. Bell, *Proc. Roy. Soc. Lond.* Ser. A **154**, 414 (1936); J. Horuiti and M. Polanyi, *Acta Physicochim. URSS* **2**, 205 (1935).
27. E. L. Mackor, A. Hofstra, and J. H. Van der Waals, *Trans. Faraday Soc.* **54**, 66 (1958).
28. A. Streitwieser, *Molecular Orbital Theory for Organic Chemists*, Wiley, New York, 1961, p. 323.
29. P. M. G. Bavin and M. J. S. Dewar, *J. Chem. Soc.* 164 (1956); M. J. S. Dewar and T. Mole, *ibid.* 1441 (1956); M. J. S. Dewar, T. Mole, and E. W. T. Warford, *ibid.* 3581 (1956).
30. M. J. S. Dewar and T. Mole, *J. Chem. Soc.* 342 (1957); S. F. Mason, *ibid.* 1233 (1959).
31. A. Streitwieser, W. B. Hollyhead, G. Sonnichsen, A. H. Pudjaatmaka, C. J. Chang, and T. L. Kruger, *J. Am. Chem. Soc.* **93**, 5096 (1971).
32. J. E. Gordon, *J. Org. Chem.* **26**, 738 (1961).
33. R. P. Bell and O. M. Lidwell, *Proc. Roy. Soc. Lond.* Ser. A **176**, 88 (1940); R. P. Bell, R. D. Smith, and L. A. Woodward, *ibid.* **192**, 479 (1948); R. P. Bell, E. Gelles, and E. Möller, *ibid.* **198**, 310 (1949); R. P. Bell and H. L. Goldsmith, *ibid.* **210**, 322 (1952).
34. C. G. Swain and A. S. Rosenberg, *J. Am. Chem. Soc.* **83**, 2154 (1961); C. G. Swain, D. A. Kuhn, and R. L. Schowen, *ibid.* **87**, 1553 (1965).
35. R. G. Pearson and R. L. Dillon, *J. Am. Chem. Soc.* **75**, 2439 (1953).
36. R. P. Bell and W. C. E. Higginson, *Proc. Roy. Soc. Lond.* Ser. A **197**, 141 (1949).

37. R. P. Bell and A. F. Trotman-Dickenson, *J. Chem. Soc.* 1288 (1949); R. P. Bell and G. L. Wilson, *Trans. Faraday Soc.* **46**, 407 (1950).
38. R. P. Bell, Ref. 3, p. 176.
39. J. A. Feather and V. Gold, *J. Chem. Soc.*, 1752 (1965).
40. R. F. Hudson and G. Klopman, *J. Chem. Soc.* **5** (1964).
41. B. D. England and D. J. McLennan, *J. Chem. Soc.* (B) 696 (1966).
42. J. F. Bunnett, *Angew. Chem.* **74**, 731 (1962).
43. N. H. Cromwell and P. H. Hess, *J. Am. Chem. Soc.* **83**, 1237 (1961).
44. (a) M. Eigen, *Angew. Chem.* **75**, 489 (1963); *Angew. Chem. Intl. Ed.* **3**, 1 (1964); (b) W. P. Jencks and M. Gilchrist, *J. Am. Chem. Soc.* **90**, 2622 (1968).
45. W. J. Boyle, F. G. Bordwell, J. A. Hautala, and K. C. Yee, *J. Am. Chem. Soc.* **91**, 4002 (1969); A. J. Kresge, *ibid.* **92**, 3210 (1970).
46. D. Turnbull and S. H. Maron, *J. Am. Chem. Soc.* **65**, 212 (1943); G. W. Wheland and J. Farr, *ibid.* **65**, 1433 (1943).
47. M. L. Bender, *Chem. Rev.* **60**, 53 (1960); M. L. Bender and B. W. Turnquest, *J. Am. Chem. Soc.* **79**, 1659 (1957).
48. J. Hine and R. D. Weimar, *J. Am. Chem. Soc.* **87**, 3387 (1965); T. N. Hall, *J. Org. Chem.* **29**, 3587 (1964).
49. K. E. Jabalpurwala and R. M. Milburn, *J. Am. Chem. Soc.*, **88**, 3224 (1966).
50. A. I. Biggs and R. A. Robinson, *J. Chem. Soc.* **388** (1961); P. Ballinger and F. A. Long, *J. Am. Chem. Soc.* **82**, 795 (1960).
51. L. A. Errede, *J. Phys. Chem.* **64**, 1031 (1960); R. S. Neale, *ibid.* **68**, 143 (1964).
52. L. P. Hammett, *Physical Organic Chemistry*, McGraw-Hill, New York, 1940; P. R. Wells, *Chem. Rev.* **63**, 171 (1963).
53. V. Gold and E. G. Jefferson, *J. Chem. Soc.* 1416 (1953).
54. M. Kilpatrick, *J. Am. Chem. Soc.* **50**, 2891 (1928); A. R. Butler and V. Gold, *J. Chem. Soc.* 2305 (1961).
55. M. L. Bender, *Chem. Rev.* **60**, 82 (1960).
56. A. Williams and R. A. Naylor, *J. Chem. Soc.* (B) 1967 (1971).
57. T. C. Bruice and R. Lapinski, *J. Am. Chem. Soc.* **80**, 2265 (1958).
58. R. F. Hudson and G. W. Loveday, *J. Chem. Soc.* 1068 (1962).
59. W. P. Jencks and J. Gerstein, *J. Am. Chem. Soc.* **86**, 4658 (1964).
60. J. F. Bunnett and G. T. Davis, *J. Am. Chem. Soc.* **82**, 665 (1960).
61. T. C. Bruice and S. J. Benkovic, *J. Am. Chem. Soc.* **86**, 418 (1964); T. C. Bruice, A. Donzel, R. W. Huffman, and A. R. Butler, *ibid.* **89**, 2106 (1967).
62. A. J. Kirby and W. P. Jencks, *J. Am. Chem. Soc.* **87**, 3209 (1965).
63. J. F. Bunnett, *Quart. Rev.* **12**, 1 (1958).
64. R. Greenhalgh and R. F. Hudson, *J. Chem. Soc.* (B) 325 (1969).
65. W. P. Jencks and J. Carriuolo, *J. Am. Chem. Soc.* **83**, 1743 (1961).
66. S. L. Johnson, *Adv. Phys. Org. Chem.* **5**, 281 (1967).
67. S. A. Khan and A. J. Kirby, *J. Chem. Soc.* (B) 1175 (1972).
68. L. R. Fedor and T. C. Bruice, *J. Am. Chem. Soc.* **87**, 4138 (1965).
69. S. M. Felton and T. C. Bruice, *J. Am. Chem. Soc.* **91**, 6721 (1969).
70. T. C. Bruice and G. L. Schmir, *J. Am. Chem. Soc.* **79**, 1663 (1957).

71. H. K. Hall, Jr., *J. Org. Chem.* **29**, 3539 (1964).
72. H. C. Brown, *J. Am. Chem. Soc.* **67**, 503 (1945).
73. L. R. Fedor, T. C. Bruice, K. L. Kirk, and J. Meinwald, *J. Am. Chem. Soc.* **88**, 108 (1966).
74. C. Brown, R. F. Hudson, V. T. Rice, and A. R. Thompson, *Chem. Comm.* 1255 (1971).
75. R. Greenhalgh and R. F. Hudson, *Chem. Comm.* 1300 (1968); *Phosphorus* **2**, 1 (1972).
76. A. L. Green, B. Saville, G. L. Sainsbury, and M. Stansfield, *J. Chem. Soc.* 1583 (1958).
77. R. F. Hudson and R. C. Woodcock, unpublished results; R. C. Woodcock, Ph.D. Thesis, University of Kent, 1972.
78. J. Epstein, R. E. Plapinger, H. O. Michel, J. R. Cable, R. A. Stephani, R. J. Hester, C. Billington, Jr., and G. R. List, *J. Am. Chem. Soc.* **86**, 3075 (1964).
79. J. Epstein, D. H. Rosenblatt, and M. M. Demek, *J. Am. Chem. Soc.* **78**, 341 (1956).
80. W. P. Jencks and A. J. Kirby, *J. Am. Chem. Soc.* **87**, 3199 (1965).
81. J. D. Aubort and R. F. Hudson, *Chem. Comm.* 938 (1970); 1342 (1969).
82. W. P. Jencks, *J. Am. Chem. Soc.* **80**, 4581 (1958).
83. O. Exner and B. Kakac, *Coll. Czech. Chem. Comm.* **28**, 1656 (1963); O. Exner, *ibid.* **29**, 1337 (1964); O. Exner and J. Holubek, *ibid.* **30**, 940 (1965); O. Exner and W. Simon, *ibid.* **30**, 4078 (1965).
84. G. F. Smith, *J. Chem. Soc.* 521 (1943).
85. M. L. Bird, E. D. Hughes, and C. K. Ingold, *J. Chem. Soc.* 634 (1954); V. Gold and D. P. N. Satchell, *ibid.* 1635 (1956); S. Winstein, E. Grünwald, and H. W. Jones, *J. Am. Chem. Soc.* **73**, 2700 (1951).
86. E. Grunwald and S. Winstein, *J. Am. Chem. Soc.* **70**, 846, 1948; R. E. Robertson, *Can. J. Chem.* **31**, 589 (1953); **33**, 1536 (1955); **35**, 1319 (1957).
87. E. D. Hughes, C. K. Ingold, and C. S. Patel, *J. Chem. Soc.* 526 (1933); C. K. Ingold, *Structure and Mechanism in Organic Chemistry*, 2nd ed., Bell, London, 1969.
88. G. Klopman and R. F. Hudson, *Helv. Chim. Acta.* **44**, 1915 (1961).
89. R. F. Hudson and G. Klopman, *J. Chem. Soc.* 1062 (1962).
90. K. Clarke and K. Rothwell, *J. Chem. Soc.* 1885 (1960).
91. M. R. Crampton, *Adv. Phys. Org. Chem.* **7**, 211 (1969).
92. R. Foster and R. K. Mackie, *Tetrahedron* **18**, 1131 (1962); P. Baudet, *Helv. Chim. Acta.* **49**, 545 (1966).
93. S. D. Ross, *Progr. Phys. Org. Chem.* **1**, 31 (1963).
94. J. E. Dixon and T. C. Bruice, *J. Am. Chem. Soc.* **94**, 2052 (1972).
95. J. Miller, A. L. Beckwith, and G. D. Leahy, *J. Chem. Soc.* 3552 (1952); J. Miller, G. D. Leahy, M. Liveris, and A. J. Parker, *Aust. J. Chem.* **9**, 382 (1956).
96. A. Ricci, M. Foà, P. E. Todesco, and G. Vivarelli, *Tetrahedron Lett.* 1935 (1965).

97. L. Cattalini, A. Orio, and A. Doni, *Inorg. Chem.* **5**, 1517 (1966).
98. L. Cattalini, A. Orio, and M. Nicolini, *J. Am. Chem. Soc.* **88**, 5734 (1966).
99. L. Cattalini, M. Nicolini, and A. Orio, *Inorg. Chem.* **5**, 1674 (1966).
100. H. G. Schuster-Woldan and F. Basolo, *J. Am. Chem. Soc.* **88**, 1662 (1966).
101. L. S. Meriwether and M. L. Fiene, *J. Am. Chem. Soc.* **81**, 4200 (1959).
102. J. O. Edwards and R. G. Pearson, *J. Am. Chem. Soc.* **84**, 16 (1962).
103. K. B. Wiberg, *J. Am. Chem. Soc.* **77**, 2519 (1955).
104. W. P. Jencks and J. Carriuolo, *J. Am. Chem. Soc.* **82**, 1778 (1960).
105. A. L. Green, G. L. Sainsbury, B. Saville, and M. Stansfield, *J. Chem. Soc.* 1583 (1958).
106. J. Epstein, V. E. Bauer, M. Saxe, and M. M. Demek, *J. Am. Chem. Soc.* **78**, 4068 (1956).
107. R. G. Pearson and D. N. Edgington, *J. Am. Chem. Soc.* **84**, 4607 (1962).
108. R. G. Pearson, D. N. Edgington, and F. Basolo, *J. Am. Chem. Soc.* **84**, 3233 (1962).
109. W. P. Jencks *Catalysis in Chemistry and Enzymology*, McGraw-Hill, New York, 1969, p. 107.
110. M. J. Gregory and T. C. Bruice, *J. Am. Chem. Soc.* **89**, 4400 (1967); S. Oae, Y. Kodoma, and Y. Yano, *Bull. Chem. Soc. Jap.* **42**, 1110 (1969).
111. M. J. Gregory and T. C. Bruice, **89**, 2327 (1967).
112. T. C. Bruice, A. Donzel, R. W. Huffman, and A. R. Butler, *J. Am. Chem. Soc.* **89**, 2106 (1967).
113. W. P. Jencks and J. Carriuolo, *J. Am. Chem. Soc.* **82**, 1778 (1960); S. L. Johnson, *Adv. Phys. Org. Chem.* **5**, 284 (1967).
114. R. Swidler and G. M. Steinberg, *J. Am. Chem. Soc.* **78**, 3594 (1956).
115. J. D. Aubort and R. F. Hudson, *Chem. Comm.* 937 (1970).
116. K. M. Ibne-Rasa and J. O. Edwards, *J. Am. Chem. Soc.* **84**, 763 (1962); G. Klopman, K. Tsuda, J. B. Louis, and R. E. Davis, *Tetrahedron* **26**, 4549 (1970).
117. S. F. Mason, *Quart. Rev.* **15**, 296 (1961); *J. Chem. Soc.* 1240 (1959).
118. E. Heilbronner and E. Haselbach, *Helv. Chim. Acta.* **53**, 684 (1970).
119. J. A. Zoltewicz, G. Grahe, and C. L. Smith, *J. Am. Chem. Soc.* **91**, 5501 (1969); T. Kauffmann and R. Wirthwein, *Angew. Chem.* **83**, 21 (1971); *Angew. Chem. Intnl. Ed.* **10**, 20 (1971).
120. W. Adam, A. Grimison, and R. Hoffmann, *J. Am. Chem. Soc.* **91**, 2590 (1969).
121. R. A. Abramovitch, G. M. Singer, and A. R. Vinutha, *Chem. Comm.* 55 (1967); H. L. Jones and D. L. Beveridge, *Tetrahedron Lett.* 1577 (1964).
122. R. Hoffmann, *Acct. Chem. Res.* **4**, 1 (1971).
123. R. J. Gillespie and R. S. Nyholm, *Quart. Rev.* **11**, 339 (1957).
124. R. Gleiter, E. Heilbronner, and V. Hornung, *Angew. Chem. Intl. Ed.* **9**, 901 (1970); M. J. S. Dewar and S. D. Worley, *J. Chem. Phys.* **51**, 263 (1969); A. J. Yencha and M. A. El-Sayed, *ibid.* **48**, 3469 (1968).
125. R. S. Mulliken, *J. Phys. Chem.* **56**, 295 (1952).
126. Y. Kadoma, S. Tamagaki, and S. Oae, *Chem. Comm.* 1115 (1972).

127. S. Wolfe, A. Rauk, L. M. Tel, and I. G. Csizmadia, *J. Chem. Soc.* (B) 136 (1971).
128. J. A. Barltrop, P. M. Hayes, and M. Calvin, *J. Am. Chem. Soc.* **76**, 4348 (1954).
129. M. A. Greenbaum, D. B. Denney, and A. K. Hoffmann, *J. Am. Chem. Soc.* **78**, 2563 (1956); R. F. Hudson, *Structure and Mechanism in Organophosphorus Chemistry*, Academic, London, 1965, pp. 168–172.
130. R. F. Hudson and F. Filippini, *J. Chem. Soc. Chem. Comm.* 726, (1972).
131. D. Sutter, *Z. Naturforsch.* **20a**, 1676 (1965).
132. O. Foss, K. Johnsen, and T. Reistad, *Acta. Chem. Scand.* **18**, 2345 (1964).
133. R. M. Stroud and C. H. Carlisle, *Acta. Cryst.* **B28**, 304 (1972).
134. H. Bock and G. Wagner, *Angew. Chem.* **84**, 119 (1972).
135. R. S. Mulliken, *J. Am. Chem. Soc.* **72**, 4493 (1950).
136. H. J. Brass and J. O. Edwards, *J. Chem. Soc.* Perk II, 627 (1972).
137. A. R. Katritzky, R. A. Y. Jones, D. L. Ostercamp, K. A. F. Record, and A. C. Richards, *J. Chem. Soc.* Perk. II, 34, 1972.
138. J. O. Edwards and R. G. Pearson, *J. Am. Chem. Soc.* **84**, 16 (1962); T. C. Bruice, A. Donzel, R. W. Huffman, and A. R. Butler, *ibid.* **89**, 2106 (1967).
139. F. Filippini and R. F. Hudson, *J. Chem. Soc. Chem. Comm.*, 522 (1972).
140. R. F. Hudson and G. Klopman, *Theoret. Chim. Acta* **8**, 165 (1967).
141. J. Hine and R. D. Weimar, *J. Am. Chem. Soc.* **87**, 3387 (1965).
142. J. Gerstein and W. P. Jencks, *J. Am. Chem. Soc.* **86**, 4655 (1964).
143. J. A. Zoltewicz and L. W. Deady, *J. Am. Chem. Soc.* **94**, 2765 (1972); J. A, Zoltewicz and H. L. Jacobson, *Tetrahedron Lett.* 189 (1972).
144. M. Dessolin and M. Laloi-Diard, *Bull. Soc. Chim.* 2946 (1971).
145. A. L. Green and B. Saville, *J. Chem. Soc.* 3887 (1956); B. E. Hackley. Ph.D. Thesis, University of Delaware, 1956.
146. J. D. Aubort and R. F. Hudson, *Chem. Comm.* 1378 (1970).
147. W. P. Jencks and M. Gilchrist, *J. Am. Chem. Soc.* **90**, 2622 (1968).
148. M. Dessolin, M. Laloi-Diard, and M. Vilkas, *Bull. Soc. Chim.* 2573 (1970).
149. J. E. Dixon and T. C. Bruice, *J. Am. Chem. Soc.* **94**, 2052 (1972).
150. J. D. Aubort and F. Filippini, private communication.
151. C. E. Moore, *Tables of Atomic Energy Levels*, National Bureau of Standards Circular 467, U.S. Govt. Printing Office, Washington, 1949–52.
152. W. P. Jencks and M. Gilchrist, *J. Am. Chem. Soc.* **90**, 2622 (1968).
153. E. C. Baughan and M. Polanyi, *Trans. Faraday Soc.* **37**, 648 (1941); M. G. Evans and M. Polanyi, *ibid.* **34**, 11 (1938); R. A. Ogg and M. Polanyi, *ibid.* **31**, 604 (1935).
154. R. F. Hudson, *Structure and Bonding* **1**, 221 (1966).
155. J. Chatt, S. Ahrland, and N. R. Davies, *Quart. Rev.* **12**, 265 (1958).
156. R. J. P. Williams and J. D. Hale, *Structure and Bonding* **1**, 249 (1966).
157. R. F. Hudson and G. Klopman, *Tetrahedron Lett.* 1103 (1967).
158. G. Klopman, *J. Am. Chem. Soc.* **90**, 223 (1968).
159. G. Klopman, *J. Am. Chem. Soc.* **86**, 4550 (1964); **87**, 3300 (1965).
160. R. H. Bathgate and E. A. Moelwyn-Hughes, *J. Chem. Soc.* 2642 (1959).

161. L. B. Young, E. L. Ruff, and D. K. Bohme, *J. Chem. Soc. Chem. Comm.* 35 (1973).

162. J. F. Bunnett, C. F. Hauser, and K. V. Nahabedian, *Proc. Chem. Soc.* 305 (1961).

163. J. Hine and A. M. Dowell, *J. Am. Chem. Soc.* **76**, 2688 (1954).

164. J. Hine and D. C. Duffey, *J. Am. Chem. Soc.* **81**, 1131 (1959).

165. R. E. Davis, R. Nehring, W. J. Blume, and C. R. Chuang, *J. Am. Chem. Soc.* **91**, 91 (1969).

166. L. C. Bateman, M. G. Church, E. D. Hughes, C. K. Ingold, and N. A. Taher, *J. Chem. Soc.* 1008 (1940).

167. R. A. Sneen, J. V. Carter, and P. S. Day, *J. Am. Chem. Soc.* **88**, 2594 (1966).

168. R. E. Robertson, R. L. Heppolette, and J. M. W. Scott, *Canad. J. Chem.* **37**, 803 (1959); R. E. Robertson, *Prog. Phys. Org. Chem.* **4**, 213 (1967).

169. A. G. Ogston, E. R. Holiday, J. St. L. Philpot, and L. A. Stocken, *Trans. Faraday Soc.* **44**, 45 (1948).

170. C. D. Ritchie and P. O. I. Virtanen, *J. Am. Chem. Soc.* **94**, 1589 (1972).

171. J. F. Bunnett and W. D. Merritt, *J. Am. Chem. Soc.* **79**, 5967 (1957); C. W. L. Bevan and G. C. Bye, *J. Chem. Soc.* 3091 (1954); A. L. Beckwith, J. Miller, and G. D. Leahy, *J. Chem. Soc.* 3552 (1952); J. F. Bunnett and G. T. Davis, *J. Am. Chem. Soc.* **80**, 4337 (1958).

172. C. W. L. Bevan and J. Hirst, *J. Chem. Soc.* 254 (1956).

173. F. M. Beringer and M. Mausner, *J. Am. Chem. Soc.* **80**, 4535 (1958).

174. G. Modena, *Acct. Chem. Res.* **4**, 73 (1971).

175. W. E. Truce and M. M. Boudakian, *J. Am. Chem. Soc.* **78**, 2748 (1956); Z. Rappoport, *Adv. Phys. Org. Chem.* **7**, 1 (1969).

176. G. Modena and P. Todesco, *Gazz. Chim. Ital.* **89**, 866 (1956).

177. J. H. Fried and W. T. Miller, *J. Am. Chem. Soc.* **81**, 2078 (1959).

178. A. J. Parker, *Quart. Rev.* **16**, 163 (1962); *Adv. Phys. Org. Chem.* **5**, 173 (1967).

179. G. Wittig, *Angew. Chem.* **74**, 479 (1962).

180. R. Huisgen, *Organometallic Chemistry*, H. Zeiss, Ed., Reinhold, New York, 1960, p. 36.

181. M. Green and R. F. Hudson, *J. Chem. Soc.* 1055 (1962).

182. I. Dostrovsky and M. Halmann, *J. Chem. Soc.* 503 (1953).

183. W. P. Jencks and J. Carriuolo, *J. Am. Chem. Soc.* **82**, 1778 (1960).

184. T. C. Bruice and L. R. Fedor, *J. Am. Chem. Soc.* **86**, 4886 (1964).

185. B. Miller, *J. Am. Chem. Soc.* **84**, 403 (1962).

186. A. J. Kirby and W. P. Jencks, *J. Am. Chem. Soc.* **87**, 3209 (1965).

187. D. W. Aubrey and M. F. Lappert, *Proc. Chem. Soc.* 148 (1960).

188. L. G. Ryss and S. L. Idel, *Russ. J. Phys. Chem.* **33**, 374 (1959).

189. R. G. Pearson, D. N. Edgington, and F. Basolo, *J. Am. Chem. Soc.* **84**, 3233 (1962).

190. J. L. Kice and G. Guaraldi, *J. Am. Chem. Soc.* **90**, 4076 (1968); J. L. Kice and G. B. Large, *ibid.* **90**, 4069 (1968).

191. J. L. Kice, G. J. Kasperek, and D. Patterson, *J. Am. Chem. Soc.* **91**, 5516 (1969).
192. O. Foss, *Acta. Chem. Scand.* **1**, 8, 307 (1947); **3**, 1385 (1949).
193. P. D. Bartlett, E. F. Cox, and R. E. Davis, *J. Am. Chem. Soc.* **83**, 103, 109 (1961); R. G. Harvey, H. I. Jacobson, and E. V. Jensen, *ibid.* **85**, 1618 (1963).
194. J. O. Edwards, *J. Am. Chem. Soc.* **84**, 22 (1964); **85**, 2263 (1963).
195. U. Belluco, L. Cattalini, and A. Turco, *J. Am. Chem. Soc.* **86**, 3257 (1964).
196. U. Belluco, L. Cattalini, F. Basolo, R. G. Pearson, and A. Turco, *J. Am. Chem. Soc.* **87**, 241 (1965).
197. G. Schwarzenbach and M. Schellenberg, *Helv. Chim. Acta.* **48**, 28 (1965).
198. J. O. Edwards, *J. Am. Chem. Soc.* **76**, 1545 (1954); A. D. Awtrey and R. E. Connick, *J. Am. Chem. Soc.* **73**, 1341 (1951).
199. K. Fukui and H. Fujimoto, *Bull. Chem. Soc. Jap.* **42**, 3399 (1969).
200. P. B. D. de la Mare and C. A. Vernon, *J. Chem. Soc.* 41 (1956).
201. J. F. Bunnett and E. Baciocchi, *Proc. Chem. Soc.* 238 (1963).
202. D. J. McLennan, *J. Chem. Soc.* B 705, 709 (1966).
203. J. F. Bunnett and J. D. Reinheimer, *J. Am. Chem. Soc.* **84**, 3284 (1962); P. Kovacic and J. J. Hiller, *J. Org. Chem.* **30**, 1581 (1965).
204. R. G. Pearson and J. Songstad, *J. Am. Chem. Soc.* **89**, 1830 (1967); C. K. Jørgensen, *Inorg. Chem.* **3**, 1201 (1964).

CHAPTER 6

Electrophilic Reactions: The General Concept of Carbocations and Their Role in Electrophilic Reactions of Alkanes (σ Bases)*

G. A. Olah
Case Western Reserve University
Cleveland, Ohio

I. THE GENERAL CONCEPT OF CARBOCATIONS

Electrophilic reactions of hydrocarbons (alkenes, alkynes, π aromatics) were long recognized to involve electron-deficient, trivalent hydrocarbon ions (carbocations, or as they were until now generally called, carbonium ions).[1]

* Based on a lecture given before the British Chemical Society annual meeting, York, September, 1971; reported in Chem. Brit. **8**, 281 (1972).

Direct observation of stable, long-lived carbocations, generally in highly acidic (superacid) systems, became possible only in the last decade through methods developed in our laboratories.[2,3] This work led us recently to the recognition of a much wider, general concept of hydrocarbon cations and to the realization of five-coordinated carbon as the key to electrophilic reactions at single bonds in saturated hydrocarbons (alkanes, cycloalkanes).

A general definition of carbocations was offered based on the realization that two distinct classes of carbocations (it seems to be the logical name for all cations of carbon compounds, since the negative ions are called *carbanions*) exist.[4]

Trivalent ("classical") carbenium ions contain an sp^2-hybridized electron-deficient carbon atom, which tends to be planar in the absence of constraining skeletal rigidity or steric interference. (It should be noted that sp-hybridized, linear oxocarbonium ions and vinyl cations also show substantial electron deficiency on carbon.)

Penta- or tetracoordinated ("nonclassical") carbonium ions, which contain five or four coordinated less electron-deficient carbon atoms bound by three single bonds, and a two-electron, three-center bond (either to two additional bonding atoms or involving a carbon atom to which they are also bound by a single bond).

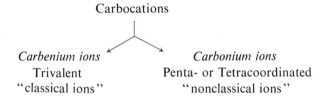

Carbocations

Carbenium ions — Carbonium ions
Trivalent — Penta- or Tetracoordinated
"classical ions" — "nonclassical ions"

Concerning the carbocation concept, it is regrettable that general usage names the trivalent, planar ions of the CH_3^+ type as *carbonium ions*. If the name is considered analogous to other *onium ions* (ammonium, sulfonium, phosphonium ions), then it should relate to the higher valency state carbocation. The higher valency state carbocations, however, clearly are not the trivalent, but the *pentacoordinated cations* of the CH_5^+ type. The German and French literature indeed frequently uses the "carbenium ion" naming for the trivalent cations. If we consider these latter ions as protonated carbenes, the naming is indeed correct,[5] and we simply could differentiate carbenium (trivalent) and carbonium (penta- or tetracoordinated) ions. It should be pointed out, however, that "carbenium ion" should be used only for the trivalent ions and not as a generic name for *all* carbocations.

Whereas the differentiation of limiting trivalent and pentacoordinated ions serves a useful purpose to establish the significant differences between these ions, it is also clear that in most specific systems there exists a continuum of charge delocalization comprising both intra- and intermolecular interactions.

A. Differentiation of Carbenium and Carbonium Ions

Experimental evidence for the differentiation of trivalent alkylcarbenium ions from pentacoordinated carbonium ions comes from nmr (^{1}H and ^{13}C), IR, Raman, and ESCA spectroscopic[6,7] study of their superacid (SbF$_5$, SbF$_5$—FSO$_3$H, Magic Acid®)[8] solutions.

Table 6-1 summarizes the ^{1}H and ^{13}C nmr parameters for the carbenium ion center in a series of carbenium ions and Fig. 6-1 shows the pmr spectra of some representative ions.

$$\begin{array}{c} R \quad \overset{+}{\underset{\diagdown}{}}{} \quad R \\ \diagdown \overset{+}{C} \diagup \\ | \\ R \end{array}$$

Trivalent
Carbenium ions

Data are characterized by substantial deshielded chemical shifts with coupling constants (J_{CH}) that indicate sp^2-hybridization.

In contrast to the highly electron-deficient trivalent carbenium centers, carbonium ions contain penta- or tetracoordinated centers, of which CH$_5$$^+$ (the methonium ion, carbonium ion) is the parent, [as CH$_3$$^+$ (methenium

$$\begin{array}{ccc} & R & \\ & | & R \\ R\!-\!\overset{|}{\underset{|}{C}}\!-\!\overset{.}{\underset{\diagdown}{<}}{}^{R} & & R \\ & R & \end{array} \qquad \begin{array}{c} R \\ \overset{.}{\underset{\diagup}{\bigwedge}} \\ R\!-\!\overset{+}{C}\!-\!-\!-\!C\!-\!R \qquad R = \text{H or allyl} \\ | \qquad | \\ R \qquad R \end{array}$$

Pentacoordinated   Tetracoordinated
Carbonium ions

ion, methyl cation, carbenium ion) is the parent for trivalent carbenium ions].

Aliphatic carbonium ions of the CH$_5$$^+$ type are indicated by the superacid chemistry of their respective alkanes[9] (σ-donors) and by mass spectrometry[10], where they are among the most abundant ions formed from alkanes. Although they are also known to exist in the gas phase in certain molecular-ion reactions, their direct observation in solution has so far not been achieved.

More rigid cycloaliphatic, particularly bicyclic, systems provide examples of directly observable, stable carbonium ions.

Table 6-1. Characteristic nmr Parameters of Alkylcarbenium Ions in SbF$_5$—SO$_2$ClF Solution at $-70°$

Ion	CH$^+$	J$_{+CH}$	J$_{+CCH}$	—CH$_2$	α—CH$_3$	β—CH$_3$	C$^+$	—CH$_3$
			δ^1$_H$				δ^{13}$_C$	
(CH$_3$)$_2$CH$^+$	13	169	3.3		4.5		-125.0	132.8
(CH$_3$)$_3$C$^+$			3.6		4.15		-135.4	146.3
(CH$_3$)$_2$C$^+$CH$_2$CH$_3$				4.5	4.1	1.94	-139.2	150.1
CH$_3$C$^+$(CH$_2$CH$_3$)$_2$				4.44	4.16	1.87	-139.4	150.8

The most disputed of nonclassical carbonium ions, the *2-norbornyl cation*, can be generated from 2-*exo*-halonorbornanes in SbF$_5$—SO$_2$ClF solution under long-lived ion conditions and at low temperature.[11] At $-156°$ it exists in a nonexchanging, static form. ^{1}H and ^{13}C nmr spectroscopy (Table 6-2) indicate that the bridging pentacoordinated methylene carbon atom is tetrahedral in nature and carries little charge. The methine carbons, to which bridging takes place, are tetracoordinated, the charge delocalized mostly into the methine bonds. This nmr spectrum (Fig. 6-2) ($-156°$) is the first experimental observation of a nonclassical carbonium ion formed by C—C σ-bond delocalization, that is, the σ route to the symmetrically delocalized ion. In other words, the process can be visualized as an intramolecular alkylation process, that is, of the C$_1$—C$_6$ σ bond by the developing electron-deficient center at C$_2$. It should also be emphasized that the same ion is observed when

generated by the alternate π route from β-Δ^3-cyclopentenylethyl halides.

Recently with Mateescu we found it also possible to apply x-ray photoelectron spectroscopy (ESCA) to the study of carbocations either in frozen superacid solutions or as isolated salts.[7] This method allows direct measurement of carbon $1s$ electron binding energies. Increasing electron-deficiency about a carbon is reflected by an increase in its $1s$ binding energy. As a result, highly electron-deficient classical alkyl and cycloalkyl carbenium centers (i.e. in the *tert*-butyl and adamantyl cations) possess carbon $1s$ binding energies approximately 4 eV greater than those of the remaining less

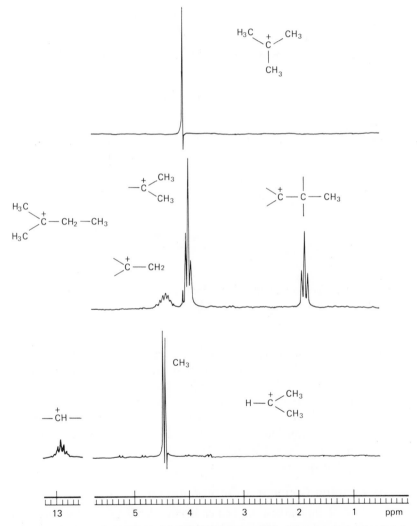

Fig. 6-1. Pmr spectra in SbF$_5$—SO$_2$ClF solution at $-60°$ of the trimethylcarbenium ion (top), dimethylethylcarbenium ion (middle) and dimethylcarbenium ion (bottom).

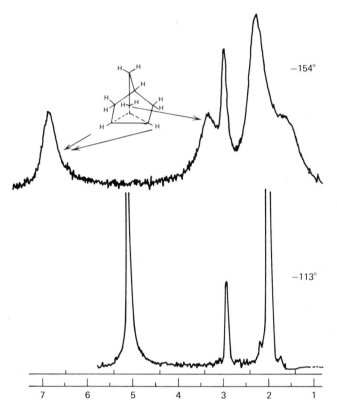

Fig. 6-2. Proton nmr spectra (100 MHz) of the "norbornyl cation" in SbF_5—SO_2ClF—SO_2F_2 solution at temperatures between $-113°$ and $-154°$.

electropositive carbon atoms. In contrast, in nonclassical carbonium ions, such as the norbornyl cation, there exists no such highly electron-deficient carbocation center and the photoelectron spectrum indicates only two modestly electron-positive carbon atoms (the tetracoordinated methine atoms; see subsequent discussion) separated by about 1.5 eV from the remaining carbon atoms, with the bridging methylene carbon indicating no detectable electron-deficient "shift" (see Figs. 6-3 and 6-4).

B. The Bonding Concept in Carbonium Ions

The bonding of the carbonium center is considered to involve three two-electron covalent bonds with the fourth bond being a two-electron three-center bond.[4] This type of bond is involved in symmetrically delocalized ions like CH_5^+ and the norbornyl cation. Thus, the interaction involves the main

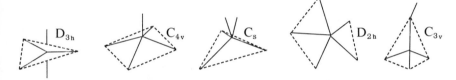

For simplicity it is suggested to use triangular dotted lines to depict the three-center bonds since full straight lines are used to symbolize two-center, two-electron bonds

$$\left[H_3C \cdots \begin{smallmatrix} H \\ \\ H \end{smallmatrix} \right]^+ \quad \text{or} \quad \left[H_3C \begin{smallmatrix} H \\ \\ H \end{smallmatrix} \right]^+$$

lobes of the covalent bonds (front-side interaction). Since an electrophile will attack the points of highest electron density, attack will occur on the covalent bonds themselves and not on the relatively unimportant back lobes.

Considering possible structures for the methonium ion with $D_{3h'}$, C_{4v}, C_s, D_{2h}, or C_{3v} symmetry,

D_{3h} C_{4v} C_s D_{2h} C_{3v}

Table 6-2. Characteristic ^1H and ^{13}C nmr Parameters of Nonclassical Carbonium Ion Centers

	δ_{1H}	J_{CH}	δ_{13C}
H_A H_A / H_B H_B	H_A 3.05 H_B 6.59	H_A 145.8 H_B 184.5	C_A 171.4 C_B 68.5
H_A / H_B H_B	H_A 3.25 H_B 7.04	H_A 218.9 H_B 193.8	C_A 159.8 C_B 67.9
H_A / H_B H_B	H_A 3.24 H_B 7.48	H_A 216.4 H_B 192.3	C_A 157.6 C_B 78.9

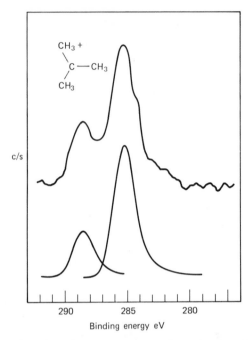

Fig. 6-3. Carbon 1*s* photoelectron spectrum of *tert*-butyl cation (lower curve computer resolved).

Olah, Klopman, and Schlosberg suggested[12] preference for the C_s front-side protonated form, based on consideration of the observed chemistry of methane in superacids (hydrogen-deuterium exchange and more significantly polycondensation indicating ease of cleavage to CH_3^+ and H_2) and also on the basis of self-consistent field calculations.[12] More extensive calculations,[13] including *ab initio* calculations utilizing an "all-geometry" parameter search,[14] confirmed the C_s symmetry structure to be favored. This structure is about 2 kcal/mole below the energy level of the C_{4v} symmetry structure, which in turn is about 8 kcal/mole favored over the trigonal bipyramidal D_{3h} symmetry structure. At the same time it should also be recognized that ready interconversion of stereoisomeric forms of CH_5^+ is possible by a pseudo-rotational type process, or as Mutterties suggested naming stereoisomer-ization processes of this type, by "polytopal rearrangements"[15] (also called sometimes "polyhedral rearrangements"). We prefer to call intramolecular carbonium ion rearrangements as "bond-to-bond rearrangements" (see subsequent discussion). Hydrogen-deuterium scrambling observed in super-acid solutions of deuteriated methane strongly indicate such processes.

It is of interest to note that the isoelectronic boron compounds can be used as model compounds for carbenium and carbonium ions. Trimethyl-boron has been used for comparison with the trimethylcarbenium ion and it is suggested that BH_5 similarly is a suitable model for $CH_5{}^+$.[4]

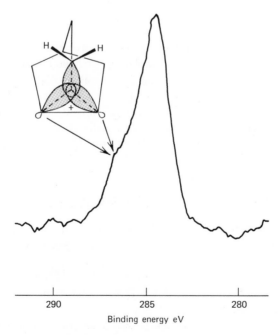

BH_5 is indicated in the acid hydrolysis of borohydrides.[16] When the hydrolysis is carried out with deuteriated acid not only HD, but also H_2 is formed. This indicates that the attack by D^+ is on the B—H bond, followed by polytopal (bond-to-bond) rearrangement, before cleavage takes place. The ease with which diborane exchanges hydrogen to deterium when treated

Fig. 6-4. Carbon $1s$ photoelectron spectrum of the norbornyl cation.

with deuterium gas also implies the formation of the three-center bonded analog, BH_3D_2

$$B_2H_6 \rightleftharpoons 2\,BH_3 \qquad BH_3 + D_2 \rightleftharpoons H{-}\overset{\overset{\displaystyle H}{|}}{\underset{\underset{\displaystyle H}{|}}{B}}{-}\kern-3pt\cdots\kern-3pt\prec\begin{smallmatrix} D \\[2pt] \\ D \end{smallmatrix} \rightleftharpoons H{-}\overset{\overset{\displaystyle D}{|}}{\underset{\underset{\displaystyle H}{|}}{B}}{-}\kern-3pt\cdots\kern-3pt\prec\begin{smallmatrix} H \\[2pt] \\ D \end{smallmatrix}$$

$$BH_2D + HD$$

$$\Updownarrow$$

$$B_2H_4D_2 \text{ etc.}$$

It should be emphasized that if steric interference (as is the case in interaction of tertiary C—H bonds in isoalkanes with tertiary carbonium ions) is substantial, the formation of the triangular three-center bond must be highly unsymmetrical, although still must not be considered to lie along an extension of the interaction bond.[17]

$$\left[R_3'C{-}\kern-3pt\cdots\kern-3pt\prec\begin{smallmatrix} H \\[2pt] \\ CR_3 \end{smallmatrix} \right]^+$$

The preferred direction of attack by the electrophile should thus not be considered to be always the same and may well vary from compound to compound (depending on the reaction conditions involved). As the various possible configurations of pentacoordinated carbon do not differ very much in stability (as in $CH_5{}^+$, with the "back-side" substituted trigonal bipyramidal form considered least favorable) it is unlikely that there is any "inherently" preferred pathway: It will depend on the individual reaction conditions.

II. ELECTROPHILIC REACTIONS AT SINGLE BONDS

Pentacoordinated carbonium ions play an important role not only in reactions of nonclassical ions, but more importantly in electrophilic reactions at single bonds, for example, aliphatic substitution of alkanes and cycloalkanes. The ability of two-electron single bonds to share their electron pairs with an electron-deficient reagent via three-center bond formation is the mechanistic key to these reactions.[17,18]

The reactivity of olefins, acetylenes, and aromatic hydrocarbons toward electrophiles is based on their π-electron donor ability. Unshared (nonbonded)

n-electron pair donation by heteroatoms represents the other major route in electrophilic reactions. Contrary to frequent textbook references to electrophilic aliphatic substitution, authenticated examples are restricted to reactions involving organometallic compounds, that is, organomercurials. The only "pure" electrophilic substitutions of alkanes reported have been the recently observed hydrogen-deuterium exchange and the protolytic cleavage reactions in superacid media.[9]

Our recent studies relating to the reactivity of hydrocarbons in superacids and in reactions with strongly electrophilic reagents resulted in the discovery of the general electrophilic reactivity of covalent C—H and C—C single bonds of alkanes (cycloalkanes).[17-19] This reactivity is due to what we consider the third major type of electron donor ability, that is, the σ-donor ability (σ basicity) of shared (bonded) electron pairs (single bonds) via two-electron three-center bond formation. It is our observation that C—C and C—H single bonds of all types (i.e., tertiary, secondary, and primary) show substantial general reactivity in electrophilic reactions such as protolytic processes (isomerization, hydrogen-deuterium exchange, protolysis), alkylations, nitration, and halogenation. These observations promise to open a new area of chemistry where alkanes and cycloalkanes are used as substrates in a wide variety of electrophilic reactions and saturated single bonds, in general, can undergo electrophilic reactions.

A. Protolysis and Hydrogen-Deuterium Exchange

With superacids, alkanes readily undergo protolytic reactions involving tertiary, secondary, and primary C—H as well as C—C bonds.[9,18] In isoalkanes, tertiary C—H bond reactivity exceeds that of secondary (and primary) C—H and C—C bonds. In n-alkanes, the C—C bond reactivity is generally found to exceed that of the C—H bonds. It is also apparent that steric factors affect σ basicity as they do π or n basicity.

Bartlett and Nenitzescu have recognized that in aluminum halide catalyzed intermolecular hydride abstractions, the electrophile (carbonium ion or proton) can remove a tertiary hydrogen atom together with its bonding electron pair[19]:

$$R_3C—H + R^+ \text{ (or } H^+) \rightleftharpoons R_3C^+ + H:R \text{ (or } H_2)$$

The assumption has been that the hydrogen atom is removed with its electron pair through what amounts to a linear transition state $\overset{\displaystyle +}{\overbrace{R\text{-}\text{-}\text{-}H\text{-}\text{-}\text{-}H}}$. Lewis, Hawthorne, and Symons[20] must be credited with having first made the suggestion that the transition state in hydride abstraction could be considered triangular instead of linear: "...electrophilic reagent attacks the C—H bond and the reasonable mode of attack is on the electrons of the bond. A triangular transition state is therefore proposed." This suggestion, however, went relatively unnoticed, probably because there were no experimental data available to substantiate it and because it was considered that steric hindrance, particularly in the interaction of a carbonium ion with an isoalkane, would allow only linear interaction.

The question of the mechanism of the hydride abstraction was not further considered until, in 1967, we reported with Lukas[9a] the protolytic ionization and hydrogen-deuterium exchange of alkanes and cycloalkanes in superacids such as FSO_3H—$SbF_5(SO_2ClF)$. Hogeveen in independent work made similar observations in HF—SbF_5 solution.[9b] These studies demonstrate that in the new superacid systems not only tertiary and secondary, but also primary C—H bonds as well as C—C bonds undergo facile protolytic cleavage. To account for the observed protolytic reactions of alkanes, we suggested with Klopman and Schlosberg[12] that protolyses preferentially involve "front-side" attack through formation of triangular transition states as shown in the case of protonation of methane to $CH_5{}^+$:

$$CH_4 + H^+ \rightleftharpoons \left[CH_3\text{--}\begin{array}{c} H \\ \diagdown \\ H \end{array} \right]^+ \rightleftharpoons CH_3{}^+ + H_2$$

Subsequently we were able to show with Shen and Schlosberg[21] that in superacids, molecular hydrogen (deuterium) readily undergoes hydrogen-deuterium exchange, even at room temperature, and considered this protolytic process to involve a triangular transition state.

$$\begin{array}{c} H \\ | \\ H \end{array} + D^+ \rightleftharpoons \left[\begin{array}{c} H \\ \diagdown \\ H \end{array}\text{--}D \right]^+ \rightleftharpoons \begin{array}{c} H \\ | \\ D \end{array} + H^+$$

Hogeveen initially rejected triangular transition states in hydride abstractions by arguing that such transition states would suffer considerable strain.[22a] He also stated that he found no evidence for exchange of molecular hydrogen (deuterium) with the superacid system used (generally SbF_5 in excess HF) which he considered further evidence against triangular transition states. At the same time he showed that carbenium ions, under stable conditions, reacted with molecular hydrogen,[22a,b] but considered the reaction to involve a linear transition state:

$$R^+ + H\text{—}H \rightleftharpoons (R\text{---}H\text{---}H)^+ \rightleftharpoons R\text{—}H + H^+$$

Brouwer and Hogeveen[23] subsequently, however, also came to the conclusion that triangular transition states are to be favored in certain hydride abstractions and substitutions of alkanes.

In our continued studies, it became obvious that protolysis of single bonds including C—H as well as C—C bonds, is a general reaction such as in the case of neopentane[18]:

$$CH_3-\underset{\underset{CH_3}{|}}{\overset{\overset{CH_3\,^2}{|}}{C}}{}^{\!1}-CH_3 \quad\xrightarrow{FSO_3H-SbF_5}\quad \left[(CH_3)_3C\!\cdot\!\!\begin{array}{c}H\\ \cdots\\ CH_3\end{array}\right]^+_{} \longrightarrow (CH_3)_3C^+ + CH_4$$

$$\Big\downarrow FSO_3H-SbF_5$$

$$\left[(CH_3)_3CCH_2\!\cdot\!\!\begin{array}{c}H\\ \cdots\\ H\end{array}\right]^+ \rightleftharpoons \left[CH_3-\underset{\underset{CH_3}{|}}{\overset{\overset{CH_3}{\searrow}}{C}}-CH_2{}^+\right] \longrightarrow (CH_3)_2\overset{+}{C}CH_2CH_3 + H_2$$

Furthermore, we suggest that three-center bond formation is characteristic of all electrophilic reactions at single bonds.

Because of the oxidizing ability of antimony pentafluoride (which is subsequently then reduced to SbF_3), the possibility of one-electron transfer redox processes must also be kept in mind when dealing with systems containing SbF_5. In superacids and particularly at low temperature the equilibrium concentration of SbF_5, for example in a HF—SbF_5 system, is low. Furthermore, the protolytic reactivity of alkanes is well demonstrated using deuteriated superacids. Electron transfer from an alkane single bond is a relatively high-energy process necessitating a 4–5 eV higher activation energy than a comparable electron-transfer process from the double-bond of an unsaturated compound and consequently is not very probable in low-temperature solution chemistry. It is significant to mention that many of the protolytic reactions discussed can also be carried out with acid systems other than those containing SbF_5. HF—TaF_5 and HF—BF_3, for example, are useful superacids in the reactions. Because the redox potentials in these systems are high, it seems to be well established that alkanes react in solutions of superacids via single-bond protolytic processes.

B. Alkylation

Despite frequent literature references to electrophilic alkylation of alkanes by olefins, from a mechanistic point of view these reactions must be considered as alkylations of the olefin by the carbenium ion derived from the isoalkane by intermolecular hydride transfer. The suggested reaction mechanism (by

Schmerling[24]) is reflected by the products formed in the reaction of propylene and isobutylene with isobutane.

$$R_2'C{=}CH_2 \xrightarrow{H^+} R_2'\overset{+}{C}CH_3 + R_3CH \longrightarrow R_2'CHCH_3 + R_3C^+$$

$$R_3C^+ + R_2'C{=}CH_2 \longrightarrow R_2'C^+{-}CH_2CR_3 \xrightarrow{H^-} R_2'CHCH_2CR_3$$

Products do not contain 2,2,3-trimethylbutane or 2,2,3,3-tetramethylbutane, which would be expected primary alkylation products in direct alkylation of isobutane with propylene and isobutylene, respectively. In the Bartlett-Nenitzescu intermolecular hydride abstraction[25] from a tertiary isoalkane by a carbenium ion, the transition state either could be considered linear or triangular (from front-side attack on the C—H bond). The latter would not be symmetrical, owing to a steric effect between the carbenium ion and the tertiary isoalkane. Even with a strongly distorted triangular transition state (i.e., the reaction taking place on the C—H bond and not at the hydrogen atom), it is obvious that cleavage can result not only in intermolecular hydride transfer, but also, via proton elimination, in direct alkylation.

It should be emphasized that there is no necessity to suggest a common transition state for hydride abstraction and alkylation, only that triangular transition states of a related nature are involved. Products depend on the

$$R_3'CH + {}^+CR_3 \rightleftharpoons \left[C{-}{<}\genfrac{}{}{0pt}{}{H}{CR_3} \right]^+ \genfrac{}{}{0pt}{}{\nearrow^1 R_3'\ CCR_3 + H^+}{\searrow_2 R_3'C^+ + CR_3H}$$

R,R' = alkyl

nature of reactants, reaction conditions, and stability of products, in addition to other factors, such as strain in the formation of transition states affecting the ratios of reactions 1 and 2.

In the case of tertiary-tertiary systems (isoalkane reacting with t-alkyl cation) a close to symmetrical trigonal three-center transition state would be highly strained, and its formation therefore is impossible. However, it is unnecessary to consider a symmetrical transition state in the reactions. The tertiary carbenium ion could easily approach the tertiary C—H bond in such a way as to form an unsymmetrical transition state, which can account for the great preference of intermolecuar hydride transfer over alkylation in the reaction. The reaction of stable, alkylcarbenium ions with alkanes can be advantageously studied as a sulfuryl chloride fluoride solution at low temperature (generally as low as $-78°$). A small amount ($\sim 2\%$ of C_8 fraction) of 2,2,3,3-tetramethylbutane was detected in the reaction of t-butyl fluoro-antimonate with isobutane [17]:

$$(CH_3)_3C^+ + HC(CH_3)_3 \underset{H^+}{\overset{-H^+}{\rightleftharpoons}} (CH_3)_3C—C(CH_3)_3$$

This observation is, however, of substantial importance since no other way than direct alkylation of the C—H bond of isobutane by the t-butyl cation can account for its formation. That direct alkylation of alkanes can indeed be carried out with ease is further shown when isobutane is reacted with the less bulky isopropyl fluoroantimonate, or when propane is reacted with t-butyl fluoroantimonate, giving in the primary alkylation product 2,2,3-trimethyl-butane up to 12%.[17] As intermolecular hydrogen transfer is faster than alkylation, and the isopropyl cation more reactive than the t-butyl cation, the alkylation reaction in both systems is considered mainly to be the propylation of isobutane:

$$(CH_3)_2CH_2 + \overset{+}{C}(CH_3)_3 \rightleftharpoons (CH_3)_2CH^+ + HC(CH_3)_3$$

$$(CH_3)_3CH + H\overset{+}{C}(CH_3)_2 \underset{H^+}{\overset{-H^+}{\rightleftharpoons}} (CH_3)_3CCH(CH_3)_2$$

The alkylation products obtained are not only those derived from the starting alkanes and an initially formed carbenium ion, but also those from the alkanes and carbenium ions formed by intermolecular hydride transfer, which generally is a faster reaction than alkylation. Since carbocations also undergo isomerization, and since intermolecular hydride transfer produces new alkanes from these ions, alkylation products obviously are increasingly complex. Generally the bulkier tertiary carbenium ions can attack the more shielded tertiary C—H bonds only with great difficulty. The processes mentioned, however, produce with ease secondary and incipient primary ions in the systems, which then alkylate C—H and C—C bonds (*alkylolysis*). The alkylation reactions of alkanes with stable carbenium ion salts in low-nucleo-philicity solvents produce a variety of products which, however, can be quantitatively analyzed with today's methods (gas chromatography, mass spectrometry). Since olefins are formed under the reaction conditions only to a minimal degree, if at all, and since at the low temperatures ($-78°$) and with short reaction times (< 30 sec), isomerizations are considered of lesser importance, we feel that indeed, for the first time, we were able to achieve *direct alkylation of alkanes*. Our claim is supported by product compositions obtained reflecting hydrogen transfer, primary alkylation, and secondary alkylation, without significant olefin formation or isomerization, and not observed in any previous acid catalyzed alkylation. Alkylation of alkanes can also be effected with the recently described $CH_3F \rightarrow SbF_5$ and

$C_2H_5F \rightarrow SbF_5$ complexes [26,27] (which show incipient methyl and ethyl cation properties) and dialkylhalonium ions.[28] It is also reasonable now to suggest that in conventional Friedel-Crafts systems, direct alkylation can play a role, although in systems where olefin formation is possible, alkene reactivity vastly exceeds that of alkanes.

Singlet carbenes are strong electrophiles and their well-known insertion reactions into C—H (and in some instances C—C) single bonds can be considered as further examples of σ-bond reactivity.

C. Nitration

The general concept of electrophilic reactivity of single bonds (σ donors) can be further demonstrated in the case of a typical electrophilic reaction such as nitration with nitronium ion. To avoid the possibility of any free-radical reaction resulting from the use of nitric acid and to avoid acid cleavage of reaction products, nitrations were carried out with stable nitronium salts, like $NO_2{}^+PF_6{}^-$, in aprotic solvents such as methylene chloride-sulfolane.[19] In the case of methane (where the nitration product—nitromethane—is not acid sensitive) anhydrous HF or FSO_3H were also used as solvents.

$$R\text{—}H \xrightarrow[\substack{CH_2Cl_3\text{-sulfolane} \\ 25° \text{ dark}}]{NO_2{}^+PF_6{}^-} R\text{—}NO_2 + H^+$$

Our studies so far have been directed primarily to mechanistic aspects, and yields in methylene chloride-sulfolane solutions (which are competing n-bases compared with hydrocarbon σ-bases, like methane) are low (ranging from 1% nitration of methane at 25° to 2–5% of nitration of higher alkanes, to 10% nitration of adamantane). The electrophilic aliphatic nitronium ion nitration of alkanes again involves attack at both C—H bonds, resulting in substitution (i.e., nitration) and at C—C bonds, causing *nitrolysis*.

$$CH_4 \xrightarrow{NO_2{}^+} \left[CH_3\text{--}\overset{H}{\underset{NO_2}{\diagup}} \right]^+ \longrightarrow CH_3NO_2 + H^+$$

$$CH_3\text{—}\overset{\displaystyle CH_3}{\underset{\displaystyle CH_3}{\overset{|}{\underset{|}{C}}}}\text{—}H \xrightarrow[\tfrac{1}{2}]{NO_2{}^+}$$

$$\left[(CH_3)_3C\text{--}\overset{H}{\underset{NO_2}{\diagup}} \right]^+ \longrightarrow (CH_3)_3CNO_2 + H^+$$

$$\left[(CH_3)_2C\text{--}\overset{\overset{H}{|}}{\underset{NO_2}{\diagup}}CH_3 \right]^+ \longrightarrow CH_3NO_2 + (CH_3)_2CH^+$$

Since tertiary and secondary nitroalkanes are cleaved with ease by strong acids, aliphatic electrophilic nitration is affected by protolytic cleavage of products and therefore generally cannot be carried out in strong acid media.

D. Chlorination

Chlorination of alkanes can also be carried out under electrophilic conditions, that is, Friedel-Crafts type acid catalyzed conditions at 25°, in the dark.

In contrast to other electrophiles, the nature of positive halogens is not yet sufficiently defined. However, we were successful in chlorinating methane, ethane, propane, and the higher alkanes with Cl_2—SbF_5 in SO_2ClF solution at $-78°$ in the dark. In this reaction substitution products (*chlorination*) are formed as well as C—C cleavage products (*chlorolysis*). The products formed can be explained again as the result of the attack of the electrophile on the corresponding bonds with the formation of a two-electron three-center bond in the transition state. In the reaction of ethane with Cl_2—SbF_5 in SO_2ClF solution at low temperature (thus under conditions favoring stable ions) the observed main reaction products are the dimethylchloronium ion (chlorolysis) and the diethylchloronium ion (chlorination with subsequent formation of the chloronium ion).

It is interesting to note that in the case of electrophilic halogenations (such as chlorination) alkanes and alkenes act as similar electron-pair donors. The difference between the paraffins and the olefins in their reactions with "positive" chlorine, in which the unshared electron pairs of chlorine are also involved, can be found in the reaction products formed. In the case of alkenes where the σ bonds are not involved in the reaction, three-membered ring cyclic halonium (chloronium) ions are formed, whereas in the case of the alkanes where the C—C σ bonds are involved directly in the reaction, open-chain dialkyl chloronium ions are formed (chlorolysis):

$$CH_2\text{—}CH_2 + \text{``}Cl^+\text{''} \rightleftharpoons \left[\begin{array}{c} CH_2\text{—}CH_2 \\ |\underline{Cl}| \end{array}\right]^+ \longrightarrow CH_2\text{—}CH_2 \underset{Cl}{\overset{+}{\diagdown \diagup}}$$

$$CH_3\text{—}CH_3 + \text{``}Cl^+\text{''} \rightleftharpoons \left[\begin{array}{c} CH_3 \quad CH_3 \\ |\underline{Cl}| \end{array}\right]^+ \longrightarrow H_3C \underset{Cl}{\overset{+}{\diagdown \diagup}} CH_3$$

In our studies we also were successful in carrying out chlorination of methane, ethane, propane, isobutane, neopentane, and so on, at 25° in the dark, generally in the presence of molar equivalent or excess chlorine (under pressure) using a variety of catalysts like aluminum chloride, ferric chloride, antimony pentachloride, stannic chloride, zinc chloride, and phosphorus pentachloride. Under the mild reaction conditions, these chlorides themselves show no chlorinating ability and are considered to act as true catalysts. (Ferric chloride, antimony pentachloride, and phosphorus pentachloride, at higher temperature and in the light or with free radical initiators, themselves are chlorinating agents.) The reaction conditions used are generally not considered to favor radical formations, although it must be pointed out that electrophilic chlorine (Cl^+ or more probably Cl_2^+) can be considered as a radical cation (i.e., the triplet state). At the same time, the chlorine atom ($Cl\cdot$) is a strong electrophile, thus the differentiation between ionic and radical chlorine is not as clear as in the case of other substituting agents. Furthermore, some catalysts, such as PCl_5 could facilitate, through coordination to Cl_2 subsequent homolytic cleavage and thus radical chlorination.

The Friedel-Crafts type chlorinations of alkanes gave reasonable yields (2–3% in the case of methane but up to 55% in the chlorination of higher alkanes, including neopentane) and fairly high selectivity. In the chlorination of methane only methyl chloride was obtained (with no methylene chloride or chloroform observed even by gas chromatography). Stronger Lewis acid catalysts ($AlCl_3$, $FeCl_3$, etc.) cause increased dehydrochlorination of secondary and tertiary alkyl chlorides with subsequent chlorine addition leading to vicinal dichlorides. Phosphorus pentachloride (as well as $ZnCl_2$, $SnCl_4$, etc.) is more selective and in the chlorination of isobutane, for example, up to 40% of products consist of t-butyl chloride (with some 50% isobutyl chloride). Neopentane yields with mild catalysts only neopentyl chloride (with some dichloride) but no amyl chlorides.

III. CONCLUSIONS

In conclusion the differentiation of *trivalent carbenium* from *pentacoordinated carbonium ions*, based on their direct observation showing marked differences, for example, in their nmr and photoelectron spectra, should put an end to much of the "myth" surrounding organic cations and to the so-called "classical-nonclassical ion controversy." In defining limiting categories of carbocations of course it must be clearly understood that there exists the possibility of a whole spectrum of ions of intermediate degrees of delocalization (or partial "nonclassical" nature) and therefore division in strictly limiting categories frequently is arbitrary. Whereas the norbornyl cation itself, for example, is a completely symmetrically delocalized nonclassical carbonium ion, the 2-methyl-norbornyl cation shows partial delocalization. More importantly, the concept of pentacoordinated carbonium ion formation via electron sharing of single bonds with electrophilic reagents in three-center bond formation promises to open up an important new area of chemistry. Whereas the concept of tetravalency of carbon obviously is not affected, carbon penta- (or tetra-) coordination as a general phenomenon seems to be well established. Trivalent carbenium ions play a major role in electrophilic reactions of π and n donors, whereas pentacoordinated carbonium ions are the key to electrophilic reactions of σ-donor single bonds.

n Donors (Nonbonded Electron Pair Donors)	π Donors (Bonded Electron Pair Donors)	σ donors
—Ö—	Olefins $\diagup C{=}C\diagdown$	—C—H
—S—	Acetylenes —C≡C—	—C—C—
—N:	Aromatics ⬡	H—H
—Halg:		

Concerning the nature of the transition state of the electrophilic reactions at single bonds,[30] the experimental product compositions strongly favor pentacoordinated carbonium ion-type transition states, involving carbocation centers with three regular covalent two-electron bonds and one two-electron three-center bond formed via "front-side" attack as contrasted with linear or back-side attack.

Front-side attack involving a trigonal transition state would lead to retention of optical activity if reactions were carried out at an optically active carbon center. Experiments in this regard are in progress in our laboratories, but there are many experimental difficulties in carrying out such reactions under controlled conditions. However, we have been able to carry out electrophilic substitution reactions (deuteration, alkylation, nitration, chlorination) at bridgehead positions of rigid systems, such as adamantane, where back-side attack and olefine formation is impossible (for back-side attack the electrophile would need to pass through the cage structure[18,19,31]). These reactions give direct experimental evidence for the direction of the electrophilic attack and the nature of the pentacoordinated carbonium ion-type transition state.

References

1. For a comprehensive review, see C. K. Ingold, *Structure and Mechanism in Organic Chemistry*, 2nd ed., Cornell Univ. Press, Ithaca, 1969.
2. For reviews, see (a) *Carbonium Ions*, G. A. Olah and P. v. R. Schleyer, Eds., Wiley-Interscience, New York, Vol. I, 1968, Vol. II, 1970, Vols. III and IV in preparation; (b) D. Bethell and V. Gold, *Carbonium Ions, An Introduction*, Academic Press, New York, 1967.
3. For reviews, see G. A. Olah, *Chem. Eng. News* **45**, 76 (March 27, 1967); *Science* **168**, 1298 (1970).
4. For a comprehensive discussion, see G. A. Olah, *J. Am. Chem. Soc.* **93**, 808 (1972), and references given therein.
5. For a review and discussion, see J. J. Jennen, *Chimia* **20**, 309 (1966); however, for a contrasting view on general use of carbonium ions, see H. Volz, *Chemie unser. Zeit.* **4**, 101 (1970).
6. For summaries, see (a) G. A. Olah and C. Pitman, Jr., in *Advances in Physical and Organic Chemistry*, Vol. IV, V. Gold, Ed., Academic, New York, 1966. (b) G. A. Olah and J. A. Olah, in *Carbonium Ions*, G. A. Olah and P. v. E. Schleyer, Eds., Wiley-Interscience, New York, 1970, Vol. II, p. 715.
7. (a) G. A. Olah, Gh. D. Mateescu, L. A. Wilson, and M. H. Gross, *J. Amer. Chem. Soc.* **92**, 7231 (1970); (b) G. A. Olah, Gh. D. Mateescu, and L. Riemenschneider, *ibid.* in press.
8. Magic Acid® is a registered trade mark of Cationics, Inc. Cleveland, Ohio, and is also available, together with other superacids and cationic reagents, from this company as well as from Aldrich Chemical.
9. (a) G. A. Olah and J. Lukas, *J. Am. Chem. Soc.* **89**, 2227, 4739 (1967); **90**, 933 (1968); (b) G. A. Olah and R. H. Schlosberg, *ibid.* **90**, 2126 (1968); (c) H. Hogeveen and A. F. Bickel, *Chem. Commun.* 635 (1967); H. Hogeveen, C. J. Gaasbeek, and A. F. Bickel, *Rec. Trav. Chim. Pays-Bas* **88**, 703 (1969); H. Hogeveen and C. J. Gaasbeek, *ibid.* **87**, 319 (1968).
10. (a) V. L. Tal'roze and A. L. Lyubimova, *Dokl. Akad. Nauk SSSR* **86**, 909 (1952); (b) F. H. Field and M. S. B. Munson, *J. Am. Chem. Soc.* **87**, 3289 (1965), and references therein.
11. G. A. Olah, A. M. White, J. R. DeMember, A. Commeyras, and Ch. Y. Lui, *J. Am. Chem. Soc.* **92**, 4627 (1970), and references given therein.
12. G. A. Olah, G. Klopman, and R. H. Schlosberg, *J. Am. Chem. Soc.* **91**, 3261 (1969).
13. (a) A. Gamba, G. Korosi, and M. Simonetta, *Chem. Phys. Lett.* **3**, 20 (1969); (b) W. Th. A. M. Van Der Lugt and P. Ros, *ibid.* **4**, 389 (1969); (c) J. J. C. Mulder and J. S. Wright, *ibid.* **5**, 445 (1970); (d) H. Kollmar and H. O. Smith, *ibid.* **5**, 7 (1970).
14. (a) V. Dyczmons, V. Staemmler, and W. Kutzelnigg, *Chem. Phys. Lett.* **5**, 361 (1970); (b) A. Dedieu and A. Veillard, presented at the 21st Annual Meeting

of the French Physical Chemical Society, Paris, September, 1970; A. Veillard, personal communication; (c) W. A. Lathan, W. J. Hehre, and J. A. Pople, *Tetrahedron Letters*, 2699 (1970); (d) W. A. Lathan, W. J. Hehre, and J. A. Pople, *J. Am. Chem. Soc.* **93**, 808 (1971).

15. E. L. Muetterties, *J. Am. Chem. Soc.* **91**, 1636 (1969).
16. G. A. Olah, P. Westerman, Y. K. Mo, and G. Klopman, *J. Am. Chem. Soc.* **95**, 3794 (1973).
17. G. A. Olah and J. A. Olah, *J. Am. Chem. Soc.* **92**, 1256 (1971).
18. G. A. Olah, Y. Halpern, J. Shen, and Y. K. Mo, *J. Am. Chem. Soc.* **93**, 1251 (1971).
19. G. A. Olah and C. H. Lin, *J. Am. Chem. Soc.* **93**, 1259 (1971).
20. (a) E. S. Lewis, and M. C. R. Symmons, *Quart. Rev.* **12**, 230 (1958); (b) M. F. Hawthorne and E. S. Lewis, *J. Am. Chem. Soc.* **80**, 4296 (1958).
21. G. A. Olah, J. Shen, and R. H. Schlosberg, *J. Am. Chem. Soc.* **92**, 3831 (1970).
22. (a) H. Hogeveen, C. J. Gaasbeek, and A. F. Bickel, *Rec. Trav. Chim. Pays-Bas* **88**, 716 (1969). (b) A. F. Bickel, C. J. Gaasbeek, H. Hogeveen, J. N. Oelderik, and J. C. Platteeuw, *Chem. Commun.*, 634 (1967).
23. D. M. Brouwer and H. Hogeveen, *Prog. Phys. Org. Chem.* **9**, 179 (1972).
24. L. Schmerling, *J. Am. Chem. Soc.* **66**, 1422 (1944); **67**, 1778 (1945); **68**, 153 (1946).
25. (a) P. D. Bartlett, F. E. Condon, and A. Schneider, *J. Am. Chem. Soc.* **66**, 1531 (1944); (b) C. D. Nenitzescu, M. Avram, and E. Sliam, *Bull. Soc. Chim. Fr.*, 1266 (1955), and earlier references given herein.
26. G. A. Olah, J. R. DeMember, and R. H. Schlosberg, *J. Am. Chem. Soc.* **91**, 2112 (1969).
27. G. A. Olah, J. R. DeMember, R. H. Schlosberg, and Y. Halpern, *J. Am. Chem. Soc.* **93**, 156 (1972).
28. (a) G. A. Olah and J. R. DeMember, *J. Am. Chem. Soc.* **91**, 2113 (1969); (b) **92**, 718 (1970); (c) **92**, 2562 (1970).
29. G. A. Olah, R. Renner, P. Schlling, and Y. K. Mo, *J. Am. Chem. Soc.*, **95**, 7686 (1973).
30. G. A. Olah, in *Symposium on the Transition State*, ed. J. E. Dubois, Gordon and Breach, 1972.
31. G. A. Olah and Gh. D. Mateescu, 159th National Meeting of the American Chemical Society, Houston, Texas, February 23, 1970, Abstr. Papers.

CHAPTER 7

Cycloaddition Reactions: A Simple Theoretical Approach

W. C. Herndon,* J. Feuer,
W. B. Giles, D. Otteson, and
E. Silber
Texas Tech University
Lubbock, Texas

* Present address, University of Texas at El Paso, El Paso, Texas.

I. INTRODUCTION

Over the past two decades several useful theoretical approaches to reactivity problems have been developing. The theoretical developments include the direct quantum-mechanical calculation of potential energy surfaces,[1-13] empirical methods to estimate activation energies based on group-additivity schemes[14-18] or classical potential functions,[19-23] and the use of correlation diagrams and orbital symmetry rules.[24-29] The latter approach, embodied in theory under the name "the Woodward-Hoffmann rules," has been extensively used to rationalize stereo- and regioselectivities of photochemical and thermal rearrangements and cycloadditions. Extensive reviews of all of these methods exist, and the applications of classical and quantum-mechanical methods have been recently summarized.[30]

None of the methods mentioned above is applicable to the entire myriad of reactivities and selectivities that are observed for these types of chemical reactions. Quantum-mechanical calculation of reaction surfaces may still be too inaccurate to compare subtle differences in rates or orientations due to a slight change in a substituent group. In any case, the obtention of all of the necessary points to define the exact energy variations as bond angles and bond lengths vary in three dimensions is a time-consuming and expensive project. However, precise details concerning the timing and nature of elementary steps of a reaction mechanism may not be obtainable in any other way. For example, recent calculations of the energy surfaces for retrocycloaddition of cyclobutane to yield ethylene[31] and geometrical isomerization of cyclopropane[32] show that intermediate biradical species (tetramethylene and trimethylene respectively) do not exist as minima in the potential energy reaction surfaces. The reliabilities of the calculations are hard to assess, but the results are still interesting and provocative.

The classical mechanical method based on the minimization of empirically obtained potential functions also has not been developed to the degree of accuracy necessary for making specific predictions about substituent and structural effects upon reaction rate. To illustrate, the calculated activation energies of Cope rearrangements differ from the experimental values by from 4 to 13 kcal/mole.[22,23] In addition, the method likewise suffers from the necessity of point-by-point plotting of a reaction surface. The group-additivity schemes for estimating heats and entropies of formation of transition states and biradicals are much more successful, and have been applied to nearly all known examples of thermal organic cycloadditions and rearrangements.[14-18] Disadvantages are that an arbitrary choice between concerted or biradical two-step mechanisms must be made before calculations, and photochemical reactions are not easily characterized by the theory.

The Woodward-Hoffmann approach has been used to delineate stereochemical features of every type of organic reaction.[24-29] After examination of orbital correlation or state diagrams, a reaction can be described as an allowed or not-allowed concerted reaction. An alternative, equivalent approach is to examine the phase relationships and symmetries of interacting orbitals to accomplish the same description.[33-41] However, it is generally not possible to make a quantitative choice between two different allowed modes of reaction, or to estimate the energy difference between a concerted and a non-concerted reaction pathway. An estimation or correlation of relative rates in a reaction series is an additional problem that cannot be treated by the Woodward-Hoffmann rules.

II. PERTURBATION THEORY

Many of the quantitative aspects of thermal and photochemical organic reactions are amenable to a simple quantum-perturbational theory of reactivity.[30,35-65] This method is especially useful for cycloaddition reactions, and in its different stages of elaboration, the perturbational approach is capable of providing either a detailed picture of a single reaction surface or a semiquantitative correlation of relative reactivities in an extensive series.

The basic idea is that one may begin with molecular wave functions for isolated, separated molecules, and then calculate the energy change resulting from the perturbing influence of one molecule upon the other. The transition complex of the reaction under consideration is the perturbed state, and its energy relative to that of the unperturbed separated molecules can be calculated more accurately by perturbation theory than by a quantum-mechanical variational method in which the interaction energy is obtained as the difference between the total isolated molecule and composite system energies.

The central points of perturbational reactivity theory are best elucidated by considering a simple interaction diagram for two molecules R and S uniting to form a transition complex, $R \cdots S$, Fig. 7-1. For the present, the discussion will be limited to the change in π-electronic energy, and the objective is to estimate the difference in energy between separated molecules and transition complex.

The energy levels of R are $F_1, F_2, F_3, \ldots, F$, the highest occupied level (HOMO) being F_{HO}. The energy levels of S are $E_1, E_2, E_3, \ldots, E$ with the lowest vacant level (LVMO) being E_{LV}. The energies and coefficients of the MOs of R and S are assumed to be known. In the transition complex, the total intermolecular interaction is a sum of the interactions of each specific molecular energy level of R with each individual level of S plus whatever Coulombic terms may be appropriate to the reaction. If we consider only the interaction between HOMO of R and LVMO of S, elementary perturbation

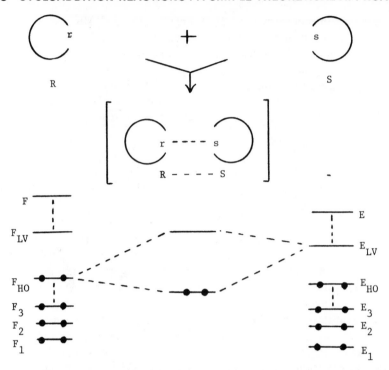

Fig. 7-1. Interaction diagram for reaction of *R* and *S*.

theory shows that the result of the interaction is a repulsion of the levels, the occupied level becomes more stable, the unoccupied level less stable.

A Hückel-type formulation of intermolecular perturbation theory leads to equations for the intermolecular perturbation energy Δ that are quite simple in form if the zero-differential overlap approximation is adopted, Eqs. (1)–(4).[55,57–59]

$$\Delta = Q + \Delta_1 + \Delta_2 + CT + P \tag{1}$$

$$Q = \frac{-Q_r Q_s \Omega}{\varepsilon} \tag{2}$$

$$\Delta_1 = N c_r c_s \gamma_{rs} \tag{3}$$

$$\Delta_2 = N \frac{(c_r c_s \gamma_{rs})^2}{F - E} \tag{4}$$

Q is a first-order Coulombic energy term that can be calculated as terms of the atomic charges. The first-order term Δ_1 arises if two partially occupied orbitals are degenerate or nearly degenerate in energy. It is a stabilizing energy term that has been called the delocalization energy or charge-transfer energy, and it can be calculated in terms of the coefficients c_r and c_s of the interacting molecular wave functions and the interaction integral γ_{rs}. The second-order term represents a stabilizing energy due to mixing of occupied orbitals of one molecule with vacant orbitals of the other molecule. N is an occupancy number, equal to unity if one or three electrons are involved in the interaction and equal to 2 if two electrons are involved.

The charge-transfer and polarization terms CT and P can be understood in terms of the diagram of Fig. 7-2. These configurations will make a contribution to the energy of the composite state if one molecule is a good donor and the other a good acceptor, or if both molecules are quite polar. The first-order part of the contribution of these terms would generally dominate over second-order terms, and would be calculated according to Eq. (3). In Eqs. (2), (3), and (4) each term is understood to be correctly summed over all relevant interactions between the two molecules. The form of the interaction integral γ is not specified, but in some applications it has been taken as proportional to overlap integrals. Its value will obviously depend upon the mechanism chosen to be calculated.

Several derivations of more complete and general theories of intermolecular perturbation theory have been published. The final result has been theory based on an SCF formalism and calculations carried out within the framework of valence-shell SCF methods.[63–65]

In our work we have only considered applications, and have left the theoretical evolution in the able hands of theorists. Generally we have looked

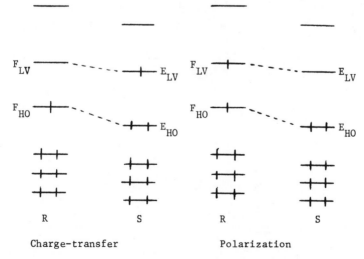

Fig. 7-2. Possible stabilizing configurations in the reaction of R and S.

for the simplest methods that can be applied to cycloaddition reactions. Two methods have been extensively tested and seem to yield consistently accurate correlations of reactivity and stereochemistry. These methods are Hückel-type perturbation calculations based on Eq. (1), and the use of molecular interaction diagrams. In the following, our approach in applying these two ideas to cycloaddition reactions will be outlined, and several illustrative examples will be given. The work of many investigators will be cited and we acknowledge our debt to the numerous workers in the field, since many of our ideas are derived from the works of other theoreticians and experimentalists.

III. INTERACTION DIAGRAMS

The construction of a molecular interaction diagram for a cycloaddition reaction can provide qualitative insights into substituent effects on rates and stereochemistry that may be more useful to organic chemists than full calculations of any type. Molecular diagrams have often appeared in the literature of molecular orbital theory and spectroscopy, and can be based upon the results of MO calculations or, preferably, on experimental values. Our viewpoint is that any reliable information, experimental or theoretical, that can be useful should be used. Photoelectron spectroscopic experiments and valence-electron SCF MO calculations are good sources for relevant information.

To illustrate, molecular diagrams for ethylene and butadiene are shown in Fig. 7-3. The values of the energies of the occupied levels were obtained from photoionization spectroscopy experiments,[66-68] and the higher energy levels are positioned on the basis of known excitation values to give the various excited states.[69] The phase of the wavefunctions for each level are, of course, obtained from calculations or by analogy to similar systems.

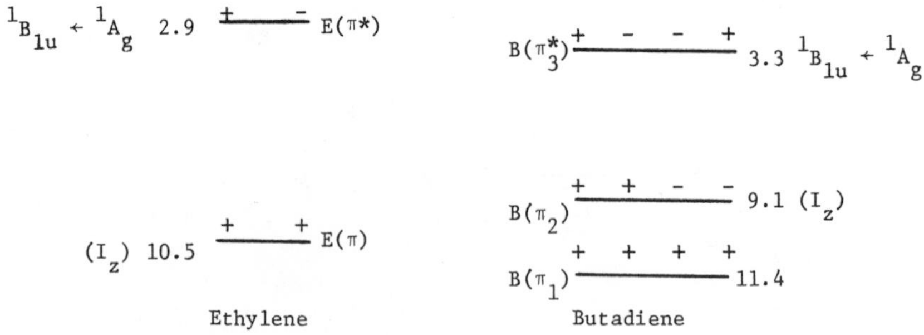

$^1B_{1u} \leftarrow {}^1A_g$ 2.9 $\underline{+\quad\quad -}$ $E(\pi*)$

$B(\pi_3^*)\underline{+\quad -\quad -\quad +}$ 3.3 $^1B_{1u} \leftarrow {}^1A_g$

(I_z) 10.5 $\underline{+\quad\quad +}$ $E(\pi)$

$B(\pi_2)\underline{+\quad +\quad -\quad -}$ 9.1 (I_z)

$B(\pi_1)\underline{+\quad +\quad +\quad +}$ 11.4

Ethylene

Butadiene

Fig. 7-3. Molecular diagrams (energies in electron volts).

The molecular interaction diagram indicates the energy-level occupancies for the two reacting molecules, and the dominating orbital interactions contributing to the perturbation energy (stabilization energy) for a particular transition state are indicated by connecting lines, or by depiction of the energy level splittings due to the interactions. Figure 7-4 shows the interaction diagram for the Diels-Alder reaction of butadiene with ethylene. The dominant interactions are $B(\pi_2)^2 \to E(\pi^*)^0$, and $E(\pi)^2 \to B(\pi_3^*)^0$, and the phase variations are suitable for maximum stabilization of a concerted $_\pi 2_s + {_\pi}4_s$ cycloaddition.

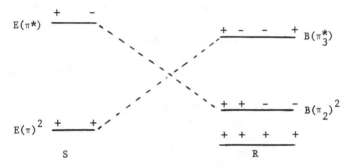

Fig. 7-4. Molecular interaction diagram for cycloaddition of butadiene and ethylene.

In the following discussion, this diagram and similar interaction diagrams for 2 + 2 photochemical and thermal cycloadditions will be considered as starting points for qualitative considerations of mechanism and stereochemistry. More quantitative aspects of the reactions will be correlated by actual calculations.

IV. APPLICATIONS

A. 2 + 4 Cycloaddition Reactions

The interaction diagram of Fig. 7-4 shows that dominating interactions in these types of reactions are those of the frontier orbitals. Electron-withdrawing or electron-donating substituents in either diene or dienophile will strengthen one interaction and weaken the other.[70] For example, an electron-donating substituent in butadiene raises both $B(\pi_2)$ and $B(\pi_3^*)$, leading to a decreased energy gap $B(\pi_2) \to E(\pi^*)$ and an increased energy gap $E(\pi) \to B(\pi_3^*)$.[71] A simple algebraic argument based on Eq. (4) shows that the strengthening effect is stronger (see Appendix I), and that both electron-attracting and electron-releasing substituents in either reactant will increase reactivity.

Further, inductive substituents at the terminal carbon atoms of the diene should have a smaller effect than identical substituents at the interior carbon atoms. This is not the same result obtained previously[71] where substituent effects were only considered to affect energies. Similarly one finds that a substituent in the dienophile should have a larger effect than it exerts in the diene (Appendix I).

Several experiments could be cited in confirmation of the deductions. Many examples are summarized in review articles.[72–75] To give one case, butadiene reacts with both acrylonitrile[76] and vinyl acetate[77] at a faster rate than with unsubstituted ethylene.[78]

Two extreme cases of Diels-Alder reactions have been differentiated. The normal reaction involves an electron-rich diene and an electron-poor dienophile, but numerous examples of inverse reactions[74,79] of electron-rich dienophiles and electron-poor dienes are also known. An interaction diagram for the normal case is given in Fig. 7-5, where only the frontier orbitals are depicted. A charge-transfer state interaction that should contribute to the stabilization of the transition state is also shown. For the normal reaction, and for the inverse reaction, the dominant interaction is only strengthened by inclusion of charge-transfer states. Since the orbital phase overlap for this interaction is suitable for concerted all-suprafacial cycloaddition, extreme cases should still exhibit stereospecificity,

The regioselectivity of reactions involving unsymmetrically substituted dienes and dienophiles is best treated by actual calculations, and several results have been published.[80–82] Rather than listing results it is sufficient to point out that the selective formations of 3,4-disubstituted and 1,4-substituted cyclohexene Diels-Alder adducts are well correlated by the calculations.

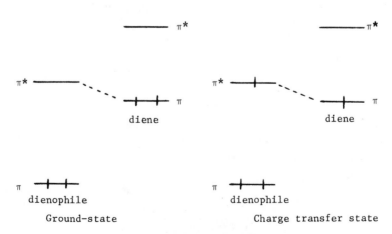

Fig. 7-5. Interaction diagram for reaction of electron-rich diene with electron-poor dienophile.

There is a renewed interest in the possibility of two-step biradical mechanisms for Diels-Alder reaction, since thermochemical calculations and experimental results[14,15,83] point to an intervention of discrete intermediates of this type for particular reactions. A recent discussion based on interaction diagrams[84] delineates those cases most likely to proceed in a nonconcerted manner. The polar Diels-Alder reactions must be concerted, but nonpolar cases might proceed by two-step cycloaddition mechanisms. In addition the possibility of competing two-step and concerted processes should not be discounted, and some perturbational calculations considering this possibility have been carried out.[85,86] Also, one should realize that the stereochemical results of retro-Diels-Alder reactions carried out at high temperature in the gas phase may not pertain to the mechanism of Diels-Alder cycloadditions studied in liquid phases at room temperature. If competition occurs, the more negative entropy requirements of concerted reactions might be cause for experimental observations characteristic of concerted reactions at lower temperatures, but at higher temperatures biradical processes might prevail.

B. 2 + 2 Photocycloaddition Reactions

The photocycloadditions of two identical olefins are allowed to be concerted all-suprafacial reactions according to the Woodward-Hoffmann criteria.[27] Experimental results indicate that such reactions are highly selective, as in the examples shown in Eqs. (5)[87] and (6).[88]

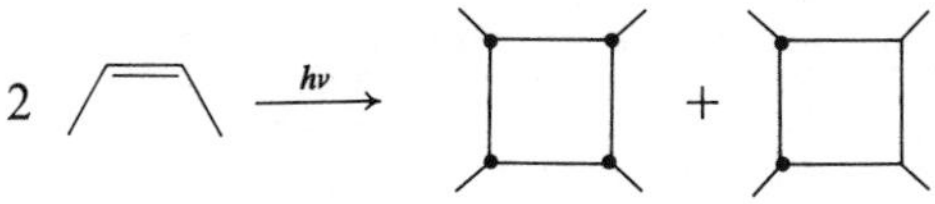

(5)

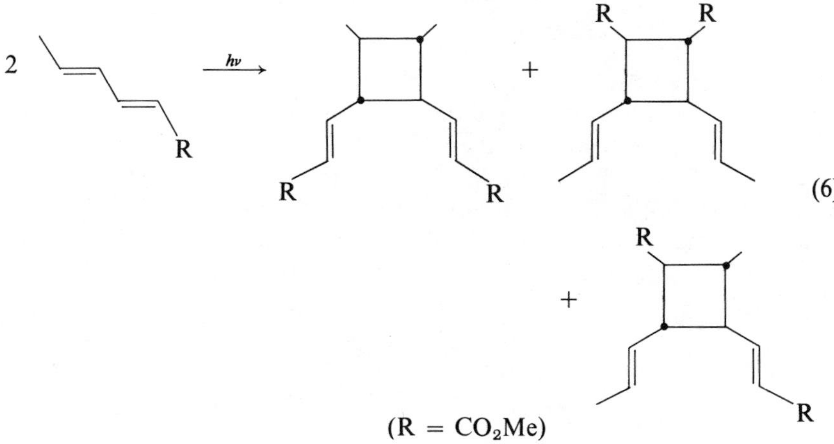

(6)

(R = CO$_2$Me)

Perturbation theory gives the same prediction, as can be seen from the interaction diagram, Fig. 7-6, since the dominant interactions are both in phase for a potential concerted reaction. Several PMO calculations have been reported, including interesting results on DNA bases.[61,86,89–91]

Photocycloadditions of two different olefins have been considered in detail and several important aspects of the reactions have been outlined.[92–96] The

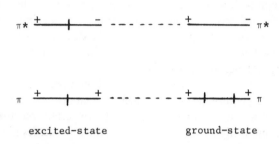

excited-state　　　　　　　ground-state

Fig. 7-6. Interaction diagram for photocycloaddition of two identical olefins.

most important conclusion pertains to the inclusion of charge-transfer states in the interaction diagram,[40,41,55,95–97] and the resultant possible stabilization of suprafacial-antarafacial (s + a) concerted reactions as compared to all-suprafacial (s + s) concerted reactions.[95] The literature should be consulted for the particulars.

An interesting photochemical cycloaddition reaction is that of an olefin and a carbonyl compound to yield an oxetane, as shown in Eq. (7).

$$\text{(7)}$$

The reaction is known as the Paterno-Büchi reaction,[98–100] and has been the subject of many mechanistic studies.[101–107] The initial step common to many Paterno-Büchi reactions involves n, π^* excitation of the carbonyl compound with a subsequent addition reaction to the olefin. Several steps, intersystem crossing, formation of exciplexes, and the like, may intervene, and the addition could be concerted or involve a biradical intermediate. Several qualitative deductions and generalizations about this type of reaction are possible after the construction of a simple interaction diagram.

Figure 7-7 shows the diagram for the cycloaddition of the n, π^* excited state of formaldehyde reacting with ethylene to yield oxetane.[70,102,108] The dominant interactions are $F(\pi^*)^1 \rightarrow E(\pi^*)^0$ and $E(\pi)^2 \rightarrow F(n)^1$. The former interaction involves a favorable orbital phase overlap for a possible all-suprafacial, concerted $_2\pi_s + _2\pi_s$ reaction. The interaction integral for this process would also have its maximum value if the two molecules approached each other in a face-to-face manner. The stereochemical consequence of such a reaction would necessarily be specific retention of substituent relative geometries.

The latter interaction would have a large interaction integral if the half-vacant nonbonding orbital of the formaldehyde and one of the π orbitals of ethylene overlapped strongly. This can occur if the ethylene π orbital approaches the formaldehyde molecule in roughly the plane of the molecule

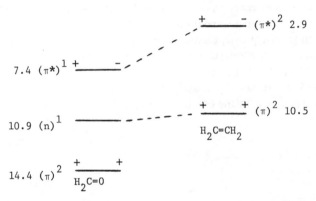

Fig. 7-7. Interaction diagram for $^1n, \pi^*$ excited formaldehyde and ethylene (energies in electron volts).

which is the nodal plane of the π system of the formaldehyde. Slight geometric distortions of the excited formaldehyde from planarity would not significantly affect this argument. Although geometries of the two reacting species can be chosen in which both new bonds are formed in concert, the most likely result of this latter interaction is to facilitate the formation of a biradical intermediate. If biradicals were formed, some loss of stereochemical specificity should be observed, the amount dependent upon the ratio of bond-rotation processes to bond-closure processes.

Actual calculations[102] show that the two mechanisms could be competing in the formaldehyde-ethylene case. Unfortunately, this particular experiment has not been carried out, but there are numerous examples of reactions with substituted carbonyl compounds and olefins, study of which has shown that the two mechanisms outlined above are quite likely to be correct descriptions of the actual molecular processes.[103–107]

The effect of substituents upon mechanism preference can be directly deduced from the interaction diagram, after tracing the effects of substituents on the molecular energy levels of the addends. Electron-attracting substituents (electronegative groups) lower all levels of a molecule, the filled levels being affected more than the vacant levels. Electron-donating groups raise both filled and vacant levels. The known values of ionization potentials for substituted aldehydes and ketones are similar to that for formaldehyde,[109,110] so only a second-order inductive effect on the energy of nonbonding electrons is apparent. Conjugative substituents lead to bands of π energy levels, the highest occupied level raised and the lowest vacant level lowered in energy. The conjugative substituent also spreads the electronic density, so that the eigenvectors at reactive sites are lowered in either π or π^* level.

Several classes of reactions can now be distinguished and are summarized in Table 7-1.[70] The table is meant to be used to predict the dominant reaction pathway, and consequently enables one to predict the stereochemical features of a particular reaction. Comparisons with some well-studied examples illustrate the utility of the theory.

Photolysis of benzaldehyde and trimethylethylene yields a mixture of cis and trans oxetanes with the two orientations shown in Eq. (8).

$$C_6H_5CHO + \qquad \xrightarrow{h\nu} \qquad \mathbf{1} \qquad \mathbf{2} \qquad (8)$$

Orientation **1** predominates and biradical intermediates generated after formation of a bond involving the lone nonbonding electron of an n, π^* excited benzaldehyde have been postulated. Table 7-1 shows that the patterns of substitution would doubly impede a concerted reaction, so that the biradical mechanism is supported for this reaction. The orientation selectivity is also correctly predicted. The dominant orbital interaction $E(\pi)^2 \to F(n)^1$ is largest for attack of the ethylene carbon with one methyl substituent on the

Table 7-1. Substituent Effects in Photoformation of Oxetanes

Mechanism	$CH_2{=}O$ Substituents			$H_2C{=}CH_2$ Substituents		
	Attracting	Donating	Conjugating	Attracting	Donating	Conjugating
Biradical	0	−	0	−	0	−
$_\pi 2_s + _\pi 2_s$	$\cdots$	+ +	−	+ +	$\cdots$	+

carbonyl nonbonding orbital, since the ground state of the olefin is polarized $(CH_3)_2C^{\oplus}\!\!-\!\!^{\ominus}CHCH_3$. This gives a larger perturbation energy, cf. Eq. (3), and implies a more stable transition state leading to the biradical which could later close to yield the major products. This argument, based on relative stabilities of transition states, gives the same results as an argument based on deduced relative stabilities of biradical intermediates. The cis-trans isomerized mixture is of course a consequence of rotation around single bonds in the biradicals.

The addition of acetone to 1,2-dicyanoethylene is found to give a completely different kind of stereochemical result.[104] Retention of configuration in the olefin moiety is observed and a concerted $_\pi2_s + _\pi2_s$ mechanism has been deduced.[104] Table 7-1 shows that the substitution pattern favors a concerted reaction. In a similar reaction, acrylonitrile plus acetone,[111] the orientation of the product in Eq. (9) is the sole orientation observed, and experiments with alkyl-substituted acrylonitriles has shown that retention of olefin geometry is again the rule.[107]

$$(CH_3)_2C\!\!=\!\!O + H_2C\!\!=\!\!CHCN \xrightarrow{\;h\nu\;} \quad\quad\quad\quad (9)$$

These results are correlated by noting that in the predominant level interaction $F(\pi^*)^1 \to E(\pi^*)^0$ the orbital coefficients of carbonyl are larger for carbon than oxygen, and those for acrylonitrile are terminal carbon larger than penultimate carbon.[70,107] Perturbation energies for concerted transition states would therefore be much larger for the observed mode of reaction than for the one with inverted stereochemistry.

Additional aspects of these photocycloadditions that can be discussed on the basis of interaction diagrams include the reactivity differences between n, π^* carbonyl triplet and singlet states, the nonreactivity of carbonyl compounds with lowest π, π^* excited states, and situations in which biradical or concerted mechanisms could be competing reactions. Our initial work in this area[70,102] confirmed the qualitative deductions that one can make from Table 7-1 with actual calculations based on HMO wavefunctions. Some stereochemical aspects of 2 + 4 photocycloadditions of dicarbonyl compounds were also correlated with calculations carried out using the same approach.[112]

About two dozen earlier theoretical treatments of excited-state additions are listed in a review,[30] and other discussions and calculations have appeared since. The importance of charge-transfer contributions to the interaction diagrams[40,41,55] has been restated in a series of articles,[84,95,108] along with arguments contending that suprafacial-antarafacial concerted cycloadditions are the logical competitors to suprafacial-suprafacial concerted reactions.[113] Our feeling is that biradical and concerted reactions of either type are the mechanistic extremes that ought to be considered in considering photocycloaddition reactions.

C. 2 + 2 Thermal Cycloaddition Reactions

The logical mechanistic extremes for reactions of this type are again concerted reactions or two-step biradical processes. The Woodward-Hoffmann rules state that thermal cycloadditions of two ethylenes must be s + a in nature if they are to be concerted reactions.[27] Most of the experimental results used in mechanistic discussions of these kinds of reactions come from kinetic and product studies of the reactions of ethylenes substituted with electron-withdrawing substituents.[114] The retrocycloaddition can also be observed,[115,116] and all of the experimental results have been interpreted in terms of biradical intermediates.

A second viable explanation of substituent effects upon stereochemistry in these cycloadditions was recently published.[113] The arguments were based upon interaction diagrams and perturbation theory, and it was demonstrated that competing s + s and s + a concerted cycloadditions could account for observed stereochemical results of several cases. The reasoning is as follows.

It is postulated that nearly all cycloadditions are concerted reactions including 2 + 2 reactions. The interaction diagram, Fig. 7-8a, shows that 2 + 2 dimerizations must be s + a in nature. Introduction of substituent asymmetry into the cycloaddends perturbs the levels so that one s + a stabilizing interaction is decreased and one increased relative to the original s + a stabilizing interactions, Fig. 7-8b. Charge-transfer states may then become important, and the incursion of an s + s concerted mechanism may

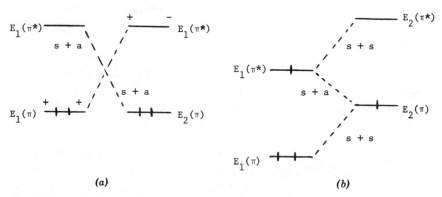

Fig. 7-8. Interaction diagram for concerted thermal cycloaddition reactions of olefins.

occur, stabilized by interactions $E_1(\pi)^2 \rightarrow E_2(\pi)^1$ and $E_1(\pi^*)^1 \rightarrow E_2(\pi^*)^0$. This happens because the interaction integral for s + s reactions is much higher than that for s + a reactions. At intermediate degrees of asymmetric substitution, competing s + s and s + a processes could lead to stereochemically mixed products previously thought to be characteristic of biradical reactions. Mixtures could also arise from the two s + a reactions that are possible.

The foregoing is a too short outline of a part of the conclusions that can be drawn from application of perturbation theory to problems in 2 + 2 cycloaddition reactions. The full article[113] should be read for a better and comprehensive presentation. Most of the conclusions seem to be sound. However, we believe that the possibility of biradical, two-step reactions for these reactions, either as competing processes or as the sole reaction pathway, cannot be excluded from consideration. Part of our reasons for endeavoring to retain biradical mechanisms in 2 + 2 cycloadditions is explained in the following paragraphs.

The only experimental criteria that have existed heretofore for biradical intermediates in 2 + 2 cycloaddition reactions have been the nonstereospecificity observed in several reactions, and the orientation selectivity (regiospecificity) correctly predicted on the basis of the most stable biradical intermediates.[114–116] At least the first of these criteria is rendered suspect as a result of the new theoretical work described above.[113] However, there are aspects of regioselectivity that can readily discriminate between a concerted process and a biradical process. These particular aspects can be illustrated by considering the thermal decomposition of a suitable substituted cyclobutane that can fragment in two different directions, Eq. (10).

$$\tag{10}$$

Is there a pattern of substitution that facilitates a concerted splitting that is counter to the predicted direction based on the estimated stability of biradical intermediates?

The substituents h_1, h_2, h_3, and h_4 are considered to be inductive substituents inducing Coulomb integrals at positions 1–4 of $\alpha + h_1$, $\alpha + h_2$, $\alpha + h_3$, and $\alpha + h_4$, respectively. A positive value of h in terms of β units (HMO theory) represents an electronegative substituent, and a negative value of h is characteristic of an electropositive substituent. The formula for the transition-state stabilizing energy of an s + a reaction given in Eq. (11) can be derived from Eq. (4) and straightforward algebra. See Appendix II for the derivation and more explanation.

$$\Delta_2 = \frac{2\gamma^2}{\beta}\left(1 - \tfrac{1}{8}[h_1 - h_2][h_3 - h_4] + \tfrac{1}{16}[h_1 + h_2 - h_3 - h_4]^2\right) \quad (11)$$

Note that the estimated perturbation energies apply to the reverse process of Eq. (10), and that the larger the energy calculated in Eq. (11), the more stabilized is the transition state. We assume the transition-state energies for the reverse processes parallel the calculated energies.

Several interesting predictions arise from application of Eq. (11). For a first example consider methyl-substituted cyclobutanes, and assign a value of $h = -0.5$ for the carbon atoms bearing a single methyl group. This assumes that the methyl substituent only exerts an inductive effect, and is one of the three methods for treating a methyl substituent that are recommended in the literature.[117] Calculated perturbation energies in units of γ^2/β (γ is the perturbation exchange integral and β is the Hückel exchange integral) are listed below, and the direction of fragmentation is signified by a dashed line. Relative rates of thermal decompositions at 700°K are given in parentheses.[118] These were estimated from known Arrhenius parameters, with the values for cis and trans isomers averaged.

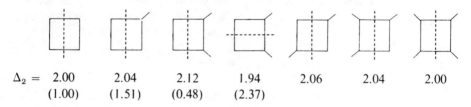

$\Delta_2 = $ 2.00   2.04   2.12   1.94   2.06   2.04   2.00
          (1.00)   (1.51)   (0.48)   (2.37)

The disubstituted cases are especially significant. The relative rates of fragmentation are not in the same order as the calculated perturbation energies. Since the calculated values are found assuming a concerted reaction, the concerted mode of fragmentation is not supported. Most workers would agree that these relatively nonpolar molecules decompose via biradical mechanisms. However, it has been argued that compounds with substituents as polar as carbonyl groups (electron-withdrawing) *must* decompose by concerted reactions.[113,119]

Only sketchy experimental evidence is presently available, and is summarized in the figures below. Carboxyl is assigned an inductive parameter of $h = +1.0$.

The stereoselectivity calculated for concerted fragmentation is again not observed, with none of the favored calculated product being observed.[120] The evidence supports a biradical reaction.

$$\Delta_2 = \begin{array}{cccc} 2.00 & 2.13 & 1.75 & 2.50 \\ (1.00) & (14.1) & (136) & (0.00) \end{array}$$

(relative rates at 683°K)

Are there any clear-cut examples of retroconcerted (2 + 2) reactions? We do find that the oxetanes shown below do fragment at 720°K as calculated for concerted reactions.[121]

$$h = +1 \quad h = +1 \qquad +1 \quad +1$$
$$h = +1 \quad h = -1 \qquad +1 \quad -1$$
$$\Delta_2 = 2.50 \qquad 2.50$$

$$+2 \quad +1 \qquad +2 \quad +1$$
$$0 \quad -1 \qquad 0 \quad -1$$
$$3.75 \qquad 1.5$$

Both directions of fragmentation are observed in about equal amounts for the first case, and the reaction to yield isobutene predominates in the second case. One might still argue that these results are derived from biradical reactions, so much additional experimental evidence is needed to clarify the issue.

An especially useful experiment would involve the decomposition of compounds with strong electronic substituents in the patterns shown below (W = electron-withdrawing, D = electron-donating, $h = \pm 2$).

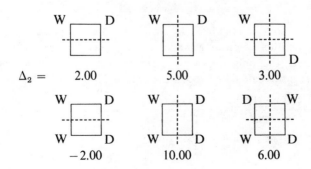

$$\Delta_2 = \begin{array}{ccc} 2.00 & 5.00 & 3.00 \\ -2.00 & 10.00 & 6.00 \end{array}$$

The relative rates would be diagnostic of mechanism, and if the stereochemical criterion was met, definite mechanism could be assigned. The stereochemistry of the reaction products and their relationship to the reactants would also help in ascertaining the possible incursion of s + s concerted reactions.

V. CONCLUSIONS

We see that there are several simple, noncomputer, theoretical methods for rationalizing and predicting the course of cycloaddition reactions. The interaction diagram and the application of easily derived perturbational formulae are particularly useful approaches. The Hückel method is a good starting point to obtain molecular eigenvalues and eigenvectors, but experimental evidence should be referred to whenever possible.

All arguments regarding whether a particular reaction proceeds via a biradical intermediate or a concerted mechanism are not settled by these theoretical considerations. However, the potential exists for these theories to make an important contribution to clarifying such questions when more experimental evidence is available.

ACKNOWLEDGMENT

The authors are very grateful to the Robert A. Welch Foundation and to the Petroleum Research Fund for financial support.

APPENDIX I. SUBSTITUENT EFFECTS ON (2 + 4) CYCLOADDITION REACTIONS

Consider the concerted cycloaddition of the π orbitals of butadiene with those of ethylene according to Eq. (1).

$$\text{(1)}$$

The transition state is stabilized by an intermolecular perturbation energy depicted in the interaction diagram shown in Fig. 7-4 (text). In the simplest approximation, the only important terms in the perturbation energy are second-order terms calculated according to Eq. (2).

$$\Delta_2 = 2\gamma^2 \frac{[c_1(HO)c_5(LV) + c_4(HO)c_6(LV)]^2}{\varepsilon_{HO}(R) - \varepsilon_{LV}(S)}$$

$$+ 2\gamma^2 \frac{[c_1(LV)c_5(HO) + c_4(LV)c_6(HO)]^2}{\varepsilon_{HO}(S) - \varepsilon_{LV}(R)} \quad (2)$$

Other terms are zero because of the neutrality and symmetry of the system.

An expression for the perturbation energy, Δ_2, is desired that takes account of the effects introduced by substituents. In the following derivation the substituents are assumed to only exert an inductive effect, represented by a change in Coulomb integral from α to $\alpha + h\beta$. Since both eigenvalues and eigenvectors appear in Eq. (2) we must first obtain the altered values for these quantities that are induced by the substituents. First-order perturbation theory can be used for this purpose.[42,43] The altered eigenvalues are given by Eq. (3) where subscript i refers to position i:

$$\varepsilon_u = \varepsilon_u{}^0 + \sum_i c_i{}^2 h_i \beta \tag{3}$$

The altered coefficients a_{iu} can be calculated using Eq. (4):

$$a_{iu} = c_{iu} + \sum_{w \neq u} \frac{[c_{iu}(c_{iw})]^2 h_i}{\varepsilon_u{}^0 - \varepsilon_w{}^0} + \sum_{w \neq u} \sum_{k \neq i} \frac{c_{ku} c_{iw} c_{kw} h_k}{\varepsilon_u{}^0 - \varepsilon_w{}^0} \tag{4}$$

In this equation, the corrections are expressed in terms of the coefficients and energies of the nonsubstituted system.

Equations (5a)–(5h) are the results of applying Eq. (4) to the ethylene and butadiene system using HMO wave functions[122]:

$$a_1(\text{HO}) = 0.60150 + 0.13011 h_1 - 0.18755 h_2 + 0.11320 h_3 - 0.05576 h_4 \tag{5a}$$

$$a_4(\text{HO}) = -0.60150 + 0.05576 h_1 - 0.11320 h_2 + 0.18755 h_3 - 0.13011 h_4 \tag{5b}$$

$$a_1(\text{LV}) = 0.60150 - 0.13011 h_1 + 0.18755 h_2 - 0.11320 h_3 + 0.05576 h_4 \tag{5c}$$

$$a_4(\text{LV}) = 0.60150 + 0.05576 h_1 - 0.11320 h_2 + 0.18755 h_3 - 0.13011 h_4 \tag{5d}$$

$$a_5(\text{HO}) = 0.70711 + 0.17678 (h_5 - h_6) \tag{5e}$$

$$a_6(\text{HO}) = 0.70711 + 0.17678 (h_6 - h_5) \tag{5f}$$

$$a_5(\text{LV}) = 0.70711 + 0.17678 (h_6 - h_5) \tag{5g}$$

$$a_6(\text{LV}) = -0.70711 + 0.17678 (h_6 - h_5) \tag{5h}$$

Equivalent expressions are given in Eqs. (6a)–(6h), where the S_i are substituent correction terms:

$$a_1(\text{HO}) = C_1(\text{HO}) + S_1 = C_1 + S_1 \tag{6a}$$

$$a_4(\text{HO}) = -C_1(\text{HO}) + S_4 = -C_1 + S_4 \tag{6b}$$

$$a_1(\text{LV}) = C_1(\text{HO}) - S_1 = C_1 - S_1 \tag{6c}$$

$$a_4(\text{LV}) = C_1(\text{HO}) + S_4 = C_1 + S_4 \tag{6d}$$

$$a_5(\text{HO}) = C_5(\text{HO}) + S_5 = C_5 + S_5 \tag{6e}$$

$$a_6(\text{HO}) = C_5(\text{HO}) - S_5 = C_5 - S_5 \tag{6f}$$

$$a_5(LV) = C_5(HO) - S_5 = C_5 - S_5 \tag{6g}$$

$$a_6(LV) = -C_5(HO) - S_5 = -C_5 - S_5 \tag{6h}$$

The energy differences that appear in the numerator terms of Eq. (4) for the altered molecules are given by Eqs. (7a) and (7b),

$$\varepsilon_{HO}(R) - \varepsilon_{LV}(S) = 1.61803\beta + \sum_{i}^{R} c_i^2 h_i - \sum_{i}^{S} c_i^2 h_i \tag{7a}$$

$$\varepsilon_{HO}(S) - \varepsilon_{LV}(R) = 1.61803\beta + \sum_{i}^{S} c_i^2 h_i - \sum_{i}^{R} c_i^2 h_i \tag{7b}$$

and are further expressed as (8a) and (b):

$$\varepsilon_{HO}(R) - \varepsilon_{LV}(S) = (\Delta\varepsilon + Y)\beta \tag{8a}$$

$$\varepsilon_{HO}(S) - \varepsilon_{LV}(R) = (\Delta\varepsilon - Y)\beta \tag{8b}$$

Equation (9) for the altered perturbation energy can now be written.

$$\Delta_2 = \frac{2\gamma^2}{\beta} \frac{[(C_1 + S_1)(C_5 - S_5) + (-C_1 + S_4)(-C_5 - S_5)]^2}{\Delta\varepsilon + Y}$$

$$+ \frac{2\gamma^2}{\beta} \frac{[(C_1 - S_1)(C_5 + S_5) + (C_1 + S_4)(C_5 - S_5)]^2}{\Delta\varepsilon - Y} \tag{9}$$

After writing the two terms over a common denominator, rearranging, and neglecting terms higher than second order in h, Eq. (10) is obtained:

$$\Delta_2 = \Delta_2{}^0 \left[1 + \left(\frac{S_1 - S_4}{2C_1} - \frac{Y}{\Delta\varepsilon} \right)^2 - \left(\frac{S_1 + S_4}{C_1 C_5} \right) S_5 \right] \tag{10}$$

Using the explicit forms for the quantities in Eq. 10, and the known values of $\Delta\varepsilon$, C_1, and C_5, a final equation for the perturbation energy in terms of the h_i can be written, Eq. (11); this equation can be used to discuss and predict substituent effects:

$$\Delta_2 = \Delta_2{}^0\{1 + [0.309(h_5 + h_6) - 0.163(h_1 + h_4) - 0.147(h_2 + h_3)]^2$$
$$- 0.077(h_5 - h_6)(h_1 - h_4) + 0.125(h_5 - h_6)(h_2 - h_3)\} \tag{11}$$

A single substituent, electron-donating or -attracting in either diene or dienophile, increases the PMO energy, since the squared term is always positive and the mixed terms are zero. Substitution in the dienophile is predicted to have four times the effect of substitution in the diene. A combination of electron-attracting groups in the diene and electron-donating groups in the dienophile, or the inverse will facilitate the reaction according to Eq. (11).

The last two terms are the source of regioselection. For two molecules each substituted with a group having opposite inductive effects, these terms will be positive for "ortho" and "para" orientations of substituents in the transition states. However, for two molecules with similar inductive substituents, Eq. (11) points to a "meta" orientation, not in accord with experimental facts.[82,123] At present, this dilemma has not been resolved.

APPENDIX II. SUBSTITUENT EFFECTS ON (2 + 2) CYCLOADDITION REACTIONS

Thermal cycloaddition involving two ethylene molecules must be suprafacial-antarafacial, and it is assumed that the geometry of the transition state is such that the overlap of orbitals at the incipient bond positions is equal ($\gamma_{1,3} = \gamma_{1,4}$).

$$
\left\| \begin{matrix} 1 \\ \\ 2 \end{matrix} \right. + \left\| \begin{matrix} 3 \\ \\ 4 \end{matrix} \right. \longrightarrow \boxed{\begin{matrix} \gamma_{1,3} \\ \\ \gamma_{1,4} \end{matrix}} \tag{1}
$$

Following the method outlined in Appendix I, Eq. (2) can be written for the intermolecular perturbation energy. Note the negative sign in the squared term for the antarafacial component of the reaction.

$$
\Delta_2 = \frac{2\gamma^2[(C_1 + S_1)(C_1 - S_3) - (C_1 - S_1)(-C_1 - S_3)]^2}{\Delta\varepsilon + Y}
$$

$$
+ \frac{[(C_1 + S_3)(C_1 - S_1) - (C_1 - S_3)(-C_1 - S_1)]^2}{\Delta\varepsilon - Y} \tag{2}
$$

Equation (3) is obtained after using the known values of the coefficients and the explicit expression for the substituent effects:

$$
\Delta_2 = \Delta_2^0[1 + \tfrac{1}{16}(h_1 + h_2 - h_3 - h_4)^2 - \tfrac{1}{8}(h_1 - h_2)(h_3 - h_4)] \tag{3}
$$

One can see that the most efficacious pattern of substitution in the olefins is to have substituents with opposing inductive effects. The squared term is then dominant with the last term positive if a head-to-head orientation is followed.

References

1. Several methods, ranging from semiempirical to *ab initio* quantum-mechanical calculations, have been used. Several hundreds of applications to organic reactions have appeared, and some will be cited later in context. Some reviews of methods and applications are listed in Refs. 2–13.

2. M. D. Newton, F. P. Boer, and W. N. Lipscomb, *J. Am. Chem. Soc.* **88**, 2353, 2361, 2367 (1966).
3. L. C. Allen and J. D. Russell, *J. Chem. Phys.* **46**, 1029 (1967).
4. K. Fukui and H. Fujimoto, *Bull. Chem. Soc. Jap.* **40**, 2787 (1967).
5. G. Blyholder and C. A. Coulson, *Theoret. Chim. Acta (Berl.)* **10**, 316 (1968).
6. K. Jug, *Theoret. Chim. Acta (Berl.)* **14**, 91 (1969).
7. S. Ehrensen, *J. Am. Chem. Soc.* **9**, 3693, 3702 (1969).
8. H. H. Jaffe, *Acct. of Chem. Res.* **2**, 136 (1969).
9. O. Sinanoglu and K. B. Wiberg, *Sigma Molecular Orbital Theory*, Yale Univ. Press, New Haven, 1970.
10. J. R. Hoyland, in *Molecular Orbital Studies in Chemical Pharmacology*, L. B. Kier, Ed., Springer-Verlag, New York, 1970, pp. 31–81.
11. G. Klopman and B. O'Leary, *Fortsch. Chem. Forsch.* **15**, 445 (1970).
12. W. C. Herndon, *Progr. Phys. Org. Chem.* **9**, 99 (1972).
13. J. N. Murrell and A. J. Harget, *Semi-Empirical Self-Consistent-Field Molecular Orbital Theory of Molecules*, Wiley-Interscience, New York, 1972.
14. S. W. Benson, *Thermochemical Kinetics*, Wiley, New York, 1968.
15. S. W. Benson, *J. Chem. Phys.* **46**, 4920 (1967).
16. H. E. O'Neal and S. W. Benson, *J. Phys. Chem.* **71**, 2903 (1967).
17. H. E. O'Neal and S. W. Benson, *J. Phys. Chem.* **72**, 1866 (1968).
18. H. E. O'Neal and S. W. Benson, *Int. J. Chem. Kinetics* **2**, 423 (1970).
19. J. B. Hendrickson, *J. Am. Chem. Soc.* **83**, 4537 (1961); **84**, 3355 (1962); **86**, 4854 (1964); **89**, 7036 (1967).
20. K. B. Wibert, *J. Am. Chem. Soc.* **87**, 1070 (1965).
21. J. E. Williams, P. J. Stang, and P. v. R. Schleyer, *Am. Rev. Phys. Chem.* **19**, 531 (1968).
22. M. Simonetta and G. Favini, *Tetrahedron Lett.* 4837 (1966).
23. M. Simonetta, G. Favini, C. Mariani, and P. Gramaccioni, *J. Am. Chem. Soc.* **90**, 1280 (1968).
24. S. I. Miller, *Adv. Phys. Org. Chem.* **6**, 185 (1968).
25. G. B. Gill, *Quart. Rev. (Chem. Soc., Lond.* **22**, 338 (1968).
26. J. J. Vollmer and K. L. Servis, *J. Chem. Ed.* **45**, 214 (1968).
27. R. B. Woodward and R. Hoffmann, *The Conservation of Orbital Symmetry*, Academic, New York, 1970.
28. S. S. Novikov, G. A. Shvekhgeimer, and A. A. Dudinskaya, *Russ. Chem. Rev.* **29**, 79 (1960).
29. T. L. Gilchrist and R. C. Storr, *Organic Reactions and Orbital Symmetry*, Cambridge Univ. Press, London, 1972.
30. W. C. Herndon, *Chem. Rev.* **72**, 157 (1972).
31. R. Hoffmann, S. Swaminathan, B. G. Odell, and R. Gleiter, *J. Am. Chem. Soc.* **92**, 7091 (1970).
32. J. A. Horsley, Y. Jean, C. Moser, L. Salem, R. M. Stevens, and J. S. Wright, *J. Am. Chem. Soc.* **94**, 279 (1972).
33. H. E. Zimmerman, *J. Am. Chem. Soc.* **88**, 1564, 1566 (1966).

34. H. E. Zimmerman, *Acct. Chem. Res.* **4**, 272 (1971).
35. M. J. S. Dewar, *Tetrahedron Suppl.* **8**, 75 (1966).
36. M. J. S. Dewar, *Spec. Pub. No. 21* (*The Chemical Society*), 177 (1967).
37. M. J. S. Dewar, *The Molecular Orbital Theory of Organic Chemistry*, McGraw-Hill, New York, 1969, Chaps. 6 and 8.
38. K. Fukui, *Tetrahedron Lett.* 2009 (1965).
39. K. Fukui, in *Molecular Orbitals in Chemistry, Physics, and Biology*, P. Löwdin and B. Pullman, Eds., Academic, New York, 1964, pp. 513–537.
40. K. Fukui and H. Fujimoto, in *Mechanisms of Molecular Migrations*, B. S. Thyagarajan, Ed., Interscience, New York, 1969, Vol. 2, pp. 117–190.
41. K. Fukui, *Fortsch. Chem. Forsch.* **15**, 1 (1970).
42. C. A. Coulson and G. S. Rushbrooke, *Proc. Camb. Phil. Soc.* **36**, 193 (1940).
43. C. A. Coulson and H. C. Longuet-Higgins, *Proc. Roy. Soc., Lond.* **A191**, 39 (1947); **A192**, 16 (1947); **A193**, 447, 456 (1948); **A195**, 188 (1948).
44. H. C. Longuet-Higgins, *J. Chem. Phys.* **18**, 265, 275, 283 (1950).
45. M. J. S. Dewar, *Proc. Camb. Phil. Soc.* **45**, 638 (1949).
46. M. J. S. Dewar, *J. Chem. Soc.* 2329 (1950).
47. M. J. S. Dewar, *J. Am. Chem. Soc.* **74**, 3341, 3345, 3350, 3353, 3355, 3357 (1952).
48. M. J. S. Dewar, *J. Chem. Soc.* 3532 (1952).
49. M. J. S. Dewar, *J. Chem. Soc.* 1617 (1954).
50. M. J. S. Dewar, *Adv. Chem. Phys.* **8**, 65 (1965).
51. K. Fukui, T. Yonezawa, and H. Shingu, *J. Chem. Phys.* **20**, 722 (1952).
52. K. Fukui, T. Yonezawa, C. Nagata, and H. Shingu, *J. Chem. Phys.* **22**, 1433 (1954).
53. K. Fukui, in *Modern Quantum Chemistry*, O. Sinanoglu, Ed., Academic, New York, 1965, Part I, pp. 49–84.
54. K. Fukui, C. Nagata, T. Yonezawa, H. Kato, and K. Morokuma, *J. Chem. Phys.* **31**, 287 (1959).
55. K. Fukui and H. Fujimoto, *Bull. Chem. Soc. Jap.* **41**, 1989 (1968).
56. K. Fukui, H. Hao, and H. Fujimoto, *Bull. Chem. Soc. Jap.* **42**, 348 (1969).
57. R. F. Hudson and G. Klopman, *Tetrahedron Lett.* 1103 (1967).
58. G. Klopman and R. F. Hudson, *Theoret. Chim. Acta* (*Berl.*) **8**, 165 (1967).
59. G. Klopman, *J. Am. Chem. Soc.* **90**, 223 (1968).
60. L. Salem, *J. Am. Chem. Soc.* **90**, 543 (1968).
61. A. Devaquet and L. Salem, *J. Am. Chem. Soc.* **91**, 3793 (1969).
62. A. Devaquet, *Mol. Phys.* **18**, 233 (1970).
63. R. Sustmann and G. Binsch, *Mol. Phys.* **20**, 1 (1971).
64. R. Sustmann and G. Binsch, *Mol. Phys.* **20**, 9 (1971).
65. A. Imamura, *Mol. Phys.* **15**, 225 (1968).
66. A. D. Baker, C. Baker, C. R. Brundle, and D. W. Turner, *Int. J. Mass. Spec. Ion Phys.* **1**, 285 (1968).
67. M. I. Al-Joboury and D. W. Turner, *J. Chem. Soc.* 4434 (1964).
68. M. J. S. Dewar and S. D. Worley, *J. Chem. Phys.* **49**, 2454 (1968).
69. D. F. Evans, *J. Chem. Soc.* 1735 (1960).

70. W. C. Herndon, *Tetrahedron Lett.* 125 (1971).
71. R. Sustmann, *Tetrahedron Lett.* 2721 (1971).
72. J. G. Martin and R. K. Hill, *Chem. Rev.* **61**, 537 (1961).
73. J. Sauer, *Angew. Chem. Intern. Ed. Engl.* **5**, 211 (1966).
74. J. Sauer, *Angew. Chem. Intern. Ed. Engl.* **6**, 16 (1967).
75. S. Seltzer, *Adv. Alicyclic Chem.* **2**, 1 (1968).
76. J. Doucet and R. Rumpf, *Bull. Soc. Chim. France* 610 (1954).
77. K. Alder and H. F. Rickert, *Ann.* **543**, 1 (1930).
78. L. M. Joshel and L. W. Butz, *J. Am. Chem. Soc.* **63**, 3350 (1941).
79. W. E. Bachmann and N. C. Deno, *J. Am. Chem. Soc.* **71**, 3062 (1949).
80. J. Feuer, W. C. Herndon, and L. H. Hall, *Tetrahedron* **24**, 2575 (1968).
81. O. Eisenstein, J. M. Lefour, and N. T. Anh, *Chem. Comm.* 969 (1971).
82. T. Inukai, H. Sato, and T. Kojima, *Bull. Chem. Soc. Jap.* **45**, 891 (1972).
83. W. von E. Doering, *Int. Union Pure and Appl. Chem.*, Twenty-third Congress, Special Lectures, Vol. 1, 235 (1971).
84. N. D. Epiotis, *J. Am. Chem. Soc.* **94**, 1924 (1972).
85. W. C. Herndon and J. Feuer, *J. Org. Chem.* **33**, 417 (1968).
86. L. Salem, *J. Am. Chem. Soc.* **90**, 553 (1968).
87. H. Yamazaki and R. J. Cretanovic, *J. Am. Chem. Soc.* **91**, 520 (1969).
88. H. P. Kaufmann and A. K. Sengupta, *Ann.* **681**, 39 (1965).
89. K. Fukui, T. Yonezawa, and C. Nagata, *Bull. Chem. Soc. Jap.* **34**, 37 (1961).
90. R. Sayre, J. P. Harlos, and R. Rein, in *Molecular Orbital Studies in Chemical Pharmacology*, L. B. Kier, Ed., Springer-Verlag, New York, 1970, pp. 207–237.
91. C. Nagata, A. Inamura, Y. Tagashira, and M. Kodama, *J. Theoret. Biol.* **9**, 357 (1965).
92. M.-H. Whangbo and I. Lee, *J. Korean Chem. Soc.* **13**, 273 (1969).
93. G. Ahlgren and B. Akermurk, *Tetrahedron Lett.* 1885 (1970).
94. P.-S. Song, M. L. Harter, T. A. Moore, and W. C. Herndon, *Photochem. Photobiol.* **13**, 521 (1971).
95. N. D. Epiotis, *J. Am. Chem. Soc.* **94**, 1941 (1972).
96. W. C. Herndon, *Fortschr. Chem. Forsch.*, in press (1974).
97. K. Fukui, K. Morokuma, and T. Yonezawa, *Bull. Chem. Soc. Jap.* **34**, 1178 (1961).
98. G. Buchi, C. G. Inman, and E. S. Lipinsky, *J. Am. Chem. Soc.* **76**, 4327 (1954).
99. E. Paterno and G. Chieffi, *Gazz. Chim. Ital.* **39**, 341 (1909); E. Paterno, *ibid.* **44**, 463 (1914).
100. D. R. Arnold, *Advan. Photochem.* **6**, 301 (1968).
101. For leading references to mechanistic studies of Paterno-Buchi reactions see Refs. 102–107.
102. W. C. Herndon and W. B. Giles, *Mol. Photochem.* **2**, 277 (1970).
103. N. J. Turro and P. A. Wriede, *J. Am. Chem. Soc.* **92**, 320 (1970).
104. J. C. Dalton, P. A. Wriede, and N. J. Turro, *J. Am. Chem. Soc.* **92**, 1318 (1970).

105. I. H. Kochevar and P. J. Wagner, *J. Am. Chem. Soc.* **92**, 5742 (1970).
106. N. C. Yang and W. Eisenhardt, *J. Am. Chem. Soc.* **93**, 1277 (1971).
107. J. A. Barltrop and H. A. J. Carless, *J. Am. Chem. Soc.* **94**, 1951 (1972).
108. N. D. Epiotis, *J. Am. Chem. Soc.* **94**, 1946 (1972).
109. S. D. Worley, *Chem. Rev.* **71**, 295 (1971).
110. K. Watanabe, T. Nakayama, and J. Mottl, *J. Quant. Spectrosc. Radiat. Transfer* **2**, 369 (1962).
111. J. A. Barltrop and H. A. J. Carless, *Tetrahedron Lett.* 3901 (1968).
112. W. C. Herndon and W. B. Giles, *Chem. Comm.* 497, (1969).
113. N. D. Epiotis, *J. Am. Chem. Soc.* **94**, 1935 (1972).
114. For leading references see P. D. Bartlett, *Int. Union Pure and Appl. Chem.* Twenty-third Congress, Vol. 4, 281 (1971); *Science* **159**, 833 (1968).
115. H. M. Frey, *Adv. Phys. Org. Chem.* **4**, 148 (1966).
116. M. R. Willcott, R. L. Cargill, and A. B. Sears, *Progr. Phys. Org. Chem.* **9**, 25 (1972).
117. A. Streitwieser, Jr., *Molecular Orbital Theory for Organic Chemists*, Wiley, New York, 1961, Chap. 4.
118. C. T. Genauk, F. Kern, and W. D. Walters, *J. Am. Chem. Soc.* **75**, 6196 (1953); M. N. Das, F. Kern, T. D. Coyle, and W. D. Walters, *ibid.* **76**, 6271 (1954); H. R. Gerberich and W. D. Walters, *ibid.* **83**, 3935, 4884 (1961).
119. W. H. Richardson, M. B. Yelvington, and H. E. O'Neal, *J. Am. Chem. Soc.* **94**, 1619 (1972).
120. W. D. Herndon and Dale Otteson, unpublished results.
121. W. C. Herndon, unpublished results.
122. C. A. Coulson and A. Streitwieser, Jr., *Dictionary of π-Electron Calculations*, Freeman, San Francisco, 1965.
123. T. Inukai and T. Kojima, *J. Org. Chem.* **36**, 924 (1971).

CHAPTER 8

Photochemical Reactions: Correlation Diagrams and Energy Barriers

J. Michl*
University of Utah
Salt Lake City, Utah

I. A SIMPLE MODEL OF PHOTOCHEMICAL REACTIVITY IN SOLUTIONS

There is remarkably little agreement on how one should go about understanding and predicting photochemical reactivity of large molecules in dense media. The quite considerable recent developments in the theory of radiationless transitions[1] have not yet advanced to a stage directly useful for prediction

* Alfred P. Sloan Fellow, 1971–1973.

of reactivity of large organic molecules. To understand which products are formed in a given reaction, many authors presently think in terms of "easy" and "difficult" motions as dictated by the corresponding change in energy of the lowest excited state of given multiplicity,[2–5] often in terms of Woodward-Hoffmann rules or of perturbation theory. It is sometimes pointed out that the nature of the lowest excited state may change during the reaction course and that one needs to look for points of easy crossing back to the ground-state surface,[6–8] so that the Woodward-Hoffmann rules based on correlation diagrams for the lowest excited state could well be irrelevant.[7,8] A useful theory of photochemical reactions must be able to distinguish between excited singlet and triplet behavior, which Woodward-Hoffmann rules[2] and many of the other approaches[4–7] do not. Also, a complete theory should do more than predict the nature of products. It should rationalize and predict phenomena such as wavelength dependence of photochemical reactivity and occurrence of hot ground-state reactions, which apparently take place even in dense media.[9] For a longer but admittedly still incomplete list of literature dealing with these problems, the reader is referred to Ref. 10.

To start our discussion it appears highly desirable to have a model for what occurs between light absorption by a ground-state molecule and formation of a thermally equilibrated ground-state product. Otherwise, it will not be clear what to calculate or estimate, let alone how to do it. The situation is much clearer in ground-state reactivity, where the absolute reaction rate theory with its concept of activated complex provides an acceptable if inaccurate framework for qualitative and semiquantitative arguments applicable to large organic molecules. We shall base our discussions on a simple model selectively synthesized from views mostly already expressed in the literature and summarized in Ref. 10.

Briefly, we postulate that there are no distinct and separate photochemical processes. Photophysical processes already well known from molecular spectroscopy sometimes take a molecule* back to the well in the ground-state potential energy hypersurface S_0 where it started from, and then they are called photophysical, and sometimes to another such well in S_0 corresponding to another conformer or isomer, and in the latter case they are called photochemical. In either case, we shall postulate the same series of events. At the beginning, the motions of the nuclei are governed by the shape of the ground-state potential energy hypersurface S_0 in the well characterizing a given starting compound. Electronic excitation upon absorption of a photon is so fast that the nuclei do not move significantly during the process (Franck-Condon principle). Because of the proximity of a large number of excited states, the Born-Oppenheimer approximation is poor for the excited molecule

* A "supermolecule" in the case of bimolecular reactions.

and the nuclear motions are probably often not simply governed by a single potential energy hypersurface (particularly in regions of avoided crossings). In an undoubtedly complicated and presently unpredictable manner, the nuclei will acquire kinetic energy at the expense of the original electronic energy, losing it again to the surrounding medium (solvent, acting as a heat bath). In a very short time, around 10^{-11}–10^{-12} sec, the excited singlet molecule will reach one of the minima in its S_1 hypersurface. If the minimum is close in energy to the S_0 hypersurface, and in particular if it corresponds to an avoided crossing with S_0 hypersurface (a "funnel" or "hole" in the S_1 surface),* the nuclear motion will take it further on down to one of the valleys in the S_0 hypersurface without further delay. Which valley is reached may depend on the direction from which the funnel in S_1 was originally reached (Fig. 8-1). Other processes, such as fluorescence or intersystem crossing to the triplet with rate constants about 10^8 sec^{-1}, will typically be unable to compete. If the minimum in S_1 reached by the molecule is separated from S_0 by a sizable energy gap, these other processes can compete efficiently.

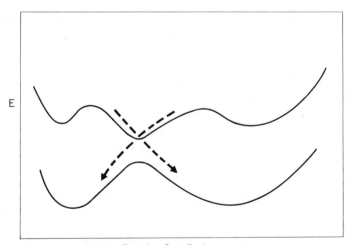

E

Reaction Coordinate

Fig. 8-1. An avoided crossing of two Born-Oppenheimer hypersurfaces, resulting in a "funnel" in the upper state. Crossover probability is high, molecules approaching the funnel from different directions may end up in different valleys of the lower state (dashed lines).

* A funnel in S_1 results from an avoided crossing between S_0 and S_1. Whether there actually is a minimum in S_1 as a result of this crossing depends on details of the slopes. Since the electronic wavefunction changes fast as a function of nuclear coordinates in such a region, transition from one surface to another can be induced easily by nuclear motion.

Also, thermal energy may allow the molecule to escape from the minimum in S_1 into another one. Such a process may then be governed by ordinary transition-state theory. In addition, there may be time for diffusion to bring close a quencher or a reaction partner while the excited molecule is in the minimum in S_1. These make available new "funnels" through which return to S_0 is possible, or new minima in S_1 of lower energy (exciplexes). Final loss of excess vibrational energy takes the molecule into one of the wells in S_0, depending on where the S_0 surface is finally reached and the photochemical process ends. In dense media, the loss of excess vibrational energy usually occurs very fast. If for some reason it does not, the excess kinetic energy of the nuclei may take the molecule over barriers in S_0 into valleys other than the one originally reached (hot ground-state reactions). In solutions, this appears probable only if the nuclei are already moving in the direction toward the barrier at the time when the S_0 surface is reached.

If the original excitation is into the triplet manifold, or if intersystem crossing into that' manifold occurs before S_0 is reached, the same kind of processes will take the molecule into one of the many minima in the T_1 hypersurface. If at that geometry the T_1 state is the ground state (e.g., in certain carbenes), the photochemical process is over and ground-state chemistry will usually follow. If not, a transition to the S_0 surface will follow. The radiationless $T_1 \rightarrow S_0$ process is relatively fast if the energy separation is small (e.g., in biradicals), particularly if heavy atoms are present. It is slower if the gap is large, in which case phosphorescence can sometimes compete. In general, however, the spin-forbidden $T_1 \rightarrow S_0$ transition is slower than otherwise similar $S_1 \rightarrow S_0$ transitions, and this also increases opportunity for approach of reaction partners and quenchers, as well as for thermal motion from one of the minima in T_1 to another (rotational equilibration, bond rearrangement, and so on).

According to the present simple model, it is relatively easy in principle to predict possible products of photochemical reactions from knowledge of the positions of minima and funnels in the S_1 and T_1 surfaces, assuming that knowledge of ground-state chemistry will allow us to estimate the fate of the molecule once it reaches the S_0 surface as long as we know just where it occurs. On the other hand, it unfortunately appears impossible to predict simply the probabilities with which the individual numerous minima in S_1 or T_1 are reached. These may in general depend on the conformation at the time of original excitation and on the wavelength of the exciting light. These determine the direction of the first push which the nuclei receive immediately after excitation, as well as the likelihood that the molecule will move toward and above any given barrier in S_1 on its way to a minimum or funnel in S_1, and whether it indeed will have enough energy to overcome the barrier. A detailed understanding appears to require a considerably better knowledge of details of the excited states hypersurfaces and of vibronic coupling than is presently available for large molecules.

Due to the breakdown of the Born-Oppenheimer approximation for higher excited states it does not appear useful to distinguish whether a molecule is in a "hot" S_1 state or in an isoenergetic "less hot" S_2 state*. In the present model, we do not attempt to classify reactions as occurring from S_1, S_2, S_3, or T_1, T_2, T_3, and so on. They all eventually proceed through funnels or minima in S_1 or T_1. As the energy of the original excitation increases, new barriers can be overcome, new minima in S_1 (or T_1) and thus new products can be reached, but the thresholds for the new processes are related to heights of the barriers rather than onsets of new vertical excited states at the starting geometry. Thus, we find it more useful to talk in terms of "extra" or "activation" energy which certain processes may require.

The rest of this chapter is devoted to the consideration of the location of minima and barriers in the S_1 and T_1 surfaces. The former should allow us to rationalize formation of particular products, the latter to predict which of the otherwise reasonable products cannot be formed unless extra energy is available, as the corresponding minimum in S_1 or T_1 is separated by a barrier from the starting geometry. If sufficient additional energy is available it may be used to overcome the barrier or it may be lost as heat before that happens. Since we cannot tell which is more likely, predictions can only be made in a negative sense.

II. LOCATION OF MINIMA AND FUNNELS IN S_1 AND T_1 HYPERSURFACES

Really reliable estimates of positions of minima in S_1 and T_1 would require complicated and expensive calculations. Nevertheless, a case can be made for the use of very simple reasoning similar to that used for discussions of ground-state reactivity. This is based on qualitative molecular orbital and valence bond theories of chemical bonding. In most cases, chemical intuition allows us to estimate positions of minima in the S_0 hypersurface almost automatically, simply using the rules of valence and not leaving any singly occupied atomic valence orbitals unable to interact with others to form bonding orbitals. In the first approximation, one can estimate the energy of the S_0 surface for a given nuclear geometry by counting the number of bonds present, making corrections for steric strain, delocalization effects, and so on.

The situation is much more complicated for the S_1 and T_1 surfaces. Two kinds of geometries at which minima in S_1 (T_1) may be present come to mind

---

* This need not be true in more sophisticated models permitting a treatment of molecular dynamics.

immediately. First, those near which there also are minima in S_0. These should occur particularly in molecules with large chromophores in which promotion of one electron changes little the overall picture of bonding, for example, aromatic molecules. The existence of these minima is well known from fluorescence and phosphorescence spectroscopy ("spectroscopic excited states"). Photochemical reactions ending by return to S_0 from such minima "lead to excited products" and have been termed X type by Dougherty.[7] The other kind of geometries near which minima often can be expected in excited state surfaces, and through which Dougherty's N- and G-type reactions occur, will be referred to as "biradicaloid" geometries. These are geometries at which the molecule has two nonbonding orbitals occupied by a total of two electrons. Such geometries are relatively unfavorable in the S_0 state, since the two electrons are not contributing to bonding, whereas a distortion which will bring the two orbitals into interaction, making one bonding and doubly occupied, the other antibonding and empty in the S_0 state, will lead to stabilization. On the other hand, in the excited S_1 and T_1 states, the same distortion is likely to destabilize the molecule, since each of the orbitals keeps one electron and the antibonding one usually goes up faster than the bonding one goes down.

From this simple argument, one might then expect minima to occur at the same or similar geometries in the S_1 and T_1 surfaces. However, adding up energies of occupied orbitals gives only a rough indication of the total energy, since electron repulsions are then treated incorrectly. A more detailed analysis[11] shows why among biradicaloid geometries those with "loose" geometry (the two radical centers far apart) should tend to be favorable energetically in the T_1 state, while those with "tight" geometry (the two radical centers in the same region of space) should tend to be favored in the S_1 state which is of "ionic" nature. Other factors, for example steric strain, will have to be considered in each case, and exceptions will undoubtedly exist; but the general tendency is in good agreement with experimental trends. The importance of this argument for the understanding of differences between products of singlet and triplet pericyclic reactions was first pointed out by Fukui[3,12] and later in more general context by other authors.[11,13] Other factors may contribute to S-T reactivity differences: the various minima in S_1 and T_1 will generally be reached at different rates after initial excitation, and more vibrational energy will typically be liberated during a final $S_1 \to S_0$ jump as compared to a final $T_1 \to S_0$ jump, which may be important in hot ground-state reactions.

A simple example will help to illustrate the situation (Fig. 8-2). Consider the excited states of the H_2 molecule, for which experimental data are available.[15] The ground-state bond length R is 0.74 Å and any geometry for which R is larger than about 1.5 Å or so is already approximately biradicaloid.

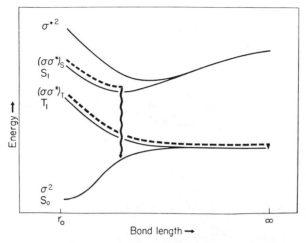

Fig. 8-2. Schematic dependence of state energies on the nature of the biradicaloid geometry (tight or loose) generated from equilibrium S_0 geometry by stretching a single bond. Only the four low-lying states which are degenerate in the first approximation are shown.

Loose geometry corresponds to $R = \infty$, tight geometry to small values of R, say 1.5–3 Å. From orbital energies alone, one might expect the $(\sigma\sigma^*)_T$, $(\sigma\sigma^*)_S$, and σ^{*2} states to be all purely dissociative (minimum at $R = \infty$), since already one electron in the σ^* orbital provides more destabilization than its bonding partner provides stabilization. In reality, only the $(\sigma\sigma^*)_T$ triplet prefers the loose geometry. The $S_1(\sigma\sigma^*)_S$ state has a minimum at $R = 1.3$ Å (even the purely "antibonding" σ^{*2} state has a minimum at $R = 2.3$ Å). This indicates the compromise between the tendency of the excited singlet to minimize the energy splitting between the σ and σ^* orbitals by increasing R, and thus minimizing the sum of energies of occupied orbitals, and at the same time to keep as tight a geometry as possible (small R) in order to minimize the electrostatic repulsion energy. In other words, T_1 (and S_0) dissociates to ground-state H atoms, but the two excited singlets behave as if they were to dissociate to H^+ and H^- and their energy increases as the bond length is stretched to infinity.*

Thus, in a dense medium, if a molecule returns to the S_0 surface via a biradicaloid minimum corresponding to a stretch in one of its bonds, it is likely to restart its life in S_0 with that particular bond stretched but not completely broken if it comes from S_1. On the other hand, if it returns from T_1, it will find itself at a geometry in which the atoms of the bond are pulled apart

---

* In reality, the picture is complicated by the existence of Rydberg states, but the basic argument remains unchanged.

completely, limited only by the solvent cage, if there is one. Recombination is clearly more likely in the former case.

Geometries of the kind described above, or "broken σ-bond minima," are reached during reactions such as photodissociation and photoabstraction. Indeed, in these reactions, singlet excitation often leads to quenching and regeneration of the starting material where triplet excitation leads to reaction.[16,17] Other important examples of biradicaloid minima are the carbene minimum and the "twisted π-bond" minimum, important in cis-trans isomerization, without much choice between "loose" and "tight" geometries, and, perhaps most important, the pericyclic biradicaloid minimum at "tight" geometries, and its "loose" counterpart, the open-chain biradicaloid minimum (Fig. 8-3).

Since from now on we shall be concerned with the last named two kinds of minima, they will be now discussed at some length. Pericyclic reactions are defined by Woodward and Hoffmann as those in which all first-order changes in bonding relationships take place in concert on a closed curve.[2] Best known examples are concerted electrocyclic, sigmatropic, and cheletropic reactions, cycloadditions, and cycloreversions. Those concerted reaction paths which

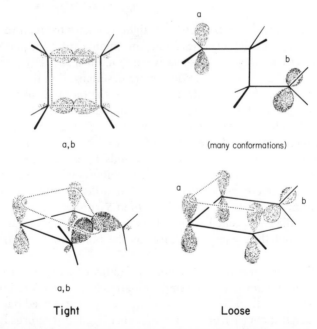

Tight   **Loose**

Fig. 8-3. Localized representation of the two non-bonding orbitals a and b encountered along the reaction path of addition of two ethylenes (top) and ring opening of cyclobutene (bottom). On the left, the tight biradicaloid geometries of the concerted reaction path (a, b in the same region of space, constructed from the same set of atomic orbitals) On the right, the loose ones of the nonconcerted reaction path (a, b separated in space).

proceed via a biradicaloid geometry are usually called (ground-state) symmetry-forbidden, those which do not, symmetry-allowed. Rules for distinguishing the two types have been discussed by many authors from various points of view.[2,3,6,18,19] In the case of forbidden reactions, there is a point along the reaction pathway at which two nonbonding orbitals are present as a result of an orbital crossing ("antiaromatic geometry"[18]). Unlike the two nonbonding orbitals of a "broken σ-bond" biradicaloid, they are delocalized through a cyclic array of orbitals, both in the same region of space (tight geometry).

The existence of a barrier in S_0 along the concerted pathway for a forbidden reaction has been recognized since the first formulation of the rules, the existence of a minimum in S_1 at the biradicaloid geometry of the activated complex has been apparently first recognized by Zimmerman[6] and by Dauben et al.,[20] investigated by van der Lugt and Oosterhoff,[8] and discussed by Dougherty.[7]

The pericyclic biradicaloid is not the only possible biradicaloid one needs to consider for a given pericyclic reaction. Any isomeric biradicaloid with the same number of bonds or bonding electrons is likely to have similar energy in its S_0 state—they all differ from the reactants or the products by the loss of one bond. Thus an obvious candidate is a loose geometry, or open-chain form of this biradicaloid: Instead of allowing the newly freed atomic orbitals obtained upon breaking a σ bond to interact with the rest of the molecule in a cyclic array of orbitals, one can twist the molecule so as to inhibit any such cyclic interaction. Then, one is either left with the original two nonbonding atomic orbitals, or perhaps one or both of them extend over larger portions of the molecule, but are still separated and not conjugated with each other. This biradicaloid is recognized as the often-postulated intermediate in the nonconcerted stepwise execution of the same reaction, and we shall refer to it as the open-chain biradicaloid. Examples are given in Fig. 8-3, which shows the tight pericyclic biradicaloids and the loose open-chain biradicaloids located along the paths of addition of two ethylene molecules and of ring opening of cyclobutene.

In accordance with the preceding arguments, one would expect the S_0 and T_1 surfaces of biradicaloids to have a minimum at the loose geometry, S_1 at the tight geometry. This is in good agreement with the general tendency of pericyclic reactions which are ground-state forbidden to proceed in stereospecific concerted manner in the excited singlet rather than stepwise (the excited molecule relaxes rapidly into the funnel in S_1 at the tight geometry). It also agrees with the general preference of triplet-state reactions for the nonconcerted path (the excited molecule relaxes into the open-chain minimum in T_1, which permits bond rotation, more of which may follow[21] after crossing to S_0). This preference cannot be explained simply by pointing out the need for spin inversion during the overall reaction and the resulting need for an intermediate: In principle, this intermediate biradical could just as well have tight concerted geometry with a cyclic array of orbitals in conjugation which does not permit loss of stereochemical information by bond rotation. Spin inversion would then lead to products directly, the reaction would follow the same pathway as the corresponding singlet reaction and would proceed with synchronous bond formation. This possibility is actually seriously considered by some authors (for references see Ref. 4), but we believe it is the exception rather than the rule. Thus, in order to account for the nonsynchronous bond formation in triplet reactions one needs more than to point out the need for an intermediate—it is important to understand why it adopts one or another geometry.

A few examples* of expected, and observed, processes follow (for references see Ref. 11):

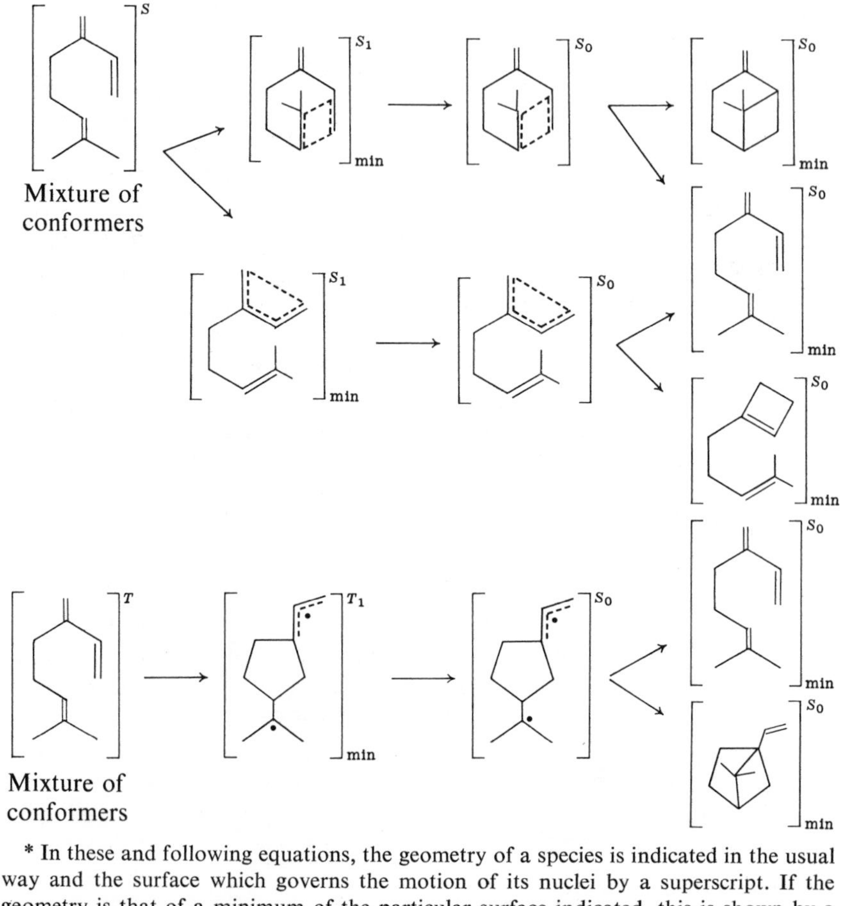

Mixture of conformers

Mixture of conformers

---

* In these and following equations, the geometry of a species is indicated in the usual way and the surface which governs the motion of its nuclei by a superscript. If the geometry is that of a minimum of the particular surface indicated, this is shown by a subscript. For an arbitrary excited singlet (triplet), the superscript is simply S (T).

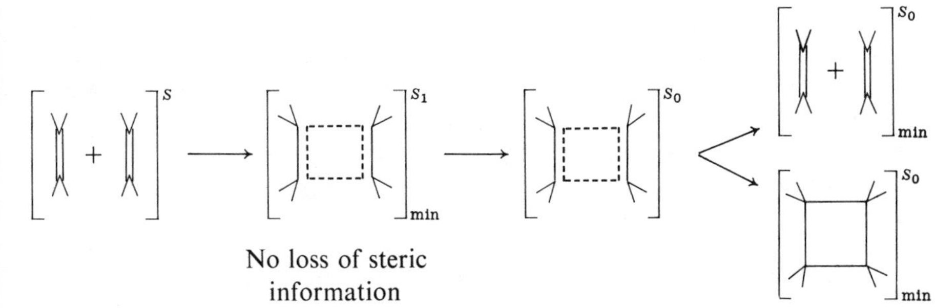

No loss of steric
information

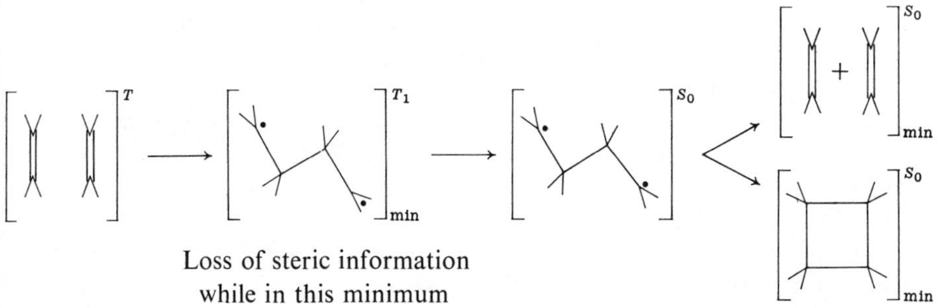

Loss of steric information
while in this minimum

Mixture of
stereoisomers

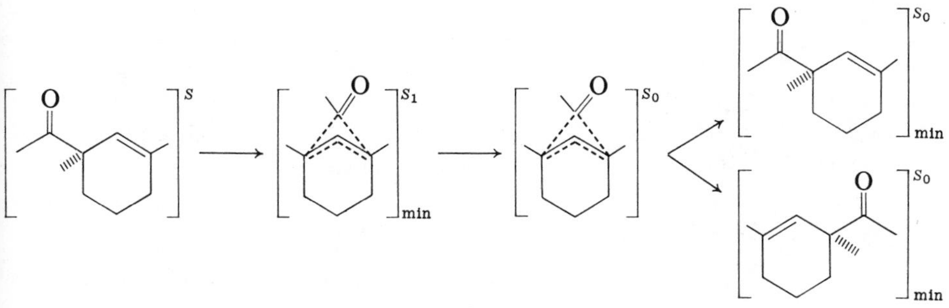

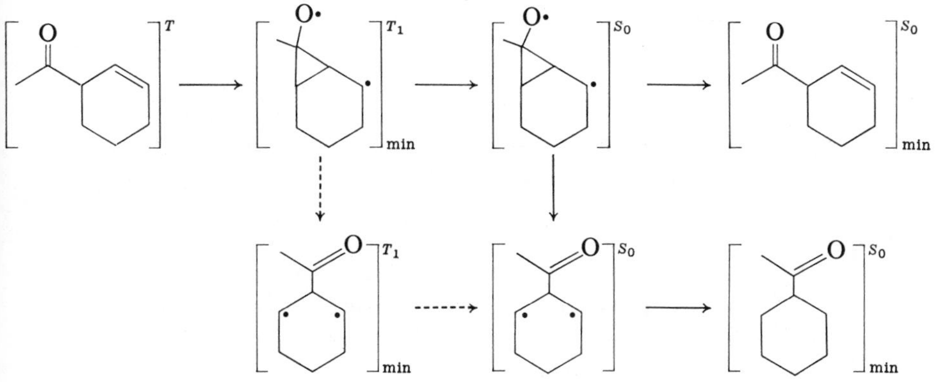

It is perhaps worth discussing one of these examples in more detail, say the last one (cf. Ref. 3). In an excited singlet, a β, γ-unsaturated ketone has a nearby funnel in S_1 corresponding to the "antiaromatic" geometry indicated in the second formula. Through this, the ground-state energy surface S_0 is reached and relaxation follows either back to the starting material or to the rearranged product. In the excited singlet, the loose geometries indicated for the triplet would be uphill from the tight "antiaromatic" one. The opposite is expected to be true for the T_1 surface. Here, the two open-chain minima indicated come to mind. It is not clear how much of a barrier separates them and if they are both reached (dotted arrow). Eventual intersystem crossing will bring the molecule to S_0 at a geometry of an open-chain biradical. The usual ground-state reactions to be expected afterwards are indicated and lead back to the starting material or on to the observed product. In order to estimate a position of a minimum in T_1, we use such simple-minded arguments as "a C—C π bond is weaker than a C—C σ bond," "delocalization of the odd electron is favorable as long as spatial separation from the other odd electron is preserved," and the like.

The model still allows for the possibility that some pericyclic reactions proceed in a concerted manner in the triplet state. While it has been pointed out that tight geometries are generally worse than loose geometries in the T_1 state, it is still possible that there is no minimum along the nonconcerted reaction path except at reactants or products. This is particularly likely if the reactant or product triplet has large delocalization energy (aromatic molecules). In such cases, we would expect the reactant or product geometry, respectively, to be reached in the T_1 state after excitation into the triplet manifold. If the S_0-T_1 separation is large and the experimental conditions suitable, it should be possible to observe phosphorescence; for example,

$$\left[\begin{array}{c}\text{◯▭}\end{array}\right]^T_{\min?} \longrightarrow \left[\begin{array}{c}\text{◯◯}\end{array}\right]^{T_1}_{\min} \longrightarrow \left[\begin{array}{c}\text{◯◯}\end{array}\right]^{S_0}_{\min} \quad (+ h\nu)$$

$$(1)$$

A nonconcerted version of this reaction would proceed in "semirotatory" manner and give ground-state naphthalene, as shown at top of page 313. However, it seems highly likely that the naphthalene triplet with one C—H bond twisted perpendicular to the molecular plane is uphill from the usual naphthalene triplet and does not correspond to a minimum, so that the concerted process (1) occurs. It would be interesting to check experimentally whether naphthalene is formed in its triplet state. On the other hand, it is

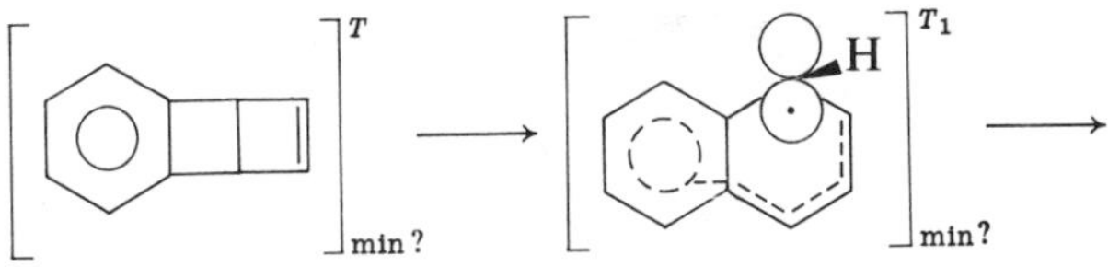

H out of plane

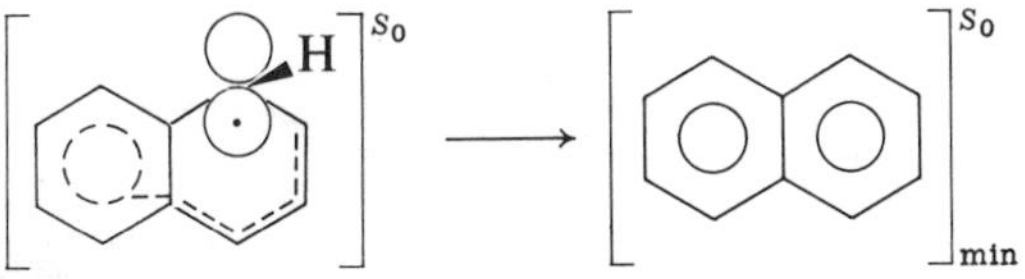

(2)

likely that the opening of cyclobutene in T_1 state proceeds in the non-stereospecific nonconcerted semirotatory manner:

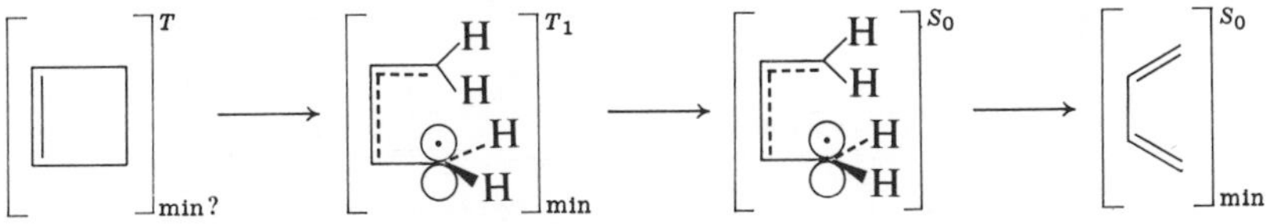

One CH_2 group twisted

III. CORRELATION DIAGRAMS AND ENERGY BARRIERS

Having pointed out some of the locations at which one is likely to encounter minima in the S_1 and T_1 surfaces, we now need to estimate whether they can be reached, given a starting geometry and excitation energy. It is in this area that the Woodward-Hoffmann rules and correlation diagrams come into their own. As far as photochemical reactivity is concerned, inaccessible minima might as well not exist. Therefore, we consider ill-advised the attempts[7,8] to dismiss the rules and correlation diagrams on which they are based as meaningless or irrelevant. The argument that the eventual conversion to S_0 proceeds through a funnel due to a state which correlates with a doubly excited state of the starting molecule rather than its lowest excited state does not show why barriers along the way should be unimportant. Such a barrier in S_1 or T_1 will result if, going along the reaction coordinate, the lowest excited state of the reactant does not correlate with one of the lowest excited states, but instead with a very highly excited state of the product (Fig. 8-4). This is entirely analogous to the barrier occurring in ground-state hypersurfaces. The existence of such barriers and their relation to correlation diagrams has been discussed by many authors (for example, Refs. 2, 14, 24–27).

STATE ENERGIES

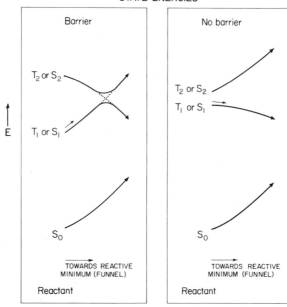

Fig. 8-4. Correlation diagram of states (schematic). Left, the $\psi_1 \to \psi_2$ configuration predominates in one of the higher excited states of the reactant. Right, it predominates in the lowest excited state.

There are undoubtedly also barriers other than those due to the course of orbital correlations, for example those due to steric hindrance, but we shall limit ourselves to the former in the following. Since the amount of energy available to the molecule will determine which barriers are unsurmountable, phenomena such as wavelength dependence or two-photon photochemistry should be particularly suited for discussion in terms of correlation diagrams and this is how the author originally became interested in the problem[14,22-24].

A. Pericyclic Reactions

Along the ground-state allowed pathways of these concerted reactions barriers generally occur in S_1 as well as T_1, since the $1 \to -1$ excited configuration of the reactant correlates with a more highly (but still singly) excited state of the product and vice versa.[2,27] Besides, these paths in general have no minima in either S_1 or T_1 along the way[8]; the only ones are expected to occur near the nuclear geometries of the reactant and the product. Thus, even if extra energy is made available, other processes involving more accessible minima will most likely occur preferentially and reactions along concerted ground-state Woodward-Hoffmann allowed pathways are seldom seen in photo-chemistry.

On the other hand, along the ground-state forbidden pathway of a peri-cyclic reaction one can generally expect a minimum in the S_1 surface [6-8] and we now need to determine whether it is separated from the starting geometry by a barrier in S_1. In the neighborhood of the minimum in S_1 there is a maximum in S_0 since in this region the molecule goes through biradicaloid geometry and effectively loses one bond. In the MO picture (Fig. 8-5), the ground-state barriers are due to the loss of one occupied bonding orbital, and an empty antibonding orbital, and their replacement by a pair of approxi-mately nonbonding orbitals half-way through the reaction. Excited states of the reactant represented by configurations in which the former is still full and the latter empty will have just as much reason to exhibit an increasing slope (they will correlate with triply excited states of the product if the ground state correlates with a doubly excited state). Those in which the former only has one electron and the latter is empty as well as the ones in which the former is full and the latter has one electron can be expected to still have roughly half of the ground-state slope (they will correlate with doubly excited states of the product). Only a state in whose predominant configuration an electron has been removed from the former and added to the latter ("characteristic con-figuration") should correlate with a singly excited state of the product. For this "characteristic" state, there is no more reason to expect a barrier. If it

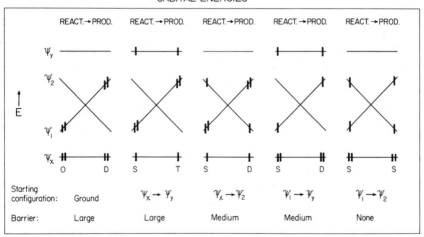

ORBITAL ENERGIES

Fig. 8-5. Correlations between the ground (0) and the excited (S, singly; D, doubly; T, triply) configurations along a typical ground-state forbidden reaction coordinate. Size of expected symmetry-imposed barrier is indicated.

happens to be the lowest excited state of the reactant, the photochemical reaction from S_1 or T_1 will have no "symmetry-imposed" barrier. If it is one of the higher states, a barrier will be present in S_1 (T_1) and its height will depend on just how high the required excited state is and how fast it comes down along the reaction coordinate (Fig. 8-4). If the characteristic configuration contributes equally to several different excited states, the qualitative argument breaks down and an actual calculation of the surfaces would be required.

Thus, the first part of the problem of constructing a correlation diagram is to identify orbitals mostly responsible for the ground-state energy barrier. In particular, these are the bonding orbital of the reactant, say ψ_1, which is increasing in energy and becoming antibonding (or at least nonbonding) in the product (the "critical orbital" of Buenker et al.[28]), and the antibonding orbital of the reactant, say ψ_2, which is decreasing in energy and becoming bonding (or nonbonding) in the product. The second part is to identify the "characteristic" excited state of the reactant as the one which has a large contribution from the $\psi_1 \to \psi_2$ singly excited "characteristic" configuration. The higher this state is above the lowest excited state, the higher the barrier to be expected. It is worth noting that a singlet double excitation $\psi_1{}^2 \to \psi_2{}^2$ would also suffice, but this configuration is usually rather high in energy in the reactant. However, it is the one which is responsible for the existence of the funnel in S_1 surface, which the molecule reaches in the course of singlet pericyclic reactions.[6-8]

In many cases, enough is known from experiments and calculations about the order of states in reactants and products to estimate how high the "characteristic" state (S or T) lies in each. Very often, ψ_1 is the highest occupied and ψ_2 the lowest free orbital of the reactant and they cross during the reaction, so that ψ_1 becomes the lowest free and ψ_2 the highest occupied orbital of the product[2,27] (the crossing may be avoided in systems of low symmetry). These cases can be straightforwardly treated by the frontier orbital theory.[3] This kind of crossing will be referred to as "normal orbital crossover"[14]; the $1 \to -1$ state of the reactant correlates with the $1 \to -1$ state of the product both in singlet and triplet. Often, for instance in polyenes, these are just the lowest excited states (both S and T) of the reactant and product and no barriers due to correlation are imposed starting on either side. Sometimes, for instance in the case of naphthalene, benzene, and their derivatives, the $1 \to -1$ singlet of reactant or product is not its lowest excited singlet and a barrier is expected. In these cases, a consideration of states rather than configurations alone is essential, since a state represented by a mixture of $1 \to -2$ and $2 \to -1$ configurations (^1L_b) is below the "characteristic" $1 \to -1$ (^1L_a) state. The difference in their energies is usually quite small and one would expect the resulting barrier to be small. This allows prediction of temperature- and/or wavelength-dependent quantum yields for photochemical reactions and emissions.[29] A good example of the former effect will be discussed at the end of this chapter. The latter effect has been observed in dense media in benzene derivatives.[22,30] In the triplet state, no barriers of this origin are expected in photochemistry of benzene or naphthalene derivatives since the $1 \rightarrow -1$ triplet is lowest even after allowance for configuration interaction.

The fact that for a particular reaction coordinate barriers are most often either absent in both S_1 and T_1, or present in both S_1 and T_1, does not imply that the products of singlet and triplet photochemistry will be the same, since that depends on the location of minima which is often different in the S_1 and T_1 surfaces. Woodward-Hoffmann allowedness in the excited state along a particular reaction coordinate is a necessary but not sufficient condition for the molecule to actually give photochemical products by following this coordinate, and if sufficient extra energy is available it need not even be a necessary condition. Even if a molecule in its T_1 state should start out on a reaction coordinate corresponding to a concerted cycloaddition reaction, which is energetically feasible, it will not stay there long and will fall into the nonconcerted biradical minimum in T_1 located nearby (due to the slope of the T_1 surface it may more likely head for it directly from the beginning).

Barriers larger than those discussed so far can be expected in cases of "abnormal orbital crossover,"[14] that is, those in which one of the crossing orbitals ψ_1 and ψ_2 is not a frontier orbital of the reactant (HOMO or LFMO).* In these cases, straightforward application of the frontier orbital theory to reactant or products or possibly both is incorrect. An example is shown in Fig. 8-6. According to simple PPP or HMO calculations,[31] the bonding orbital ψ_1 of the reactant I, which increases in energy during the concerted disrotatory opening of the four-membered ring and eventually becomes antibonding in the product II, turns out to be the bonding orbital of the ethylene chromophore, and not HOMO of naphthalene which is clearly higher in energy.[32] On the other hand, the orbital ψ_2 into which the electron needs to be placed is the LFMO of naphthalene; the even higher located LFMO of ethylene would also do since it behaves similarly. The "characteristic" singlet and triplet states of the reactant thus are highly energetic charge-transfer states in which an electron has been transferred from the ethylene chromophore to the naphthalene chromophore (or ethylene $\pi\pi^*$ states which are probably comparable in energy). The same situation has been reported[14] for the ring-opening of the benzologue III to pleiadene IV, where the "characteristic"

---

* However, the barrier found in the photochemical cycloreversion of heptacyclene to two acenaphthylene molecules is rather small. It was rationalized by Chu and Kearns[26] as due to an "abnormal" orbital crossover.

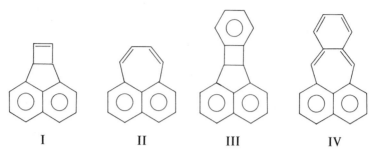

I II III IV

states are charge-transfer states in which an electron is transferred from benzene to naphthalene chromophore. In both I and III, a high-energy barrier to ring opening is expected to result from the abnormal orbital crossover. This should prevent the molecules from reaching the pericyclic minimum for ring opening in their S_1 energy surface—the singlet states of both have indeed been reported to be unreactive.[23,33] Assuming the lowest triplet states of the products to be planar like triplet states of other large

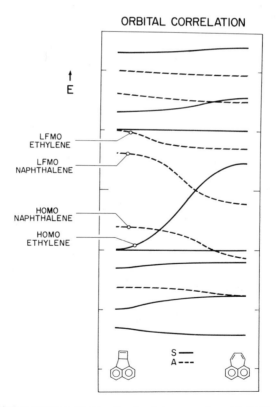

Fig. 8-6. Orbital correlation diagram for disrotatory opening of I. [31]

aromatic molecules (and unlike triplet states of unsaturated molecules with terminal methylene groups), and considering geometrical constraints imposed on the molecules, we conclude that reaction in triplet state, in which the molecule would not go through a pericyclic funnel but should reach S_0 surface by efficient intersystem crossing (small energy gap) after reaching the lowest triplet state of product, is equally forbidden. Indeed the lowest triplets of both reactants, I and III, are unreactive. In both cases, extra energy provided by a second photon takes the molecule over the barrier and leads to product formation.[23,33]

Using similar logic based on orbital correlation diagrams obtained from repeated approximate calculations (PPP and HMO) for points along the reaction coordinate, one can obtain predictions for new molecules.[24] For example, opening of the four-membered ring in V should not require correlation-imposed activation energy (Fig. 8-7), while in the azulene derivative VI it should (the $\psi_1 \rightarrow \psi_2$ state is again of charge-transfer nature).

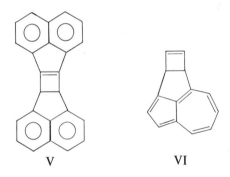

V   VI

It should perhaps be remarked that in cases of abnormal crossing in which reaction is induced by supplying additional energy so that the barrier can be overcome, the initial excitation need not be into the high-energy "characteristic" state. All that is needed is to supply enough energy for the motion of nuclei toward the barrier that they can pass above it. Thus, in cases of reactions such as I → II and III → IV, we do not claim that it is actually necessary to first effect a charge transfer before the reaction can proceed. Actually, experimentally, the state from which the reaction starts is one of the locally excited higher triplets of naphthalene. It seems fruitless to speculate at present about the detailed course of the conversion of electronic energy into the kinetic energy of nuclei which follows.

B. Other Reactions

Most known photochemical processes are not pericyclic reactions and this has led to skepticism concerning the relevance of Woodward-Hoffmann rules to photochemical problems on the part of some investigators. Yet, many

ORBITAL CORRELATION

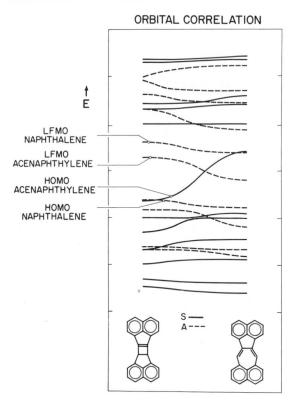

Fig. 8-7. Orbital correlation diagram for disrotatory opening of V. [31]

photochemists have been drawing potential energy curves for excited states very closely related in spirit to those which Woodward and Hoffmann drew for pericyclic reactions, albeit sometimes without specific reference to molecular symmetry, orbitals, and so on (e.g., see Refs. 17 and 25); similar barriers appear in elaborate calculations.[34] The spirit of the argument remains the same, but it is usually even harder to estimate magnitudes of the barriers. As before, one inquires about the presence of a barrier between the starting geometry and the geometry of the minimum or funnel through which return to S_0 is to occur (depending on multiplicity, a loose or tight geometry may be favored). As before, it is necessary to identify the singlet or triplet "characteristic" state which correlates smoothly with a low-lying excited state at the biradicaloid geometry. To do this, we again need to identify the orbitals ψ_1 and ψ_2. For instance, for the cis-trans isomerization of ethylene, the "characteristic" configuration $\psi_1 \to \psi_2$ is the $\pi \to \pi^*$ state, and for dissociation of a singly bonded diatomic molecule, it is the $\sigma \to \sigma^*$ state. In a large molecule it may be difficult to identify any one excited state as the characteristic state: for example, many of the $\sigma \rightarrow \sigma^*$ excited states of toluene will have some contribution from the $\sigma \rightarrow \sigma^*$ bond orbital excitation in say one of the three C—H bonds in the methyl group. However, the characteristic state can still be said to have $\sigma \rightarrow \sigma^*$ rather than $\pi \rightarrow \pi^*$ character. Similarly, the characteristic state for α-cleavage in a ketone triplet is of $\sigma \rightarrow \sigma^*$ nature, not $n \rightarrow \pi^*$. For additional examples of such arguments, see Ref. 25.

In both of the above examples, the characteristic configuration, although it cannot be clearly identified in terms of delocalized molecular orbitals, is not important in the lowest excited state of the molecule at the starting geometry. Since in both cases the lowest excited configurations ($\pi\pi^*$ and $n\pi^*$, respectively) can be expected to go up in energy as the σ bond is stretched, rather than down (or at least remaining constant), a barrier would be expected (Fig. 8-4). The question now is, how high will it be? If there were no interaction between the orbitals ψ_1 and ψ_2 of the characteristic configuration (in the first approximation, σ and σ^* bond orbitals of the bond to be broken) and the orbitals participating in the excitation involved in the lowest excited state ($\pi\pi^*$ or $n\pi^*$ in our examples), it should be high indeed—the crossing between the rising lowest excited state and descending "characteristic" state would not even be avoided. This would be the situation if we inspected the reaction coordinate say for the dissociation of one of the aromatic C—H bonds in toluene or a C—C bond distant from the carbonyl in our ketone. Such reaction coordinates undoubtedly lead to local minima in the S_1 and T_1 hypersurface, but these are separated by high barriers from the starting geometry. There is indeed no experimental evidence for such dissociations upon excitation into the lowest excited states. On the other hand, during cleavage of one of the methyl C—H bonds in toluene and of the C—CO bond in ketones, the characteristic orbitals ψ_1 and ψ_2 will interact with the orbitals participating in the lowest excited state, the crossing will be avoided, and the barrier reduced. It is unfortunately not easy to estimate the amount of the interaction without a calculation. Recent *ab initio* results show that the barrier is largely removed in the dissociation of CH_2O into CHO and H in the triplet state.[34] Experimentally, the barriers for α cleavage of ketones are small and can be overcome using thermal energy.[17] On the other hand, the barriers in the S_1 and T_1 surfaces of toluene are higher. The former can be overcome if short-wavelength radiation is used, the latter by absorption of a second photon by the triplet state.[35]

Some qualitative arguments can nevertheless be made and a few examples follow. It should be possible to relate the height of barriers for dissociation of methylanthracenes into anthrylmethyl radicals with the MO coefficients of the HOMO and LFMO of anthracene, and one can argue that ketones with lowest triplets of $\pi\pi^*$ type should have higher barriers for the α-cleavage reaction than those with $n\pi^*$ lowest triplets. Similarly, the opening of benzo-cyclobutene to o-xylylene in the triplet state will be expected to have a barrier basically similar to that for dissociation of ethylbenzene triplet to benzyl and methyl radicals, only lowered by relief of steric strain. Additional substitution can be expected to help to reduce the barrier: $\alpha,\alpha,\alpha',\alpha'$-tetraphenylbenzo-cyclobutene, analogous to 1,1,1,2,2-pentaphenylethane, is known to open in a sensitized photochemical reaction,[36] whereas low-temperature irradiation of benzocyclobutene itself under similar conditions results in no change.[37] In this manner, one can understand the seemingly contradictory tendency of photochemical reactions to give some of the least stable products imaginable, given the reactants, and yet to be often facilitated by substitution which stabilizes those products. The former is due to the fact that minima in S_1 and T_1 usually occur just where S_0 has "maxima," the latter to the fact that suitable substitution often helps to reduce barriers along the way, so that those minima in S_1 and T_1 can be reached. We may conclude this section by re-marking that photochemical reactions for which the "characteristic" state is not the lowest excited state are likely to involve barriers. Although it is very hard to predict how high these will be, qualitative comparisons between re-lated molecules should often be possible. Better answers require calculations. In this respect, the situation is not very different from that outlined for peri-cyclic reactions, as described so far, where calculations, albeit of a simpler kind, were also required to distinguish between "normal" and "abnormal" crossings. The connection between the two cases is actually more funda-mental, as will become clear in the following section.

IV. DERIVATION OF CORRELATION DIAGRAMS BY INSPECTION OF HMO ORBITALS

It would be rather desirable to be able to predict when orbital crossover for a pericyclic reaction will be "normal" and when "abnormal" without having to perform calculations. This can indeed be done, but it is first necessary to look at the Woodward-Hoffmann rules in somewhat unusual light.[24] We shall concentrate on the case of electrocyclic reactions, but the approach can be generalized to others and a case of a cycloreversion reaction is worked out at the end.

Figure 8-8 shows a derivation of the well-known orbital and state correla-tion diagrams [2,8,27] for a disrotatory and a conrotatory opening of cyclo-butene in two steps: first, on the left-hand side, breaking the σ-bond, but not allowing any interaction of the resulting two radicals with the rest of the molecule (the ethylene orbital energies thus remain unaffected by the reaction, the σ and σ^* orbital energies behave similarly as in dissociation of ethane into two methyl radicals). The exact way in which the σ-bond breaking is actually

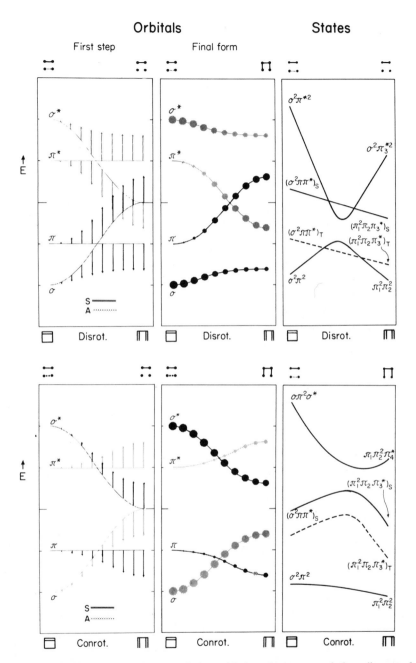

Orbitals

First step   **Final form**

States

Fig. 8-8. "Backwards" derivation of the orbital and state correlation diagram for concerted ring opening of cyclobutene. In symbols above the figures, dots indicate positions of the interacting orbitals, full lines represent π bonds, and dotted lines σ bonds. Details of the state correlation are compatible with approximate calculations. [8] Only a few crucial states are shown. For additional information see text.

performed has no important effect on the course of the correlation; Fig. 8-8 shows results of a very approximate calculation (HMO[31]) assuming that it occurs by simultaneous rotation of the two methylene groups, the original σ interaction of the two atomic orbitals forming the σ bond gradually decreasing, while the π interaction increases but is small and is neglected in the calculation as in butadiene itself. In the second step, the σ and σ^* orbitals are allowed to interact with the ethylene double bond; the interaction is still zero at the cyclobutene geometry when they are of wrong symmetry, but it gradually increases as the rotation increases the π component of the AOs from which the σ and σ^* orbitals are constructed. The calculated effect of the interaction is indicated by arrows and the resulting curves are shown to the right of the original ones. The size of the circles indicated on these curves shows the weight which the originally bonding σ (dark circles) or antibonding σ^* (light circles) combination of the two rotating AOs has in the individual MOs during the course of the reaction. It is easy to understand qualitatively the contents of this figure: from left to right the original σ bonding orbital of the single bond moves up in energy, its two AOs interacting less and less with each other as the disrotation proceeds; at the same time, as the π component of this originally σ bonding orbital increases, it interacts more and more with π orbitals of the rest of the molecule, but only with those which are reasonably close in energy, have the right nodal properties (symmetry), and have large coefficients at atoms adjacent to those of the original single bond. As a result, it "pushes up" those orbitals with which it interacts most strongly in those regions where it has energy comparable to theirs (only one such orbital in our case). In an entirely analogous fashion, the original σ^* antibonding orbital decreases in energy and "pushes down" those orbitals which have appropriate nodal structure. Again, only one is available since our original π system is ethylene and has a total of only two π orbitals. The bonding one, π, is without a node across the bond and is suitable for interaction with the newly generated π component of the original bonding σ combination if the opening is disrotatory and with the π component of the original antibonding σ^* combination if it is conrotatory. These are drawn in dark lines in Fig. 8-8. The reason why the nodal character of the newly formed π component of the original σ and σ^* combinations is just the opposite for disrotatory and conrotatory opening is clear from Fig. 8-9, which shows the σ and σ^* combination of the AOs on the now partially rotated methylene groups of the original σ bond, and also their projection into σ and π components. Similarly, since the ethylene π^* orbital has a node across the bond, its symmetry is suitable for interaction with the original σ^* combination if the opening is disrotatory and with the original σ orbital if it is conrotatory. These are drawn as light lines in Fig. 8-8. Accordingly, already a purely qualitative consideration of the nodal properties of the original σ and σ^* bond orbitals at the bond

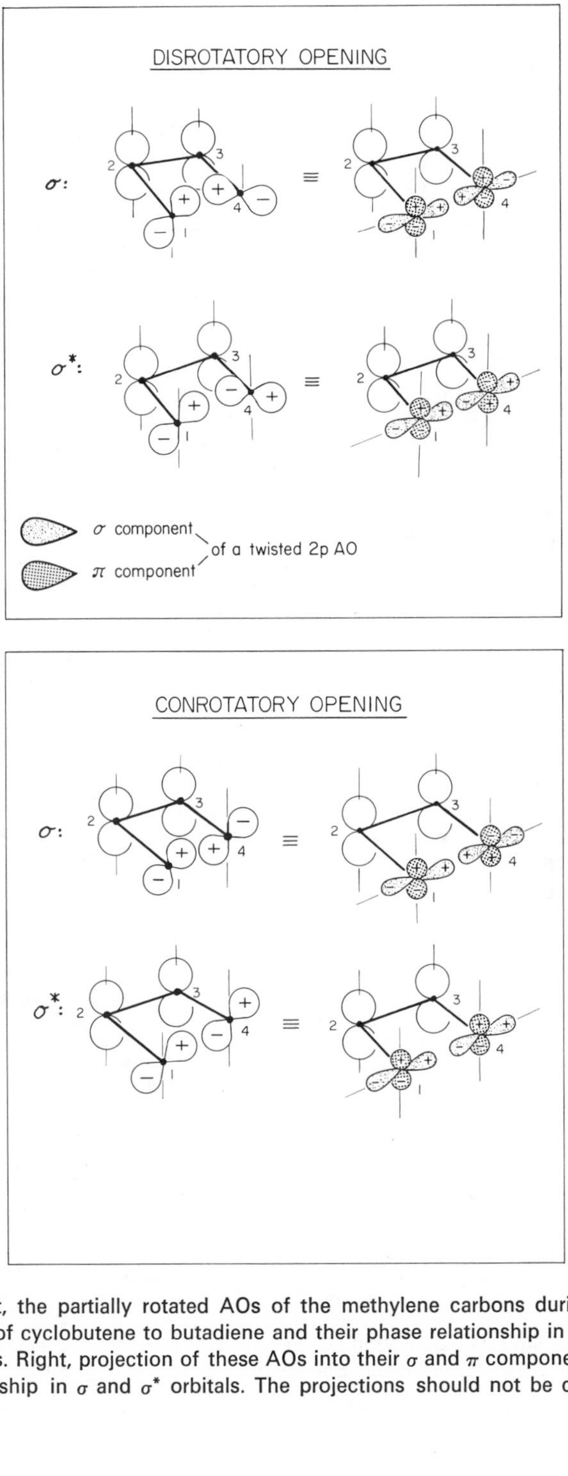

Fig. 8-9. Left, the partially rotated AOs of the methylene carbons during concerted ring opening of cyclobutene to butadiene and their phase relationship in the original σ and σ^* orbitals. Right, projection of these AOs into their σ and π components, and their phase relationship in σ and σ^* orbitals. The projections should not be confused with d orbitals.

which is being broken, and of nodal properties of orbitals of the π system at the sites at which it interacts with the newly freed AOs, allows the conclusion that ground-state opening of cyclobutene is allowed in the conrotatory mode, forbidden in the disrotatory mode. In this simple case (one each of σ, σ^*, π, π^* orbitals) our derivation of the correlation diagram is merely a translation into lines of the verbal description given in Fukui's frontier orbital approach.[3] Figure 8-10 shows a similar stepwise derivation of the orbital correlation diagram for disrotatory opening of 1,3-cyclohexadiene. Again, simple consideration of the nodal properties of the orbitals involved allows this diagram to be constructed without any calculations. Since the properties of the σ and σ^* orbitals are easily remembered once and for all, all that is needed is the form of HMO orbitals for the π electron system in the electrocyclic component with fewer π electrons. For this reason we call this a "backward" derivation,

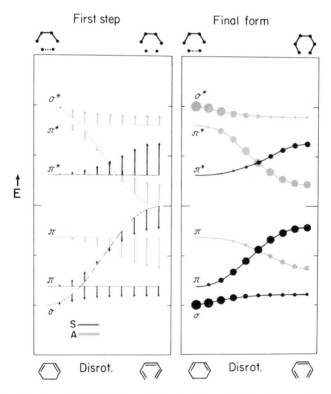

Fig. 8-10. "Backwards" derivation of the orbital correlation diagram for disrotatory ring opening of 1,3-cyclohexadiene. See text and caption to Fig. 8-8.

the normal way[2] being to look at the molecular orbitals of the species with more π electrons. One can ask the question[27] as to why the stereochemistry of the more conjugated product should be determined by nodal properties of orbitals in the product alone rather than those of the reactant; Fig. 8-8 shows that one can start from either side, the complete picture being, of course, best represented by the whole correlation diagram. The "backward" way has its pitfalls since in more complicated cases it is hard to estimate how much which curves will interact and "repel" when interaction is turned on. If the order of orbital energies for both reactant and product is established by calculation, the difficulty disappears. However, for molecules of sufficient symmetry, it is then just as easy to draw correlation diagrams using the noncrossing rule. For allowed reactions and those ground-state forbidden reactions involving a normal orbital crossover, identical results are readily obtained by Fukui's frontier orbital method.[3] This method will, however, fail if applied blindly to cases of abnormal orbital crossover, in which the initial orbitals ψ_1 and ψ_2 are not the frontier orbitals (HOMO and LFMO). We see the value of our "backward" method not so much in providing yet another derivation of Woodward-Hoffmann rules as in the insight it provides for a particular course the correlation of orbitals takes in the diagram (normal or abnormal) and for easy identification of ψ_1 and ψ_2.*

Using the above approach, we can find the "characteristic" orbitals ψ_1 and ψ_2, and thus the nature of orbital crossover, simply from knowledge of nodal properties and approximate size of coefficients of the molecular orbitals of the electrocyclic reactant with fewer π electrons (in our experience, HMO and PPP orbitals give identical answers). The least bonding MO in any of the chromophores present in the starting molecule which interacts strongly with the bonding combination (σ) of the AOs on the two carbons originally joined by a single bond will be ψ_1 and the least antibonding MO in any of the chromophores present which interacts strongly with the antibonding combination (σ^*) of the AOs on the same two carbons will be ψ_2. If ψ_1 is HOMO of the original molecule and ψ_2 is its LFMO, the orbital crossover is normal; otherwise it is abnormal. To determine which orbitals will interact strongly with the bonding combination of the AOs on the two carbons of the bond being broken, which has no nodal plane across this bond for a disrotatory opening but has such a nodal plane if the opening is conrotatory (Fig. 8-9), we need to inspect the nodal properties of these orbitals, which should match those of the bonding combination on the originally single bond, as well as the

---

* This is of no great consequence for thermal reactions for which one only needs to know whether the reaction coordinate involves an orbital crossover, i.e., a biradicaloid geometry, somewhere along the way. This is perhaps most easily derived from the rules of Dewar and Zimmerman, which avoid any construction of correlation diagrams altogether.[6,18]

approximate size of expansion coefficients in these MOs at atoms adjacent to those of the originally single bond. One proceeds similarly for the anti-bonding combination of the AOs on the bond being broken, remembering that it has opposite nodal properties compared with the bonding combination.

There is an unavoidable degree of arbitrariness in the demand for "large" coefficients on the appropriate atoms in the MOs under inspection. The smaller these coefficients, the "less avoided" will be the crossing with the σ bonding combination, the later in the reaction it will be that the interaction will start, and the more of a barrier one can expect.

It is now possible to take another look at the results for the disrotatory ring opening in I, V, and VI. The HMO coefficients in the pertinent positions of naphthalene are collected in Table 8-1; the coefficients in the positions 1 and 2 of acenaphthylene are 0.51 and 0.51 in HOMO and 0.32 and −0.32 in LFMO; the coefficients for azulene are given in Fig. 8-11; and the coefficients of ethylene, of course, are 0.71 and 0.71 in HOMO and 0.71 and −0.71 in LFMO. It is clear that the course of correlation diagrams in Figs. 8-6 and 8-7 could have been predicted from an inspection of HMOs. Figure 8-11 illus-

Table 8-1. HMO Coefficients of Naphthalene and Benzene

Naphthalene HMOs								
−3		−1.30β	0.17	0.17	0.17	−0.40	0.40	0.40
−2		−1.00β	−0.41	0.41	0.41	0.0	0.0	0.0
−1	(LFMO)	−0.62β	0.26	0.26	−0.26	0.43	0.43	−0.43
1	(HOMO)	0.62β	0.26	−0.26	0.26	0.43	0.43	−0.43
2		1.00β	0.41	0.41	0.41	0.0	0.0	0.0
3		1.30β	−0.17	0.17	0.17	0.40	0.40	0.40

Benzene HMOs			
−2	−1.0β	0.29	0.29
−1	−1.0β	0.50	−0.50
1	1.0β	0.50	0.50
2	1.0β	0.29	−0.29

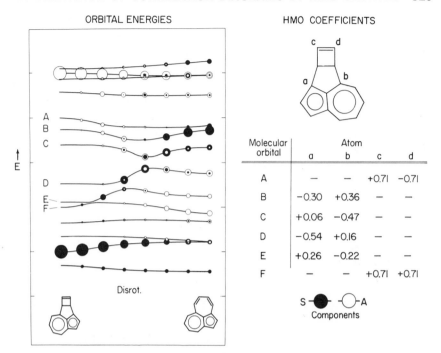

Fig. 8-11. Relation between the course of orbital correlation during the disrotatory opening of VI and the HMO coefficients at the crucial atoms. See text.

trates this for the case of VI. It shows the signs and magnitudes of coefficients at the crucial atoms in azulene and ethylene. Due to the lack of suitable symmetry, a symmetric and an antisymmetric component can be found for each of the azulene orbitals. The top two bonding orbitals in azulene are clearly unsuitable for strong interaction, so the ethylene π orbital is ψ_1. The second lowest antibonding orbital of azulene is ψ_2 but the π^* orbital of ethylene would perform even better. Thus, as noted previously, a charge-transfer state or ethylene π-π^* state is the "characteristic" state, depending on which is lower, and need for activation energy is predicted. An actual HMO calculation gives the orbital correlation diagram, also shown in Fig. 8-11, in good agreement with expectations.

The role of the π-electron system is clear from preceding examples and the course of correlations in Figs. 8-8 to 8-10, and is seen to be closely related to our previous discussion of photodissociation reactions. In the absence of assistance by the π system, the states which correlate smoothly with low-lying excited states of products involves highly energetic $\sigma \rightarrow \sigma^*$ excitation, whereas if the π system assists, only a less energetic $\pi \rightarrow \pi^*$ excitation is needed. An energy barrier along the reaction coordinate will result if the excitation needed does not correspond to the lowest excited state of the initial molecule; this is much more likely in the former case than the latter. Such assistance is also important in ground-state reactions, but in a different way: It removes the biradicaloid character of the broken bond at all points along the "ground-state allowed" electrocyclic reaction coordinate (see Fig. 8-8, conrotatory opening of cyclobutene) but not along the "ground-state forbidden" one (Fig. 8-8, disrotatory opening of cyclobutene).

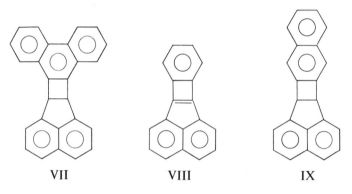

Knowledge of HMO orbitals of simple chromophores, combined with knowledge of the nature of their lowest excited singlet and triplet states in terms of configurations, allows numerous easy predictions. For triplet reactions, one also needs to estimate whether the reaction will proceed through a biradicaloid minimum at loose geometry or all the way to the excited triplet of the product [cf. processes (1) and (2) on pp. 312, 313]. For example, in the phenanthrene derivative VII ψ_1 is the HOMO of phenanthrene (coefficients in positions 9 and 10 are both 0.42), and ψ_2 is the LFMO of phenanthrene (coefficients in 9 and 10 are 0.42 and -0.42). The lowest triplet is HOMO $\rightarrow$ LFMO (^3L_a) and no symmetry-imposed barrier would be expected (except that the MO coefficients are somewhat too small for efficient interaction, but at present only an experiment can show how serious this is). The singlet HOMO $\rightarrow$ LFMO state (^1L_a) is only second in energy, but is not very high above the ^1L_b singlet, so that a small barrier is expected for the singlet reaction.

This result suggests that there may be a simpler way of predicting the presence or absence of barriers in the ring-opening of molecules such as I and III, and that it is perhaps not at all necessary to derive any correlation diagrams: Perhaps all that is necessary for an easy reaction is to localize the lowest excited state in the cyclobutene rather than the naphthalene portion of the molecule. After the publication of our first experimental results on the two-photon ring-opening III → IV,[23] this was indeed suggested by other authors.[7,18] They pointed out that during the reaction, the molecule is isoconjugate with the cyclobutadiene derivative VIII, in which the naphthalene moiety is merely an inconsequential appendage on the antiaromatic benzocyclobutadiene system. The main objection to this rationalization is based on experimental data of which the above authors were unaware.[38] The reaction III → IV occurs upon excitation to a state in which the naphthalene moiety is excited to a higher triplet, but does not occur upon direct excitation of the benzene chromophore to its first excited singlet. Also, IX behaves just like III and requires two photons although the excited states of the two naphthalene chromophores are very much alike, both are excited during broad-band irradiation, and it is indeed hard to speak of strictly localized excitation. Both examples show that the "localization" of excitation does not simply determine reactivity by itself.

The approach described here also permits other interesting predictions. Inspection of the molecular orbitals of naphthalene and benzene (Table 8-1) shows that 1,4-Dewar anthracene X should have a barrier for opening to

$$\text{X} \qquad\qquad \text{XI}$$

anthracene in S_1 and T_1, since the "characteristic" configuration is $2 \to -2$, corresponding to the high-energy B_a band of naphthalene. On the other hand, the 9,10 isomer XI has no barrier imposed by orbital correlation in its T_1 state and only a small barrier is expected in the S_1 state. To see this, it is perhaps appropriate to discuss the role of the benzene chromophore in some detail. Because of orbital degeneracy, benzene (Table 8-1) has two HOMOs, one with and one without a node across the bond between the atoms of interest (the two which form part of the four-membered ring), the latter with larger coefficients. The same holds for the two LFMOs but now the orbital with a node has larger coefficients. Thus, one or the other of the two HOMOs must be ψ_1 and will be labeled 1, while one or the other of the LFMOs must be ψ_2 (labeled -1). Then $1 \to -1$ will be the label of the characteristic configuration, whether the opening is disrotatory or conrotatory, but it should decrease in energy faster in the former case (larger coefficients at the crucial atoms). It is reasonable to label the orbitals as stated since as soon as the ring opening is in progress, the "characteristic" orbital ψ_1 goes up and ψ_2 down in energy so that they will indeed become the HOMO and LFMO of the system, respectively, and the other two will then be appropriately labeled 2 and -2.

The numbering $1, -1, 2, -2$ used in Table 8-1 refers to disrotatory opening.* The $1 \to -1$ (and $2 \to -2$) transitions are polarized horizontally in the drawing in Table 8-1 and correspond to the L_a (and B_a) bands of disubstituted benzene. The ^3L_a state is the lowest triplet and therefore no barrier is imposed by state correlation along the concerted pathway. This is ordinarily unimportant since the nonconcerted semirotatory pathway will be followed as discussed earlier. In a case like XI, however, the concerted pathway is of interest, since we would expect excited triplet anthracene to be the product.

On the other hand, the ^1L_a state is not the lowest singlet. Transition into the lowest singlet (^1L_b) is polarized vertically in the drawing in Table 8-1, corresponds to a mixture of $1 \to -2$ and $2 \to -1$ excitations, and thus correlates with a highly excited state of products. Since the ^1L_a state is about 10,000 cm^{-1} higher in energy in benzene itself, and coming down along the reaction coordinate, a small barrier is to be expected on the disrotatory concerted path toward the biradicaloid pericyclic minimum.

Photochemical electrocyclic reactions in the opposite sense, from the more conjugated species to the one with fewer π electrons and one more σ bond, may be hard to effect for many of the molecules discussed here simply because of the energetics. For example, the first absorption band of IV lies in the infrared region[39] and undoubtedly provides an efficient return path to S_0. In other cases, the first excitation energy may be sufficient to make the overall process exothermic. Inspection of the correlation diagrams or of nodal properties of the MOs (e.g., Fig. 8-6) shows that the orbital crossover for this reverse reaction is mostly normal, that is, a HOMO $\to$ LFMO excitation is required. However, if the orbital crossover is abnormal in the usual (ring-opening) reaction sense, the HOMO $\to$ LFMO state of the open form correlates with one of the higher singly excited states of the closed form, rather than with its lowest excited state. Depending on just how high the corresponding excited state lies, a smaller or larger barrier can be expected even for the ring-closing reaction. It should be generally lower than that for the ring-opening reaction, first because the uphill correlation of the HOMO $\to$ LFMO state is now only with a higher singly excited state rather than a doubly or even triply excited state; and second, because it now takes less motion along the reaction coordinate to reach the minimum in S_1 at the biradicaloid geometry (HOMO and LFMO) of the more conjugated open form are usually already less bonding and antibonding, respectively, than those of the closed form). In the triplet state, we would not expect the reverse ring-closing reaction to proceed at all since T_1 has a minimum at the geometry of the open species, say II or IV. In species in which T_1 has a minimum at an open-chain nonconcerted geometry, other reactions may proceed via that minimum.

* During conrotatory opening, the energies of orbitals ψ_1 and ψ_2 change as indicated, but not enough to actually cross. Therefore, labels ψ_1 and ψ_2 are not really appropriate.

Other kinds of pericyclic reactions can be discussed in similar ways as electrocyclic reactions and we shall show this on the example of the cyclo-reversion reaction of XII to give phenanthrene and acetylene. We have recently studied this reaction experimentally.[40] The experimental data are most easily interpreted as implying a barrier in S_1 which can be overcome using shorter irradiation wavelengths and a barrier in T_1 which can be over-come by absorption of a second photon. It has been pointed out that an opening of the cyclobutene ring in XII to butadiene involves an "abnormal" orbital crossing and thus should require extra energy.[24] This is seen from the signs of the frontier orbitals of biphenyl on atoms 2 and 2′, which are 0.30,

XII

-0.30 in HOMO and 0.30, 0.30 in LFMO—the nodal properties are just the opposite of those required. This reaction has not been detected.

It is not known experimentally whether the fragmentation of XII is con-certed. Following our previous discussion, we shall assume that it is concerted in the singlet state, and proceeds through the usual pericyclic minimum (anti-aromatic geometry), and that it is stepwise in the triplet state, proceeding through an open-chain biradicaloid minimum. Figure 8-12 shows a sketch of the state energy diagram for the two pathways, drawn approximately to scale, using reasonable estimates of strain energies. A barrier in the T_1 surface along the nonconcerted pathway is indicated for reasons rather similar to those discussed earlier in connection with the dissociation of the C—H bond of toluene. In the present case, T_1 of XII is of $\pi \rightarrow \pi^*$ nature while the characteristic configuration for eaching the open-chain minimum is of $\sigma \rightarrow \sigma^*$ nature. Just like in benzyl radical, the newly freed atomic orbital on the α carbon will interact with the HOMO and LFMO orbitals involved in T_1, but the interaction should be even smaller than in the case of benzyl, since the expansion coefficient of these orbitals at position 2 of biphenyl is only 0.30. The resulting barrier may well be higher than indicated in Fig. 8-12. The figure also shows that motion along the concerted path would be endothermic and the molecules would sooner or later again collect in the minimum in T_1 at the starting geometry.

It is now necessary to show how the course of the orbital correlation along

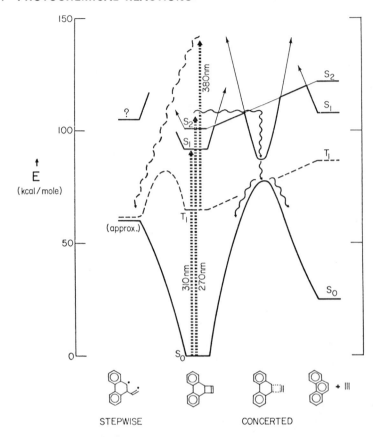

Fig. 8-12. State correlation diagrams for the fragmentation of XII. The dotted arrows indicate absorption of a photon. The wavy lines indicate how the barriers in S_1 (thick line) and T_1 (dashed line) can be overcome when sufficient energy is available.

the concerted pathway can be derived, and this is shown in Fig. 8-13. Using the same approach as above for electrocyclic reactions, we first imagine that there are no interactions between the biphenyl chromophore and cyclo-butene moiety anywhere along the reaction path. In such a hypothetical case, the correlation diagram would look as shown in light lines in Fig. 8-13. The energies of the π and π^* orbitals of biphenyl would not be affected at all, and those of the two bonding σ and two antibonding σ^* orbitals involved in the fragmentation of the cyclobutene ring to ethylene and acetylene would change with the reaction coordinate as if the biphenyl part were not present, and thus would have the form familiar[2] from the concerted fragmentation of cyclobutane to two ethylenes. The extra double bond on the acetylene is ignored in the present crude treatment. The nodal properties of these four orbitals are shown in the margin in Fig. 8-13. (The bonds which break are vertical.) As is well known, the ψ_1 orbital has a node between atoms labeled A and B in formula XII, and the ψ_2 orbital does not. Next, we imagine that the interactions between the orbitals of the four-membered ring and the biphenyl chromophore are introduced. These will occur between orbitals on atoms A and B and the positions 2 and 2′ in biphenyl. Remembering the nodal properties of the HOMO and LFMO orbitals of biphenyl, we expect the interactions to take the form indicated schematically in Fig. 8-13 by arrows and by the heavy line. The interaction is probably not very strong due to the relatively small size of the coefficients of the π orbitals at positions 2 and 2′. The final diagram need not be worked out in any further detail. It is

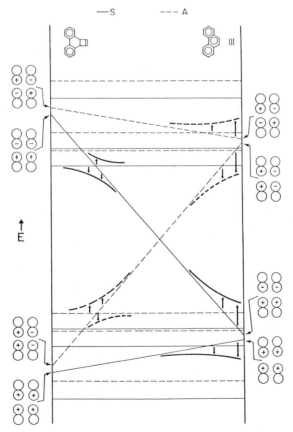

Fig. 8-13. Derivation of the correlation diagram for concerted fragmentation of XII (schematic; see text).

already clear that the reaction will be ground-state forbidden and the orbital crossover is the "normal" kind (HOMO exchanges place with LFMO). The "characteristic" configuration for the pericyclic reaction is $1 \rightarrow -1$, starting from either side. On both sides, this corresponds to the lowest triplet (^3L_a) as shown in Fig. 8-13. Again on both sides, the $1 \rightarrow -1$ excitation corresponds to the second singlet S_2, however, and a barrier is therefore to be expected as is shown in Fig. 8-12. This picture is in excellent agreement with the experimental data available at present.

In conclusion, we wish to point out again that the developments described above, although exciting to the author, represent only a very small dent in the mountain of our present ignorance of the detailed nature of photochemical processes in organic chemistry. Even in the field to which this chapter has been limited, namely location of minima and barriers in S_1 and T_1, much more remains to be done, hopefully in a more quantitative manner.

ACKNOWLEDGMENT

The author is indebted to the editor of *Molecular Photochemistry* for permission to reproduce figures, and to the donors of the Petroleum Research Fund, administered by the American Chemical Society, for partial support of this work.

References

1. J. Jortner, S. A. Rice, and R. M. Hochstrasser, *Advan. Photochem.* **7**, 149 (1969); R. Englman and J. Jortner, *Mol. Phys.* **18**, 145 (1970).
2. R. B. Woodward and R. Hoffmann, *Angew. Chem., Int. Ed. Engl.* **8**, 781 (1969).
3. K. Fukui, *Acct. Chem. Res.* **4**, 57 (1971).
4. W. C. Herndon and W. B. Giles, *Mol. Photochem.* **2**, 277 (1970).
5. H. E. Zimmerman, *Angew. Chem., Int. Ed. Engl.* **8**, 1 (1969); N. D. Epiotis, *J. Am. Chem. Soc.* **94**, 1941, 1946 (1972).
6. H. E. Zimmerman, *J. Am. Chem. Soc.* **88**, 1566 (1966); *Acct. Chem. Res.* **4**, 272 (1971).
7. R. C. Dougherty, *J. Am. Chem. Soc.* **93**, 7187 (1971).
8. W. Th. A. M. van der Lugt and L. J. Oosterhoff, *J. Am. Chem. Soc.* **91**, 6042 (1969).
9. E. F. Ullman, *Acct. Chem. Res.* **1**, 353 (1968).
10. J. Michl, *Mol. Photochem.* **4**, 243 (1972); *Topics in Current Chemistry*, **46**, 1 (in press).
11. J. Michl, *Mol. Photochem.* **4**, 257 (1972).
12. S. Kita and K. Fukui, *Bull. Chem. Soc. Jap.* **42**, 66 (1969).

13. Footnote 11 in Ref. 14, and footnote 6 in H. E. Zimmerman and G. A. Epling, *J. Am. Chem. Soc.* **94**, 3649 (1972).

14. J. Michl, *J. Am. Chem. Soc.* **93**, 523 (1971).

15. G. Herzberg, *Molecular Spectra and Molecular Structure. I. Spectra of Diatomic Molecules*, Van Nostrand, Princeton, N. J., 1950, 2nd ed., pp. 373, 532.

16. P. J. Wagner, *J. Am. Chem. Soc.* **89**, 2503 (1967).

17. J. C. Dalton, K. Dawes, N. J. Turro, D. S. Weiss, J. A. Barltrop, and J. D. Coyle, *J. Am. Chem. Soc.* **93**, 7213 (1971).

18. M. J. S. Dewar, *Angew. Chem., Int. Ed. Engl.* **10**, 761 (1971).

19. T. F. George and J. Ross, *J. Chem. Phys.* **55**, 3851 (1971); R. G. Pearson, *Acct. Chem. Res.* **4**, 152 (1971); C. Trindle, *J. Am. Chem. Soc.* **92**, 3251 (1970); W. A. Goddard III, *ibid.* **94**, 793 (1972).

20. W. G. Dauben, R. L. Cargill, R. M. Coates, and J. Saltiel, *J. Am. Chem. Soc.* **88**, 2742 (1966).

21. L. M. Stephenson and J. I. Brauman, *J. Am. Chem. Soc.* **93**, 1988 (1971).

22. R. S. Becker and J. Michl, *J. Am. Chem. Soc.* **88**, 5931 (1966).

23. J. Michl and J. Kolc, *J. Am. Chem. Soc.* **92**, 4148 (1970).

24. J. Michl, *Mol. Photochem.* **4**, 287 (1972).

25. V. G. Plotnikov, *Opt. Spectr.* **27**, 322 (1969); R. J. Cox, P. Bushnell, and E. M. Evleth, *Tetrahedron Lett.* 207 (1970).

26. N. Y. C. Chu and D. R. Kearns, *J. Phys. Chem.* **74**, 1255 (1970).

27. H. C. Longuet-Higgins and E. W. Abrahamson, *J. Am. Chem. Soc.* **87**, 2045 (1965).

28. R. J. Buenker, S. D. Peyerimhoff, and K. Hsu, *J. Am. Chem. Soc.* **93**, 5005 (1971).

29. For simple benzene derivatives studied in gas phase, even at medium pressures, wavelength dependence of this kind seems to be the rule rather than the exception. For references see W. A. Noyes, Jr., and C. S. Burton, *Ber. Bunsenges. Phys. Chem.* **72**, 146 (1968); T. L. Brewer, *J. Phys. Chem.* **75**, 1233 (1971); C. S. Parmenter, *Adv. Chem. Phys.* **22**, 365 (1972).

30. R. S. Becker, E. Dolan, and D. E. Balke, *J. Chem. Phys.* **50**, 239 (1969).

31. The calculations were performed for a series of points along the assumed disrotatory reaction coordinate using standard HMO or PPP methods and a basis set of only those atomic orbitals directly involved in the transformation (π systems of both chromophores plus the two orbitals of the σ bond to be broken). Twisting of atomic orbitals was simulated by appropriate changes in the values of β. Arbitrarily, $\beta_{\sigma\sigma}$ was set equal to $2\beta_{\pi\pi}$. The resulting correlation diagrams are insensitive to details of the calculation and are the same for both methods of calculation.

32. In the approximation used to estimate relative orbital energies in Fig. 8-6, ϕ_1 is degenerate with the second highest bonding orbital of naphthalene, but in better treatments it will be even below this one.

33. J. Meinwald, G. E. Samuelson, and M. Ikeda, *J. Am. Chem. Soc.* **92**, 7604 (1970).

34. D. M. Hayes and K. Morokuma, *Chem. Phys. Lett.* **12**, 539 (1972).

35. F. P. Schwartz and A. C. Albrecht, *Chem. Phys. Lett.* **9**, 163 (1971).

36. G. Quinkert, W.-W. Wiersdorff, M. Finke, K. Opitz, and F.-G. von der Haar, *Chem. Ber.* **101**, 2302 (1968).

37. Irradiation in a rigid glass. Triplet is definitely populated since phosphorescence can be observed. See G. Quinkert, M. Finke, J. Palmowski, and W.-W. Wiersdorff, *Mol. Photochem.* **1**, 433 (1969); C. Flynn and J. Michl, *J. Am. Chem. Soc.* **95**, 5802 (1973) have found that irradiation with very intense light (254 mμ) produces *o*-xylylene and other products, possibly in a two-photon process. This is presently under investigation.

38. J. Kolc and J. Michl, presented at the Rocky Mountain Regional Meeting of A. C. S., Colorado State University, Fort Collins, Colorado, June 30–July 1, 1972; see also Ref. 39b.

39.(a) J. Kolc and J. Michl, *J. Am. Chem. Soc.* **92**, 4147 (1970); (b) J. Kolc and J. Michl, *J. Am. Chem. Soc.* **95**, 7391 (1973).

40. J. Kolc and J. Michl, presented at the Fourth International IUPAC Symposium on Organic Photochemistry, Baden-Baden, Germany, July 16–22, 1972.

CHAPTER 9

Ionic Reactions in the Gas Phase: Study by Ion Cyclotron Resonance

R. C. Dunbar
Case Western Reserve University
Cleveland, Ohio

I. INTRODUCTION

Theoretical description of the mechanisms and rates of gas-phase ion-molecule reactions presents a challenge which has to date been at best poorly met. The theoretical approach advanced and elaborated in the early 1950s and often known as the Gioumousis-Stevenson theory has remained essentially the only theoretical model which has found agreement with any significant amount of experimental observations for systems of sufficient complexity to be of interest to most chemists. Various ingenious and elaborate theoretical models advanced since then have been troubled by severe quantitative and usually qualitative failures to predict or explain a significant body of experimental results.

The experimental aspects of this field have advanced to the point that theoretical models at several levels of sophistication could be extensively tested, and if successful could be of enormous benefit to the field. Rather than detailing past theoretical efforts, my principle aim here is to outline broadly some of the qualitative and quantitative chemistry which such theories must deal with, in the hope of stimulating theoretical interest in this chemistry. The selective nature of the coverage is designed to emphasize some of the experimental findings which appear most promising and provocative for future theoretical analysis.

The study of gas-phase ions and their reactions by ion cyclotron resonance (ICR) spectroscopy is a field which has attracted a great deal of interest. Since Baldeschwieler et al.[1] first described the technique in currently popular form, more than 150 publications and several review articles[2-4] have appeared, and it is not my purpose to give a comprehensive review of the field, but rather to outline some of its capabilities and some of the chemical problems which it has been (and will be) applied to. In the course of doing this, I hope to illuminate some of the peculiar features of gas-phase ionic chemistry which distinguish it so sharply from the liquid phase. The reader should realize that there are several other major instrumental methods actively in use for the study of gaseous ions and ion-molecule reactions, and these can be (in numerous cases have been) brought to bear on many of the same problems whose study by ICR methods is described here. The interested reader may refer to a number of excellent reviews[5-8] for more comprehensive and balanced coverage of this expansive field of chemistry.

II. PROPERTIES OF IONS

A. The ICR Technique

The basis for the ion cyclotron resonance technique[9] is the well-known cyclotron motion which an ion carries out in the plane perpendicular to a magnetic field at an angular frequency

$$\omega_c = \frac{eB}{mc} \qquad (1)$$

where e and m are the charge and mass of the ion, B is the magnetic field strength, and c is the speed of light. The marginal oscillator detection circuitry of the spectrometer applies to the ion an alternating rf electric field in the cyclotron plane, and when the phase and frequency of this field match the phase and cyclotron frequency of the ion, a resonant transfer of kinetic energy to the ion is observed by the detector. As the technique is normally used, the rf field frequency is fixed, so that the magnetic field strength at which a peak appears is characteristic of the mass of the ion. With a typical rf frequency of 307 kHz, the peak for N_2^+ (mass 28) appears at 5600 G; masses up to 200 are routinely studied.

In the most commonly used experimental configuration[10] (see Fig. 9-1), ions are generated by electron impact in the source region of the ICR cell. Since motion along the magnetic field is otherwise unconstrained, a repelling voltage is applied to the trapping plates to prevent ion loss. A voltage applied to the drift plates has the effect (the principle is the same as that of the Hall

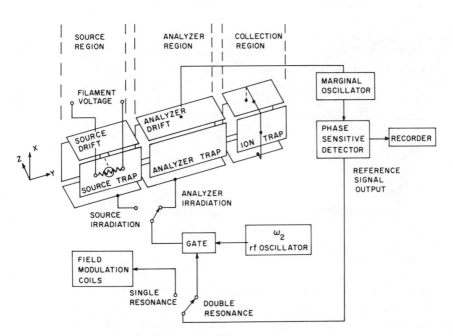

Fig. 9-1. Experimental arrangement for ICR spectroscopy.

effect) of driving the ions down the cell and into the analyzer region, and ultimately out to the ion collector. As they pass through the analyzer region in a time of the order of a few msec, the ions are detected by the marginal oscillator detector, and the signal is processed and recorded as a function of magnetic field. The ICR cell must be maintained under high vacuum to allow free motion of the ions, and the range of pressures used is from 10^{-8} torr (where isolated ions are desired) to about 10^{-3} torr (where frequent collisions with neutral molecules are desired).

The predominant use of the ICR technique has been the study of ion-molecule reaction chemistry using the double-resonance technique.[1] As an illustration, suppose we are working with diborane (B_2H_6) for which the primary electron-impact-produced ions are[11] $B_2H_2^+$, $B_2H_4^+$, and $B_2H_5^+$. As the pressure is raised, so that reactive collisions become possible, a peak corresponding to $B_4H_6^+$ will appear, and we ask which of the primary ions is responsible for its production. To answer this question by double resonance, a second (irradiating) rf field is applied to the analyzer region of the cell, while simultaneously the $B_4H_6^+$ peak is continuously monitored with the observing (marginal) oscillator. When the frequency ω_2 of the irradiating oscillator passes through the cyclotron frequency of one of the primary ions (say $B_2H_5^+$), the $B_2H_5^+$ ions will be accelerated by the irradiating field, and some of them will attain sufficient velocity to escape from the cell. *If* the product $B_4H_6^+$ peak is being produced by reaction of these $B_2H_5^+$ ions, a dip in the $B_4H_6^+$ signal will be observed. If the $B_4H_6^+$ ion is produced in a slightly endothermic reaction, the increased energy of the $B_2H_5^+$ ions due to irradiation may actually increase the reaction rate, giving a *rise* in $B_4H_6^+$ signal. If, however, these two ions are not coupled reactively, irradiation of $B_2H_5^+$ will result in no change in $B_4H_6^+$ signal intensity. In the present instance,[11] irradiation of $B_2H_2^+$ has no effect, of $B_2H_4^+$ decreases $B_4H_6^+$, and of $B_2H_5^+$ increases $B_4H_6^+$, showing the chemistry here to be

$$B_2H_2^+ \quad \xrightarrow{\quad /\!/\quad}$$

$$B_2H_4^+ \quad \xrightarrow{\ B_2H_6\ } \quad B_4H_6^+ + \text{neutral fragment} \qquad (2)$$

$$B_2H_5^+ \quad \xrightarrow[\ B_2H_6\ ]{\ \text{Endothermic}\ }$$

The appearance of these spectra is indicated in Fig. 9-2. The double-resonance technique thus provides a way of identifying and checking the energetics of ion-molecule reaction pathways, and has proven in practice to be a particularly powerful technique.

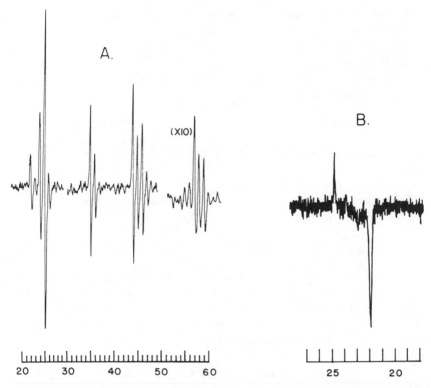

Fig. 9-2. Typical ICR spectra, $^{10}B_2H_6$ at 17 eV ionizing energy and about 10^{-5} torr pressure. (a) Single resonance spectrum with field sweep. The derivative presentation of the peaks arises from the modulated phase sensitive detection. The peaks at masses 22, 24, and 25 are primary ions $B_2H_2^+$, $B_2H_4^+$, and $B_2H_5^+$ generated at the electron beam, while the peaks at higher masses are ion-molecule reaction products. (b) Double-resonance spectrum obtained by observing mass 46 ($B_4H_6^+$) and sweeping the double resonance oscillator through the primary-ion region of the spectrum.

B. Thermochemistry of Gaseous Ions

The determination and application of various thermodynamic quantities play the same kind of central role in gas-phase chemistry that they do in other fields of chemistry. However, it is still extremely hard to find accurate reliable data for gaseous ions, uncertainties in thermodynamic quantities of several kcal being the rule rather than the exception, and a great deal of effort is devoted to attempts to improve and extend the available body of data. It will be useful to begin our discussion by defining some of the quantities of principal concern.[12,13]

The *ionization potential* of a molecule is the energy required to remove an electron from it. Unlike the other thermodynamic quantities discussed here, ionization potentials have been accurately determined and tabulated for a large variety of molecules.

The *proton affinity* (or basicity) of a neutral molecule is the negative of the enthalpy of the process

$$A + H^+ \longrightarrow AH^+ \qquad -\Delta H^0 = PA(A) \qquad (3)$$

Proton affinities thus defined are positive numbers for all known molecules, and since the gaseous (unsolvated) proton is extremely reactive, usually have values of more than 100 kcal/mole.

The proton affinity or *basicity* of an anion is defined in a similar way as the negative of the enthalpy of the reaction

$$A^- + H^+ \longrightarrow AH \qquad -\Delta H^0 = PA(A^-) \qquad (4)$$

The *acidity* of a neutral AH (or a cation BH^+) is just the negative of the basicity of A^- (or B). As an indication of how difficult it is to dissociate a gaseous molecule or ion into H^+ plus the conjugate base, the acidity of H_2Cl^+ in the gas phase is about[14] -140 kcal, corresponding to a pK_A of about $+23$. However, since we are normally concerned here with *differences* in acidities, this very low acidity is of no real concern.

The *electron affinity* of a molecule, EA(A), is the negative of the enthalpy of the reaction

$$A + e^- \longrightarrow A^- \qquad -\Delta H^0 = EA(A) \qquad (5)$$

and may range from zero to over 100 kcal/mole.[13]

Finally, the *hydrogen affinity* of an ion $HA(A^+)$ is the negative of the enthalpy of the reaction

$$A^+ + H \longrightarrow AH^+ \qquad -\Delta H^0 = HA(A^+) \qquad (6)$$

We will see that the hydrogen affinity provides a good measure of the A^+—H bond strength in the ion.

Relationships among the quantities: Two equations relating the above quantities are easily derived[15]:

$$PA(A) + 1P(A) = HA(A^+) + 1P(H) \qquad (7)$$

and

$$PA(B^-) = D(BH) + 1P(H) - EA(B) \qquad (8)$$

where the bond dissociation energy D(B—H) is a known quantity for most molecules of interest. Because of these relationships, it is not necessary to measure all of the above quantities experimentally.

1. Proton Affinities. Because it is difficult to measure absolute proton affinities, most effort has gone into comparing proton affinities of molecules by examining reactions typified by

$$CH_5{}^+ + CH_3OH \underset{k_{-1}}{\overset{k_1}{\rightleftharpoons}} CH_4 + CH_3OH_2{}^+ \qquad (9)$$

A reaction such as this will normally proceed most rapidly in the direction corresponding to a decrease in free energy. By ICR double resonance, it is easily found that reaction (9) proceeds rapidly to the right, indicating that methanol is more basic than methane; if we assume, as is often done, that the ΔS^0 change for a proton transfer reaction is negligible, so that the basicity is the same as the proton affinity, then $PA(CH_3OH) > PA(CH_4)$. By repeating this procedure for a large number of pairs of compounds, and using already known proton affinities as standards, it has been possible to establish the proton affinities of a very large number of compounds, some of which are shown in Table 9-1.

More recently,[22-24] the direct measurement of the equilibrium constant in proton transfer reactions has been used to yield very accurate thermochemical data. In this approach, a mixture of neutral and protonated molecules, such as the four species involved in reaction (9), is introduced into a modified ICR cell, and the system is given time to come to equilibrium (in perhaps 100 msec). The equilibrium constant is then found from the equilibrium concentrations of the ions, giving ΔG^0 for the reaction according to $\ln K = -\Delta G/RT$. The temperature dependence of the equilibrium constant is also readily found, so that both ΔH^0 and ΔS^0 for the reaction can be found. This promising technique yields data of very high precision, with differences as small as 0.1 kcal readily observable.

In comparison with solution-phase values, the gas-phase basicities shown in the tables are enormous, reflecting the strong tendency of the bare proton to become "solvated" by a neutral molecule. Of more interest are the trends observed in the basicities of related molecules:

1. The addition of a methylene group or other polarizable group to a neutral molecule normally has the effect of increasing the basicity, and examination of series of alcohols, ethers, amines, and halides in Table 9-1 suggests that for small molecules this increase is of the order of 5–20 kcal/mole per methylene group. This effect is undoubtedly a reflection of the increasing polarizability of the molecule, allowing it to solvate the proton more efficiently.

Table 9-1. Thermochemical Values (kcal/mole)

Species	Proton Affinity	Hydrogen Affinity	Ref.
N_2	116	162	14
CO	143	152	14
Methane	126	106	14
Ethane	143	97	16
Ethylene	160	87	15,17
Propylene	179	90	15,17
2-Butene	181	78	15,17
1-Butene	183	90	15,17
Isobutylene	195	94	15,17
Methanol	180	117	15
Ethanol	186	114	15
Methyl chloride	160	107	18
Ethyl chloride	167	107	18
Methyl bromide	163	93	18
Ethyl bromide	170	93	18
Methyl iodide	170	77	18
Ethyl iodide	175	77	18
SiH_4	≤ 146	≤ 105	4
NH_3	207	128	4
PH_3	185	102	19,20
AsH_3	175	93	4
H_2O	164	143	17
H_2S	170	97	17
H_2Se	170	85	4
HF	131	182	4
HCl	140	121	21
HBr	141	96	21
HI	145	71	21

2. The basicities of the anions in the alkoxy and amino series (Table 9-2) are decreased by the addition of methylene groups. This effect is in striking contrast to solution behavior, where the electron-donating character of the alkyl groups increases the basicity. For example, in solution the order of basicities is $CH_3O^- < C_2H_5O^- < C_3H_7O^-$, while the gas-phase order is seen to be the reverse. This effect must be due to the stabilization of the negative charge of the anions by the more polarizable alkyl groups, a stabilization which in gas phase predominates over the inductive effects seen in solution.

2. Hydrogen Affinity and Bond Strength. As we have seen above, the hydrogen affinity of an ion $HA(A^+)$ (a quantity easily calculated from the proton affinity) corresponds to the energy involved in cleaving the $A—H^+$ bond while leaving the charge on the A portion of the ion. Thus for ions AH^+

Table 9-2. Several Series of Proton Affinity Sequences

1. Order of acidities of ROH[25]:

$C_6H_5OH > (CH_3)_3CCH_2OH > (CH_3)_3COH$
$> (CH_3)_2CHOH > CH_3CH_2OH > CH_3OH > H_2O$

2. Order of acidities of R_2NH[26]:

$(C_2H_5)_2NH > (CH_3)_3CNH_2 > n\text{-}C_3H_7NH_2 > C_2H_5NH_2 > CH_3NH_2 > NH_3$

3. Order of acidities of binary hydrides[27]:

$H_2S > AsH_3 > PH_3 > SiH_4 > H_2O > NH_3 > CH_4$

in which the positive charge does not reside to a significant extent in the A—H bonding region, the breaking of the A—H bond should differ very little from the breaking of a similar bond in a neutral molecule. We can see several series of ions in Table 9-1 for which this prediction is borne out: methanol and ethanol show approximately equal hydrogen affinities, while the three series of alkyl halides, the chlorides, bromides, and iodides, show this effect very strikingly. It thus appears that the hydrogen affinity of an ion is a very useful measure of the bond strength by which a hydrogen atom binds to the ion, and provides us with an illuminating way of comparing such bonds with bonds of neutral molecules.

C. Determining Ion Properties by Chemical Reactivity

There is often considerable uncertainty about the structure of gas-phase ions. In a mass spectrometer, for example, the mass and atomic composition of an ionic species can be determined, but further information about it, particularly its structure, is generally difficult to obtain, and there are few experimental tools which can help. One of the most useful approaches is to examine the reactions of the ion with one or more neutral species. We will look at a few of the many types of ions for which reactive structure studies have been successfully applied.

1. $C_3H_6O^+$. In an elegant early application of this approach,[28,29] Diekman et al. distinguished between keto and enol forms of the acetone parent ion. They showed that the keto form of this ion (from ionization of acetone) undergoes the characteristic ion-molecule reactions

$$CH_3\overset{\overset{\displaystyle O^+}{\displaystyle \|}}{C}CH_3 + CH_3COCH_3 \longrightarrow (CH_3)_2C\overset{+}{=}O—COCH_3 + CH_3 \quad (10)$$

(a condensation reaction), and

$$\overset{\overset{\displaystyle O^+}{\|}}{CH_3CCH_3} + CH_3CO(CH_2)_3CH_3 \longrightarrow$$

$$CH_3CO(CH_2)_3CH_3{}^+ + CH_3COCH_3 \quad (11)$$

(a charge transfer reaction).

Enol forms of this ion were generated from three different sources:

$$(1) \quad \xrightarrow{-C_2H_4} \quad CH_3\overset{\overset{\displaystyle OH}{|}}{C}{=}CH_2{}^+ \quad (12)$$

$$(2) \quad \xrightarrow{-} \quad CH_3\overset{\overset{\displaystyle OH}{|}}{C}{=}CH_2{}^+ \quad (13)$$

(a McLafferty rearrangement)[30]

and $\quad (3) \quad \xrightarrow{-2} \quad \quad (14)$

(a double McLafferty rearrangement).

These enol ions did not undergo reactions (10) or (11), but instead showed a different group of characteristic reactions, typical of which was

$$CH_3\overset{\overset{\displaystyle OH}{|}}{C}CH_2{}^+ + CH_3CO(CH_2)_3CH_3 \longrightarrow$$

$$CH_3C(OH)(CH_2)_3CH_3{}^+ + CH_3COCH_2 \quad (15)$$

a proton-transfer reaction. In addition to providing a clear indication that the keto and enol structures of $C_3H_6O^+$ are distinct and not rapidly interconverting in the gas phase, this study gave the very interesting result that the three enol ions generated by reactions (12), (13), and (14) were not distinguishable in the ICR spectrometer. Diekman et al. concluded[28] that one likely explanation of this was that these three ions all rearrange to a common form, or interconvert rapidly among several different forms common to all of them.

2. $C_2H_5O^+$. Beauchamp and Dunbar[31] studied the $C_2H_5O^+$ ions from a variety of neutral precursors in another ICR study. They found that it is possible to identify at least two different $C_2H_5O^+$ structures on the basis of their reactivities, and considered it likely that these structures were $CH_3CH{=}$

OH^+ and CH_3—O^+=CH_2; with the possibility that the second structure rearranges to or interconverts with

$$\begin{array}{c} H \\ O^+ \\ H_2C - CH_2 \end{array}$$

Two of several characteristic reactions found for these ionic structures were

$$CH_3CH\overset{+}{=}OH + C_2H_5OH \longrightarrow CH_3CHO + C_2H_5OH_2^+ \quad (16)$$

(proton transfer) which was characteristic of the CH_3CH=OH^+ structure, and

$$CH_3—\overset{+}{O}=CH_2 + CH_3OC_2H_5 \longrightarrow \overset{CH_3}{\underset{+}{CH_3O}}—C_2H_5 + CH_2O \quad (17)$$

(methyl cation transfer), which was characteristic of the CH_3O^+=CH_2 structure.

Structures were examined in this way for the $C_2H_5O^+$ ions from numerous sources. As an example, the $C_2H_5O^+$ ion obtained from dimethyl carbonate,

$$\begin{array}{c} CH_3O \\ \diagdown \\ \diagup \\ CH_3O \end{array} C=O \xrightarrow[-e^-]{-CO_2H} C_2H_5O^+ \quad (18)$$

was found to show a characteristic methyl cation transfer reaction, and was assigned as having the $CH_3O^+CH_2$ structure.

3. $C_7H_8^+$. As a final illustration of this powerful technique, a recent study of Hoffman and Bursey[32] provided some rather unexpected results in the structure of $C_7H_8^+$ ions. It has long been known that the mass spectral fragmentation patterns of toluene and cycloheptatriene are so similar as to strongly suggest that the $C_7H_8^+$ parent ions from these two sources are identical.[33] Moreover, the hydrogen and carbon atoms of toluene isotopically labeled at specific sites are known to show extensive scrambling,[34] and these facts have been taken as suggesting that when toluene is ionized the parent ion rearranges very rapidly to a seven-membered ring form.[35]

However, Hoffman and Bursey found that the nondecomposing $C_7H_8^+$ cations from toluene and cycloheptatriene showed entirely dissimilar ion-molecule reaction patterns. $C_7H_8^+$ from toluene showed the reaction

$$C_6H_5CH_3^+ + RONO_2 \longrightarrow CH_3C_6H_5NO_2^+ + RO \quad (19)$$

whereas the $C_7H_8^+$ ions formed from cycloheptatriene were entirely non-reactive under similar conditions. Hoffman and Bursey considered this as showing that ring expansion in toluene ions occurs only for those ions having enough energy to decompose subsequently.

D. Determining Ion Properties Spectroscopically

In looking at structures of gas-phase ions, it is of course attractive to look for more direct methods such as spectroscopy or resonance techniques. However, the difficulty of obtaining concentrations of gas-phase ions greater than perhaps 10^{-14} M rules out direct application of traditional techniques. Recently an indirect approach to optical spectroscopic study of ions has been explored and may have useful possibilities.[36]

In this approach, the photodissociation of the ion in question is observed as a function of wavelength to produce a photodissociation spectrum characteristic of the ion. An example[37] is the determination of the structures of $C_7H_8^+$ cations arising from different precursors:

$$\text{(toluene)} \longrightarrow C_7H_8^+ \xrightarrow{h\nu} C_7H_7^+ + H \qquad (20a)$$

$$\text{(cycloheptatriene)} \longrightarrow C_7H_8^+ \xrightarrow{h\nu} C_7H_7^+ + H \qquad (20b)$$

$$\text{(norbornadiene)} \longrightarrow C_7H_8^+ \xrightarrow{h\nu} C_7H_7^+ + H \qquad (20c)$$

In observing these reactions, the $C_7H_8^+$ cations are trapped in the ICR cell, while light is beamed into the cell. Dissociation is followed directly by following the disappearance of $C_7H_8^+$ versus wavelength using the marginal oscillation detector. The resulting spectra for reactions (20a), (20b), and (20c) are shown in Fig. 9-3.

We are not yet in a position to interpret the spectra in Fig. 9-3 directly in terms of ion structure. It is immediately evident, however, that the $C_7H_8^+$ ions from toluene, cycloheptatriene, and norbornadiene are not identical, and that upon ionization these three C_7H_8 molecules do not rearrange to a common symmetrical ion, as has been suggested.[33] Figure 9-3 leaves open the possibility that some rearrangement of norbornadiene cation to the cyclo-heptatriene structure occurs, but rules out any significant rearrangement of either of these ions to the most stable toluene structure.

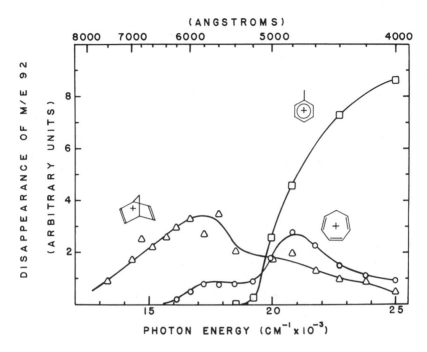

Fig. 9-3. Extent of photodissociation of $C_7H_8^+$ ions as a function of wavelength for toluene (□), cycloheptatriene (○), and Norbornadiene (△) as a function of wavelength.

An excellent illustration of how this technique provides an indirect route to the absorption spectrum of cations is the *t*-butylbenzene dissociation

$$\text{(structure)} \xrightarrow{h\nu} \text{(structure)} + CH_3 \qquad (21)$$

This dissociation is energetically possible for wavelengths as long as 10,000 Å, but in fact is observed only for wavelengths shorter than 7500 Å (1.6 eV). This evidently reflects the inability of the parent ion to absorb photons beyond this wavelength. The first excited state of the *t*-butylbenzene ion is known[38] to lie near 1.8 eV above the ground state from photoelectron spectroscopy, and it is not at all surprising that the absorption band of the ion would commence at about 1.6 eV as indicated by the photodissociation spectrum.

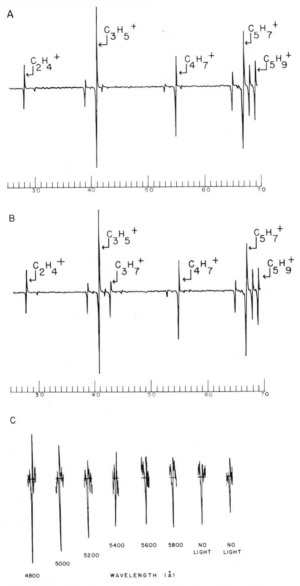

Fig. 9-4. The photon-induced reaction $C_3H_5^+ + C_2H_4 \xrightarrow{h\nu} C_3H_7^+ + C_2H_2$. (a) ICR single-resonance spectrum of ethylene with the light off, under conditions where many reaction products are formed. (b) Spectrum with identical conditions to (a), but with irradiation at 400 nm. Note the appearance of the peak at mass 43 due to the reaction of photoexcited $C_3H_5^+$. (c) $C_3H_7^+$ peak heights for irradiation at the wavelengths indicated. This is a kind of absorption spectrum for the allyl cation.

352

Another example of interest is provided by photodissociation spectra of butene cations.[39] The spectra for the reactions

$$cis\text{-}2\text{-}C_4H_8^+ \xrightarrow{\ h\nu\ } C_4H_7^+ + H \tag{22a}$$

$$trans\text{-}2\text{-}C_4H_8^+ \xrightarrow{\ h\nu\ } C_4H_7^+ + H \tag{22b}$$

are essentially identical, with the spectrum rising gradually and continuously from a wavelength onset at 4800 Å. The spectrum for the reaction

$$1\text{-}C_4H_8^+ \xrightarrow{\ h\nu\ } C_4H_7^+ + H \tag{23}$$

is, however, distinctly different from those for the 2-butene cations, showing an onset at 5200 Å, and rising more rapidly from the baseline. These spectra provide strong evidence that the exothermic rearrangement

$$1\text{-}C_4H_8^+ \longrightarrow 2\text{-}C_4H_8^+ \quad -16 \text{ kcal} \tag{24}$$

if it occurs at all, has a rate much less than 1 sec^{-1}, and indicates that the 1-butene cation is in fact a long-lived structure with a lifetime of at least several seconds.

A final example of the use of this new approach to obtaining optical spectra of ions arises from the observation that the reaction

$$C_3H_5^+ + C_2H_4 \longrightarrow C_3H_7^+ + C_2H_2 \tag{25}$$

which is very slow in the dark, is greatly accelerated by optical irradiation.[40] ICR double-resonance methods can be used to establish that this effect arises through absorption of a photon by the $C_3H_5^+$ ion (very likely having the allyl structure) and a spectrum for this photon-accelerating effect on the reaction can be produced as in Fig. 9-4. The striking feature of these results is that they indicate an optical absorption for $C_3H_5^+$ beginning about 5200 Å (2.3 eV). This implies an excited state of the $C_3H_5^+$ cation not far above 2.5 eV, which is in striking disagreement with recent theoretical estimates placing the lowest singlet state of allyl cation near 4.0 eV.

III. KINETICS OF ION REACTIONS

A. Theoretical Considerations

In a more detailed look at the reactive behavior of ions in the gas phase, we will start by considering some of the simple features of the kinetics of such reactions. Under low-pressure gaseous conditions the reaction kinetics of ions have several characteristic features which sharply differentiate these conditions from gas-phase neutral-neutral reactions and from solution phase reaction kinetics:

a. Ion-molecule reactions are extremely fast, with a great number having bimolecular rate constants between 10^{-9} and 10^{-10} cc/molecule-sec.

b. Ion molecule reactions never or almost never have rates above about 3×10^{-9}, which would appear to be a diffusion controlled upper limit on the rates.

c. These reactions apparently always proceed without activation energy.

All of these features are predicted by a simple and successful theoretical approach known as the Gioumousis-Stevenson theory, which we will begin by outlining.

The simplicity and predictive success of the Gioumousis-Stevenson theory[41,42] arise from the fact that to a good approximation the interaction of a free ion and a neutral molecule is dominated (except when the two species are very close) by the long-range electrostatic attraction between the ion and the transitory dipole moment which it induces in the molecule.[41,42]

$$V(r) = -\frac{e^2 \alpha}{2r^4} \tag{26}$$

where $V(r)$ is the potential energy at separation r, e is the ion charge, and α is the polarizability of the neutral. The interesting feature of this potential term is that it permits only two classes of collisions: "grazing" collisions in which the two particles never approach closer than a certain fairly large distance, so that no reaction is likely; and "orbiting" collisions, in which the two particles spiral toward one another until the strongly repulsive short-range forces cause them to bounce apart. By the time the ion and the neutral strike each other in an orbiting collision, their mutual potential energy of attraction will have been converted to tens of kilocalories of kinetic energy, and it is the presence of this large excess energy in the collision complex which undoubtedly overwhelms the activation energy which the reaction might otherwise require. The assumption that every orbiting collision leads to reaction leads to a bimolecular rate constant

$$K = 2\pi e \left(\frac{\alpha}{M_r}\right)^{1/2} \tag{27}$$

for the reaction, where M_r is the reduced mass of the system. Since it is not necessarily true that all orbiting collisions lead to reaction, Eq. (27) can really only be an upper limit on the reaction rate, but for many reactions it has been found to give a useful first approximation to the rate. For typical small and moderate sized molecules (mass of the order of 2–200 AMU), Eq. (27) gives rate constants of the order of 1×10^{-9} cc/molecule-sec. Thus the success of this very elementary theory is that it both predicts the correct order of magnitude for many reaction rates, and accounts readily for the observation of a rather sharp upper limit to such rates, and shows that this upper limit can in fact be thought of as a kind of diffusion-controlled limit.

B. Proton-Transfer Reactions

As much as any generalization about ion-molecule reactions is possible, one may say that in a system where proton transfer is thermodynamically favorable, this reaction is likely to be the predominant reaction channel. In fact, proton transfer reactions typically have rates near the Gioumousis-Stevenson value, and have been often cited as giving strong support for the Gioumousis-Stevenson approach to kinetics.[43] As discussed above, hundreds of such reactions have been observed in the course of finding proton affinity values.

The experimental approach to measuring such reaction rates by ICR is straightforward, and was first described in detail by Buttrill.[44] If we write a typical reaction in the form

$$A^+ + B \xrightarrow{k} C^+ + D \tag{28}$$

the primary A^+ ions will be generated at the electron beam, and will then drift down the length of the cell with the drift velocity

$$V_d = \frac{cE}{B} \tag{29}$$

where E and B are the electric and magnetic field strengths. As the A^+ ions drift down the cell they will be converted according to simple quasi-first-order kinetics into C^+ ions. Then it is easy to write expressions for the average concentrations of A^+ and C^+ ions in the analyzer region of the cell (which are the experimentally accessible numbers), and to solve for the rate constant K, giving the expression

$$K = 3m_A{}^2 I_C [nm_C{}^2 I_A (2\tau_A + \tau_A') + nm_A{}^2 I_C(\tau_A + 2\tau_A')]^{-1} \tag{30}$$

where m_A and m_C are the ion masses, I_A and I_C are the observed ICR signals, τ_A and τ_A' are the times at which A^+ ions enter and leave the ICR cell [calculated by Eq. (29)], and n is the density of B molecules. This expression can of course be extended to cover more elaborate cases,[45] but in principle at least it is always easy to find the rate constant of a reaction from measuring the ICR signals of the various ions involved.

Reaction rates have been measured in this way for a number of proton transfer reactions, as illustrated in Table 9-3. The rates are indeed near the Gioumousis-Stevenson prediction. It is interesting that several proton

Table 9-3. Proton-Transfer Rates Measured by ICR Methods

Reaction[a]	Rate Constant ($\times\ 10^{-10}$ cc/mol.-sec)	Theory	Ref.
$CH_3F^+ + CH_3F \rightarrow CH_3FH^+$	12.8 (13.6)	9	18
$C_2H_5F^+ + C_2H_5F \rightarrow C_2H_5FH^+$	15.0	10	18
$CH_3Cl^+ + CH_3Cl \rightarrow CH_3ClH^+$	12.5	10	18
$C_2H_5Cl^+ + C_2H_5Cl \rightarrow C_2H_5ClH^+$	9.4	10	18
$CH_4^+ + CH_4 \rightarrow CH_5^+$	9.5 (11.5)	13	44(46)
$CH_2D_2^+ + CH_2D_2 \rightarrow CH_3D_2^+$	5.1	13	44
$CH_2D_2^+ + CH_2D_2 \rightarrow CH_2D_3^+$	4.1	13	44
$SeH_2^+ + SeH_2 \rightarrow SeH_3^+$	4.0	—	4
$AsH_3^+ + AsH_3 \rightarrow AsH_4^+$	5.6	—	4
$PH_3^+ + PH_3 \rightarrow PH_4^+$	10.5	—	4
$NH_3^+ + NH_3 \rightarrow NH_4^+$	19	9	47
$NH_2^+ + NH_3 \rightarrow NH_4^+$	11	9	47
$D_2^+ + Ar \rightarrow ArD^+$	16(9.9)	16.3	48
$H_3^+ + Ar \rightarrow ArH^+$	3.65	19.1	49
$H_2^+ + N_2 \rightarrow N_2H^+$	19.5	22.8	49
$H_3^+ + N_2 \rightarrow N_2H^+$	10.3	19.0	49

[a] Neutral products omitted.

transfer reactions to dipolar molecules are *higher* than the Gioumousis-Stevenson rate, for example in methanol, methyl chloride, and so on, and this has often been taken as indicating the operation of a "dipole-locking" type of ion-dipole interaction.[50,51] However, we will see below that ion mobility studies provide much more conclusive evidence for this mechanism in dipolar systems.

C. More Complex Reactions

We would now like to make an effort to chart the large field of chemical reactions of ions with neutral molecules which involve more complex re-arrangements or condensations in the course of the reaction than was the case for the simple proton reactions considered above. We will see that the double-resonance techniques of ICR spectroscopy are indispensible in un-raveling the complex patterns of consecutive and competing reactions in many ionic systems. Moreover, even after a reaction has been identified by such techniques its mechanism is often not obvious, and we will see that the techniques of isotopic labeling of the reactants are also important tools in characterizing ion-molecule reactions. It will, of course, not be possible here to survey even a small fraction of the total number of known ion-molecule reactions. Our goal will simply be to cover in some detail enough different types of reactive systems to give an idea of the kinds of chemistry encountered in these perhaps unfamiliar systems.

Reactions in simple hydrocarbons have received a great deal of attention. The ionic chemistry of these molecules is often very rich and it is the rule rather than the exception that several reaction channels are observed for the reaction of each ion-neutral pair.

Among the saturated hydrocarbons methane and ethane have been investigated in some detail. The principal reactions in these systems are shown in Tables 9-4 and 9-5. One may note the general tendency of these reactions to form the highly stable product cations CH_5^+, $C_2H_5^+$, $C_3H_5^+$, $C_3H_7^+$, and $C_4H_9^+$. The butyl cation $C_4H_9^+$ particularly seems to be the end point of many reaction sequences in saturated hydrocarbons.

Among olefins ethylene has been most extensively studied and the following scheme showing the qualitative reaction pattern indicates the complexities of this reactive system[55]:

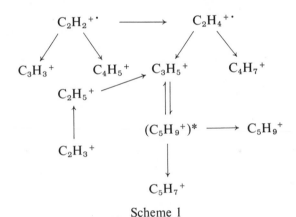

Scheme 1

The formation of $C_5H_9^+$ product in this system illustrates the collisional stabilization of an excited reaction complex. At low pressure such an ionic product which is formed with no neutral leaving group to carry off excess energy cannot be formed; however, it is found that as the pressure is raised species such as this may be stabilized before their decomposition by a non-reactive collision with another ethylene molecule. Bowers, Elleman, and Beauchamp[55] found that the excited $C_5H_9^+$ collision complex has a lifetime of $\geq 1.1 \times 10^{-4}$ sec, so that collision of the excited complex with a neutral molecule in a time much less than this can result in a collision-stabilized $C_5H_9^+$ cation.

Unsaturated hydrocarbons also undergo a variety of reactions with molecules or ions containing heteroatoms. Examples of such reactions are the

Table 9-4. Principal Ion-Molecule
Reactions of Methane
(Mass Spectrometric
Results).[52]

Reactant	Product
CH_3^+	$C_2H_3^+$
	$C_2H_5^+$
CH_4^+	CH_5^+
$C_2H_3^+$	$C_3H_5^+$
	$C_3H_6^+$

following reactions involving hydrogen sulfide and ethylene or acetylene[56]:

$$H_2S^+ + C_2H_4 \longrightarrow CH_2SH^+ + CH_3 \qquad (31)$$

$$H_2S^+ + C_2H_2 \longrightarrow HCS^+ + CH_3 \qquad (32)$$

That both of these reactions probably proceed through an interesting four-membered cyclic transition state was indicated by the results of deuterium labeling experiments, for which the results are indicated in Scheme 2 for the ethylene system and Scheme 3 for the acetylene system.

Table 9-5. Principal Ion-Molecule
Reactions of Ethane at 13 eV.[16,53,54]

Reactant	Product	Rate Constant ($\times 10^{-10}$ cc/molecule-sec)
$C_2H_2^+$	$C_3H_5^+$	Fast
	$C_4H_7^+$	Fast
$C_2H_4^+$	$C_2H_5^+$	0.10
	$C_3H_5^+$	0.14
	$C_3H_3^+$	0.10
$C_2H_5^+$	$C_4H_9^+$	0.40
$C_2H_6^+$	$C_3H_8^+$	0.23
	$C_4H_9^+$	0.10
	$C_3H_9^+$	0.17
	$C_2H_5^+$	0.33
	$C_3H_7^+$	0.14
$C_3H_6^+$	$C_4H_8^+$	0.30
$C_3H_9^+$	$C_4H_9^+$	1.00

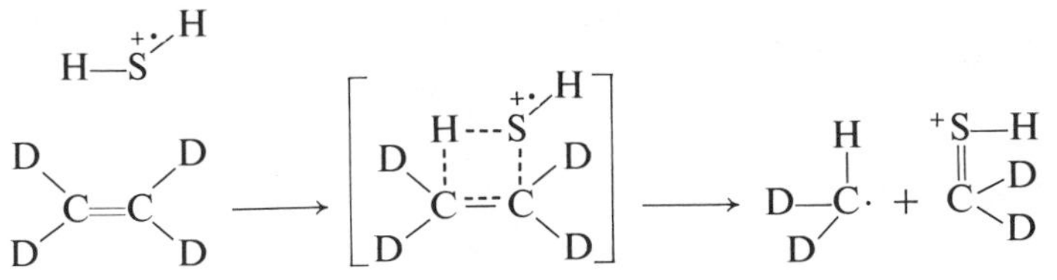

Scheme 2

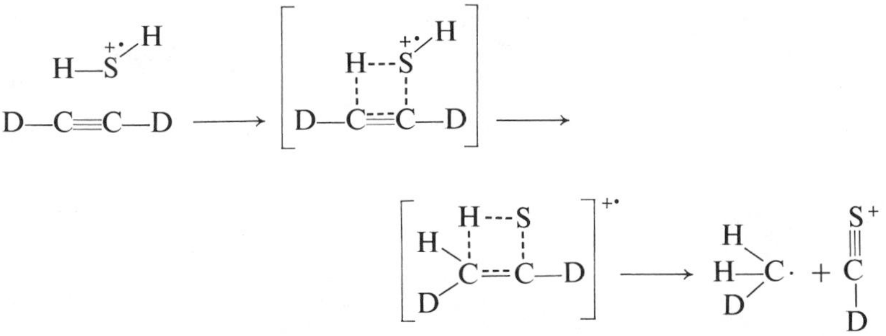

Scheme 3

In striking contrast to the complex reaction systems found for hydrocarbon molecules are the reactions of boron hydride cations with boron hydride neutrals.[11,57] A number of such reactions occur between various boron hydride cations and the simplest boron hydride neutral, diborane. The unifying feature common to all of these reactions is that each of them may be formulated as involving the transfer of a B_2H_2 group from diborane to the cation with the elimination of two molecules of hydrogen:

$$B_mH_n{}^+ + B_2H_6 \longrightarrow B_{m+2}H_{n+2}^+ + 2\,H_2 \tag{33}$$

This is suggestive of the possibility that the B_2H_2 unit in diborane consisting of the borons plus the two bridging hydrogen atoms:

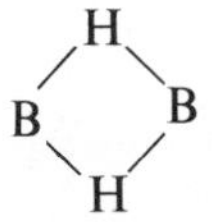

$$\tag{34}$$

is retained intact during the course of the reaction, while the terminal hydrogen atoms are eliminated. Similar behavior is found in reactions of higher boron hydrides, and it was found in a study of tetraborane, pentaborane-9, pentaborane-11, and hexaborane that the stoichiometry of the ion molecule reactions observed suggested preferential retention of the atoms involved in the boron-bridging-hydrogen frameworks of the molecule and the ion. An interesting and as yet unexplained point emerges from a comparison of the boron hydride ionic systems and the hydrocarbon systems: A survey of ion molecule reactions in hydrocarbon systems suggests that those reactions are favored in which the ionic product contains an even number of electrons with the neutral product carrying off the odd electron; by contrast the boron hydride systems show a marked preference for reactions in which the unpaired electron is retained on the ionic reaction product with the neutral product being an even electron species.

There is a large class of reactions apparently proceeding by a common mechanism which we may classify under the general heading of nucleophilic displacement reactions.[4,58] These reactions involve displacement of a small molecular fragment from an ionic substrate by a nucleophilic neutral molecule. Three examples show the elimination of HF, HCl, and BH_3 by attacking nucleophiles[4,58,59]:

$$Xe + CH_3FH^+ \longrightarrow CH_3Xe^+ + HF \tag{35}$$

$$H_2O + CH_3ClH^+ \longrightarrow CH_3OH_2^+ + HCl \tag{36}$$

$$CH_3OH + H_2B\!-\!BH_3^+ \longrightarrow CH_3OHBH_2^+ + BH_3 \tag{37}$$

It is noteworthy that rare-gas molecules like xenon can act as nucleophiles in these reactions. The leaving group may be hydrogen halide, water, halogen neutral, borane, and presumably any other small stable neutral species. Beauchamp[58] formulates all these reactions as involving initial formation of a strong hydrogen bond between the reactive ion and nucleophile, as, for example,

$$H_2O + CH_3ClH^+ \longrightarrow \begin{bmatrix} CH_3\!-\!Cl \\ \vdots \\ H\!-\!O\!-\!-\!-\!H \\ | \\ H \end{bmatrix} \longrightarrow \begin{matrix} CH_3 \\ | \\ H\!-\!O \\ | \\ H \end{matrix} + \begin{matrix} Cl \\ | \\ H \end{matrix}$$

He further proposes two general conditions which must be met for such nucleophilic displacements to occur: (a) The reaction process must be exothermic; (b) proton transfer from the ionic substrate to the entering nucleophile must not be exothermic. If the proton transfer is exothermic then proton transfer is expected to be the dominant mode of reaction.

A reaction type which apparently is closely related to such nucleophilic displacements is the ionic dehydration of alcohols by a variety of ionic reagents.[4,60] The general reaction type is exemplified by the dehydration of sec-butyl alcohol:

$$C_4H_9\overset{+}{O}H_2 + C_4H_9OH \longrightarrow (C_4H_9)_2\overset{+}{O}H + H_2O \qquad (39)$$

Beauchamp and Caserio[60] postulate that such dehydrations proceed via formation of a strong hydrogen bond, displacement presumably involving a four centered transition state:

$$C_4H_9\overset{+}{O}H_2 + C_4H_9OH \longrightarrow \left[\begin{array}{c} \overset{CH_3}{\underset{|}{}} \\ C_2H_5-CH-O{\nearrow}^{H} \\ \underset{H{\nearrow}}{O}{\cdots}H^+ \\ \underset{C_4H_9}{|} \end{array} \right] \longrightarrow$$

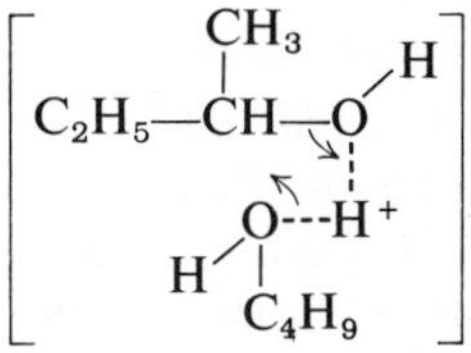

$$(C_4H_9)_2\overset{+}{O}H + H_2O \quad (40)$$

It is interesting to note that the ionic dehydration reaction becomes formally equivalent to the preceding nucelophilic displacement reactions at the time of formation of the strong hydrogen bond, so that the transition state (40) is formally equivalent to the transition state (38) postulated for nucleophilic displacements.

Carbonyl compounds of transition metals undergo a variety of ion-molecule reactions to give both binuclear complex ions and a variety of displacements of ligands on the metal atom.[61,62] Two condensation reactions giving binuclear complex ions are

$$Cr(CO)_4^- + Cr(CO)_6 \longrightarrow Cr_2(CO)_8^- + 2\,CO \qquad (41)$$

$$Fe(CO)^+ + Fe(CO)_5 \longrightarrow Fe_2(CO)_4^+ + 2\,CO \qquad (42)$$

Iron pentacarbonyl has also been found to undergo ligand displacement reactions with a number of neutral species, including water, as shown in the following scheme[61]:

$$
\begin{array}{lll}
Fe(CO)^+ \longrightarrow Fe(H_2O)^+ & & \\
Fe(CO)_2^+ \longrightarrow Fe(H_2O)_n(CO)_{2-n} & n = 1, 2 & \\
Fe(CO)_3^+ \longrightarrow Fe(H_2O)_n(CO)_{3-n} & n = 1, 2 & (43) \\
Fe(CO)_4^+ \longrightarrow Fe(H_2O)_n(CO)_{4-n} & n = 1, 2, 3 &
\end{array}
$$

Two exciting prospects for further studies of this kind are (1) the elucidation of factors affecting the stability of binuclear transition metal complex ions, and (2) determination of relative binding energies of different ligands on a given metal atom.

D. Nonreactive Collisions

The intimate connection between the behavior of ions in the ICR spectrometer and the nature and frequency of their interactions and collisions with neutral molecules leads naturally to experimental approaches aimed at obtaining quantitative information about nonreactive as well as reactive collisions. In rigorous terms one should of course distinguish between elastic collisions (no change of internal state of the ion) and nonreactive inelastic collisions (involving changes in internal state, but not in chemical composition), but in the experiments we will discuss in this section, this distinction is unimportant, and all such collisions may be termed nonreactive.

ICR studies of nonreactive collisions exploit the damping of the ion motion by random collisions with neutral molecules. In a classic study,[63] Beauchamp established in a fairly general way that the random nonreactive and resonant charge-transfer collision processes could be correctly included in the ion equation of motion in the form of a viscous damping term, giving an equation of motion

$$\frac{d\mathbf{V}}{dt} = e\left(\mathbf{E} + \frac{1}{c}\mathbf{V} \times \mathbf{B}\right) - \zeta\mathbf{V} \tag{44}$$

where $\mathbf{V}$ is the ion velocity, e its charge, $\mathbf{E}$ and $\mathbf{B}$ are the applied electric and magnetic fields, and ζ is the scalar damping coefficient. The approximation of the stochastic collision processes by a time-independent damping factor has been the central assumption of subsequent treatments,[4,64,65,66] and more recent work has indicated that such a term also accounts for the effects of limited ion lifetime in the ICR cell.[65]

For the case of the magnetic field in the z direction, the motion in the z direction in Eq. (44) separates out and is of no concern. The two coupled equations obtained for motion in the x and y directions may be solved, and motion in the x-y plane is characteristically that of a damped two-dimensional harmonic oscillator with frequency equal to the cyclotron frequency, and amplitude which decays with time constant $1/\zeta$. The ICR lineshape is then found to be

$$A(w) = \frac{C}{(w - w_c)^2 + \zeta^2} \tag{45}$$

where $A(w)$ is the (steady-state) power absorption at frequency w and w_c is the cyclotron frequency. This is a familiar Lorentzian lineshape, and the measurement of the half-width at half-height yields a value for ζ which is a measure of the collision frequency.

Before describing the results of such linewidth measurements, we may note that two alternative ways of measuring ζ have been described based on non-steady-state solutions of Eq. (44): (1) A transient ICR technique has been used exploiting the damped heterodyne interference pattern between the ions and the marginal oscillator[64]; (2) the damped power emission from initially accelerated ions has been measured.[67] Both of these techniques involve measurement of rapid transient phenomena as a function of time, and appear to suffer from very severe intensity problems, so that we consider the line-width technique described below as a much more promising source of accurate, convenient data. This technique has recently been exploited by Beauchamp[4] and in our laboratory in the experiments which we will describe.

We have measured the linewidth for the Cl^- ICR resonance line with a number of different neutral gases, with which there was no possibility of chloride ion reacting.[68] Some results are shown in Fig. 9-5, in which we have

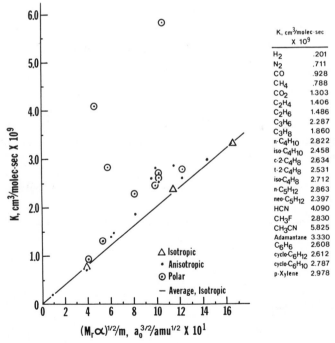

K, cm^3/molec·sec $\times 10^9$	
H_2	.201
N_2	.711
CO	.928
CH_4	.788
CO_2	1.303
C_2H_4	1.406
C_2H_6	1.486
C_3H_6	2.287
C_3H_8	1.860
$n\text{-}C_4H_{10}$	2.822
$iso\text{-}C_4H_{10}$	2.458
$c\text{-}2\text{-}C_4H_8$	2.634
$t\text{-}2\text{-}C_4H_8$	2.531
$iso\text{-}C_4H_8$	2.712
$n\text{-}C_5H_{12}$	2.863
$neo\text{-}C_5H_{12}$	2.397
HCN	4.090
CH_3F	2.830
CH_3CN	5.825
Adamantane	3.330
C_6H_6	2.608
$cyclo\text{-}C_6H_{12}$	2.612
$cyclo\text{-}C_6H_{10}$	2.787
$p\text{-}Xylene$	2.978

Fig. 9-5. Line broadening of the Cl^- ICR peak due to nonreactive collisions with a variety of molecules. K is the collision rate (proportional to linewidth), and $(M_r\alpha)^{1/2}/m$ is the (square root of the) molecular polarizability with a correction term to account for the transition from the center of mass to the laboratory coordinate system.

shown the rate of momentum relaxation due to inelastic collisions as a function of $(\alpha/M_r)^{1/2}$. As we saw in Section III.A, this should give a linear plot if the collision rate is determined solely by the ion-induced dipole potential, and indeed this is seen to be the case for nonpolar molecules. For the neutral molecules having permanent dipoles, the collision rates are seen to lie greatly above the line, and it is apparent that the ion-permanent dipole potential is acting to increase the collision rate. However, comparison with the "locked dipole" potential[50,51] shows that the effect of the dipole is much less than would be expected for complete dipole locking. The increase in rate is about 50% of that which would be expected for a fully locked dipole. From these results it is clear that the attractive force between the ion and the dipolar molecule is important in determining the nonreactive collision behavior, and that rapid rotation of the dipolar molecule does not average out this attractive force. One can expect, of course, that this same effect would show up in an enhancement of reaction rates above the Gioumousis-Stevenson rate in reactive systems, an enhancement which we noted for several reactions in Section III.B.

(Recent results show that the rate constants in Figure 5 are all too high by about a factor of 2. The new, lower numbers are in excellent agreement with the Gioumousis-Stevenson prediction for nonpolar molecules.)

E. Conclusions

It should have become evident even from this brief and fragmentary survey why ICR spectroscopy has enjoyed such a rapid growth of interest—it has shown itself to be applicable to a remarkably wide range of problems in the chemistry and physics of ions and their interactions with neutral molecules. There seems no reason to suspect that the possibilities of gaseous ionic chemistry will soon be exhausted. Each of the areas of ionic chemistry discussed here is immediately suggestive of the need for theoretical description at a higher level than the experimentalist has been able to provide. The thermochemical and spectroscopic properties of gas-phase ions, the rates and mechanisms of reactions, and the nature of nonreactive collision processes are all areas in which experimental capabilities are in danger of outstripping theoretical understanding.

ACKNOWLEDGMENTS

The research results described in this chapter were obtained with the invaluable collaboration of Dr. J. M. Kramer, Mr. P. P. Dymerski, and Mr. E. W. Fu, and were supported in part by the National Science Foundation, the Research Corporation, and the Petroleum Research Fund of the American Chemical Society.

References

1. L. R. Anders, J. L. Beauchamp, R. C. Dunbar, and J. D. Baldeschwieler, *J. Chem. Phys.* **45**, 1062 (1966).
2. G. A. Gray, *Adv. Chem. Phys.* **19**, 141 (1971).
3. J. D. Baldeschwieler and S. S. Woodgate, *Acct. Chem. Res.* **4**, 114 (1971).
4. J. L. Beauchamp, *Ann. Rev. Phys. Chem.* **22**, 527 (1971).
5. E. W. McDaniel *et al.*, *Ion-Molecule Reactions*, Wiley-Interscience, New York, 1970.
6. F. H. Field, *Acct. Chem. Res.* **1**, 42 (1968).
7. L. Friedman, *Ann. Rev. Phys. Chem.* **19**, 273 (1968).
8. "Ion-Molecule Reactions in the Gas Phase," *Advan. Chem. Ser.* **58** (1966).
9. J. D. Baldeschwieler, *Science* **159**, 263 (1968).
10. J. L. Beauchamp and J. T. Armstrong, *Rev. Sci. Instr.* **40**, 123 (1969).
11. R. C. Dunbar, *J. Am. Chem. Soc.* **90**, 5676 (1968).
12. Some useful background discussion of ion thermochemistry can be found in S. W. Benson, *J. Chem. Ed.* **42**, 502 (1965).
13. The principal tabulated source of ion thermochemical data is J. L. Franklin et al., "Ionization Potentials, Appearance Potentials, and Heats of Formation of Gaseous Positive Ions," NSRDS-NBS 26, 1969.
14. D. Holtz, J. L. Beauchamp, S. D. Woodgate, *J. Am. Chem. Soc.* **92**, 7484 (1970).
15. J. L. Beauchamp, Ph.D. Thesis, Harvard University, 1968.
16. R. C. Dunbar, J. Shen, and G. A. Olah, *J. Chem. Phys.* **56**, 3794 (1972).
17. J. L. Beauchamp and S. E. Buttrill, Jr., *J. Chem. Phys.* **48**, 1783 (1968).
18. J. L. Beauchamp, D. Holtz, S. D. Woodgate, and S. L. Patt, *J. Am. Chem. Soc.* **94**, 2798 (1972).
19. D. Holtz, J. L. Beauchamp, and J. R. Eyler, *J. Am. Chem. Soc.* **92**, 7045 (1970).
20. D. Holtz and J. L. Beauchamp, *J. Am. Chem. Soc.* **91**, 5913 (1969).
21. M. A. Haney and J. L. Franklin, *J. Phys. Chem.* **73**, 4328 (1969).
22. M. T. Bowers, D. H. Aue, H. M. Webb, and R. T. McIver, *J. Am. Chem. Soc.* **93**, 4314 (1971).
23. E. M. Arnett, F. M. Jones, III, M. Taagepera, W. G. Henderson, J. L. Beauchamp, D. Holtz, and R. W. Taft, *J. Am. Chem. Soc.* **94**, 4724 (1972).
24. R. T. McIver and J. E. Eyler, *J. Am. Chem. Soc.* **93**, 6334 (1971).
25. J. I. Brauman and L. K. Blair, *J. Am. Chem. Soc.* **92**, 5986 (1970).
26. J. I. Brauman and L. K. Blair, *J. Am. Chem. Soc.* **91**, 2126 (1969).
27. J. I. Brauman, J. R. Eyler, L. K. Blair, M. J. White, M. B. Comisarow, and K. C. Smyth, *J. Am. Chem. Soc.* **93**, 6360 (1971).
28. J. Diekman, J. MacLeod, C. Djerassi, and J. D. Baldeschwieler, *J. Am. Chem. Soc.* **91**, 2069 (1969).
29. G. Eadon, J. Diekman, and C. Djerassi, *J. Am. Chem. Soc.* **91**, 3986 (1969).
30. A common type of mass spectral rearrangement-fragmentation process.
31. J. L. Beauchamp and R. C. Dunbar, *J. Am. Chem. Soc.* **92**, 1477 (1970).

32. M. K. Hoffman and M. M. Bursey, *Tetrahedron Lett.* 2539 (1971).

33. H. M. Grubb and S. Meyerson, in *Mass Spectrometry of Organic Ions*, F. W. McLafferty, Ed., Academic Press, New York, 1963.

34. K. L. Rinehart et al., *J. Am. Chem. Soc.* **90**, 2983 (1968).

35. I. Howe and F. W. McLafferty, *J. Am. Chem. Soc.* **93**, 99 (1971).

36. R. C. Dunbar, *J. Am. Chem. Soc.* **93**, 4354 (1971).

37. R. C. Dunbar and Emil W. Fu, *J. Am. Chem. Soc.*, **95**, 2716 (1973).

38. D. W. Turner et al., *Molecular Photoelectron Spectroscopy*, Wiley-Interscience, New York, 1970.

39. J. M. Kramer and R. C. Dunbar, *J. Chem. Phys.*, in press.

40. J. M. Kramer and R. C. Dunbar, *J. Am. Chem. Soc.* **94**, 4346 (1972).

41. G. Gioumousis and D. P. Stevenson, *J. Chem. Phys.* **29**, 294 (1958).

42. E. W. McDaniel, *Collision Phenomena in Ionized Gases*, Wiley, New York, 1964.

43. A. G. Harrison, J. J. Myher, and J. C. J. Thynne, in Ref. 8, p. 150.

44. S. E. Buttrill, Jr., *J. Chem. Phys.* **50**, 4125 (1969).

45. A. G. Marshall and S. E. Buttrill, Jr., *J. Chem. Phys.* **52**, 2752 (1970).

46. M. B. Comisarow, *J. Chem. Phys.* **55**, 205 (1971).

47. W. T. Huntress, *J. Chem. Phys.* **54**, 843 (1971).

48. R. P. Clow and J. H. Futrell, *Int. J. Mass Spectrum. Ion Phys.* **4**, 165 (1970).

49. M. T. Bowers and D. D. Elleman, *J. Chem. Phys.* **51**, 4606 (1969).

50. J. V. Dugan and J. L. Magee, *J. Chem. Phys.* **47**, 3103 (1967).

51. L. P. Theard and W. H. Hamill, *J. Am. Chem. Soc.* **84**, 1134 (1962).

52. F. P. Abramson and J. H. Futrell, *J. Chem. Phys.* **45**, 1925 (1966).

53. S. Wexler and L. G. Pobo, *J. Am. Chem. Soc.* **93**, 1327 (1971).

54. M. S. B. Munson, J. L. Franklin, and F. H. Field, *J. Phys. Chem.* **68**, 3098 (1964).

55. M. T. Bowers, D. D. Elleman, and J. L. Beauchamp, *J. Phys. Chem.* **72**, 3599 (1968).

56. S. E. Buttrill, Jr., *J. Am. Chem. Soc.* **92**, 3560 (1970).

57. R. C. Dunbar, *J. Am. Chem. Soc.* **93**, 4167 (1971).

58. See Ref. 14.

59. R. C. Dunbar, *J. Phys. Chem.*, **76**, 2467 (1972).

60. J. L. Beauchamp and M. C. Caserio, *J. Am. Chem. Soc.* **94**, 2638 (1972).

61. M. S. Foster and J. L. Beauchamp, *J. Am. Chem. Soc.* **93**, 4925 (1971).

62. R. C. Dunbar, J. Ennever, and J. P. Fackler, in press.

63. J. L. Beauchamp, *J. Chem. Phys.* **46**, 1231 (1967).

64. R. C. Dunbar, *J. Chem. Phys.* **54**, 711 (1971).

65. M. B. Comisarow, *J. Chem. Phys.* **55**, 205 (1971).

66. A. G. Marshall, *J. Chem. Phys.* **55**, 1343 (1971).

67. C. A. Lieder, R. W. Wien, and R. T. McIver, Jr., *J. Chem. Phys.* **56**, 5184 (1972).

68. P. P. Dymerski and R. C. Dunbar, *J. Chem. Phys.* **57**, 4049 (1972).

Index